2025 IFIP/IEEE 33rd International Conference on Very Large Scale Integration (VLSI-SoC 2025)

Puerto Varas, Chile
12-15 October 2025

IEEE Catalog Number: CFP25LSI-POD
ISBN: 979-8-3315-9813-6

Copyright © 2025, IEEE

All Rights Reserved

Copyright and Reprint Permissions:

Abstracting is permitted with credit to the source. Libraries are permitted to photocopy beyond the limit of U.S. copyright law for private use of patrons those articles in this volume that carry a code at the bottom of the first page, provided the per-copy fee indicated in the code is paid through Copyright Clearance Center, 222 Rosewood Drive, Danvers, MA 01923.

For other copying, reprint or republication permission, write to IEEE Copyrights Manager, IEEE Service Center, 445 Hoes Lane, Piscataway, NJ 08854. All rights reserved.

*** This is a print representation of what appears in the IEEE Digital Library. Some format issues inherent in the e-media version may also appear in this print version.

IEEE Catalog Number: CFP25LSI-POD
ISBN (Print-On-Demand):   979-8-3315-9813-6
ISBN (Online):   979-8-3315-9812-9
ISSN: 2324-8432

Additional Copies of This Publication Are Available From:

Curran Associates, Inc
57 Morehouse Lane
Red Hook, NY 12571 USA

Phone: (845) 758-0400
Fax: (845) 758-2633
E-mail: curran@proceedings.com
Web: www.proceedings.com

TABLE OF CONTENTS

Noise and Quantization Parameterization of Photonic Convolution Accelerator 1
Mateus Vidaletti Costa, Mauricio Gomes de Queiroz, Raphael Cardoso, Ian O'Connor, Arnan Mitchell

Mission Profile-Driven Transistor Aging Modeling and Simulation Flow 6
Firas Ramadan, Maayan Ella, Freddy Gabbay

Trojan Attacks on Graph Convolution Neural Networks for Circuit Analysis 11
Rupesh Raj Karn, Ozgur Sinanoglu

Exploring MRAM for On-Chip Texture Storage in Rendering Applications 16
Nicolás Villegas, Stefano Romanini, Moritz Scherer, Warren Hunt, Syed Shakib Sarwar, Barbara De Salvo
Chiao Liu, Francesco Conti, Davide Rossi, Luca Benini, Jorge Gómez

Open-Source 4 K CMOS Calibration: Integrating IceMOS and Sky130 PDK 21
Mauricio Montanares, V.H. Arzate Palma, Kevin G. McCarthy, Gerardo Molina Salgado

Implementation of a 16:1 Multiplexer and 1:16 Demultiplexer on a Single Chip Using Sky130 PDK and
Open-Source EDA Tools for Silicluster 26
Uriel Jaramillo-Toral, Susana Ortega-Cisneros, Emilio Isaac Baungarten-Leon, Erick Jaramillo-Toral
Héctor Emmanuel Muñoz Zapata

Inter-chip Clock Network Synthesis on Passive Interposer of 2.5D Chiplet Considering Transmission
Line Effect 31
Tai Yan, Yiyu Wang, Zhan Li, Ning Xu, Yuanqing Cheng

Fault Modeling and Testing of Spin-Orbit Torque-Based Multipillar Memory Cell 36
Arshid Nisar, Lorena Anghel, Gregory Di Pendina

Exploring Enhancements to 1T1C FeMFET Bitcells with a Versatile DTCO Methodology 41
Rosario Pronsat, Antoine Cauquil, Pascal Vivet, Jean Coignus, Damien Deleruyelle, Cédric Marchand
Lioua Labrak, Ian O'Connor

Non-Volatile Ferroelectric-AND (FeAND) Memory Cell Design 46
Basile Darne, Miqueas Filsinger, Alberto Bosio, Damien Deleruyelle, Ian O'Connor, Bertrand Vilquin
Cédric Marchand

On the Possibility of Relying Solely on FeMFET Variability for PUF Implementations 51
Miqueas Filsinger, Antoine Cauquil, Damien Deleruyelle, David Navarro, Ian O'Connor, Cédric Marchand

TSPC-Based Low-Power High-Resolution CMOS Phase Frequency Detector 56
Dhandeep Challagundla, Venkata Krishna Vamsi Sundarapu, Ignatius Bezzam, Riadul Islam

EmFIA: A Novel Emulation-based Fault Injection Vulnerability Assessment Framework at RTL Level 61
Tanvir Rahman, Shuvagata Saha, Sujan Kumar Saha, Farimah Farahmandi, Mark Tehranipoor

Work in Progress: Exploring Ferroelectric Oscillators for Solving NP-Hard Problems Using Mallick's
Coupling Mechanism 66
Joaquin Welch, Jorge Gomez, Jaime Cisternas

Lightweight Congruence Profiling for Early Design Exploration of Heterogeneous FPGAs 71
Allen Boston, Biruk Seyoum, Luca Carloni, Pierre-Emmanuel Gaillardon

ML4FPGA: An LLM Framework for Electronic Design Automation and Verification on FPGA 76
Uchechukwu Leo Udeji, Martin Margala

Scalability analysis of multi-bank near-memory computing in low-power SoCs 81
Luigi Giuffrida, Pasquale Davide Schiavone, Michele Caon, Guido Masera, Maurizio Martina
David Atienza

LiC: Low-Cost Cache Replacement Algorithm for All Cache Levels 85
Varun Venkitaraman, Tejeshwar Thorawade, Mitul Tandon, Keerthisagar Kokkiligadda, Virendra Singh
Janak Patel

Automated Generation of Microfluidic Netlists using Large Language Models 90
Jasper Davidson, Skylar Stockham, Allen Boston, Ashton Snelgrove, Valerio Tenace
Pierre-Emmanuel Gaillardon

PVT-Robust Analog Control Stage for Buck DC-DC Converters in Open-Source SKY130 95
Giordano De Moraes Rossa, Henrique Beque, Iuri Tinti, Jorge Marín, Juan Pablo Martínez Brito

From Secure Storage to Compute-in-Memory: A Versatile Memory System using 1T-nC FeRAM 100
Rakesh Acharya, Rudra Biswas, Jiahui Duan, Prapti Panigrahi, Kai Ni, Vijaykrishnan Narayanan

A Low-Power 4-bit Tracking-Type Analog-to-Digital Converter in SKY130 Process 105
Esteban Astudillo, EduardoHolguin, Esteban Garzón, Luis Miguel Prócel

Swift Synthesis of Approximate Hardware Accelerators Using Generative Adversarial Networks 109
Muhammad Awais, Hassan Ghasemzadeh Mohammadi, Marco Platzner

Deus Ex LLMs: AI vs Humans in Post-Quantum Cryptographic Hardware Code Generation 114
Ethan Cornett, Rahul Magesh, Sharath Pendyala, Elif Bilge Kavun, Aydin Aysu

Open-Source Approach to IC Development: Validation Against Measurements of Selected Devices
from the IHP-Open-PDK 119
Krzysztof Herman, Dietmar Warning

Toward Multi-Person Breath Rate Estimation via mmWave Radar 123
Cristian Turetta, Christian Farina, Chiara Bozzini, Morteza Varasteh, Graziano Pravadelli

Application and Detection of Hardware Trojans Applied to Valid Data States of NCL Combinational
Circuits 128
João Pedro Pereira Magalhães, Tales Cleber Pimenta, Diogo Leonardo Ferreira Silva

Accelerating Machine Learning using RISC-V Vector Extension in a Manycore Platform 133
Willian Analdo Nunes, Antônio Vinicius Corrêa Dos Santos, César Marcon, Fernando Gehm Moraes

TLGLock: A New Approach in Logic Locking Using Key-Driven Charge Recycling in Threshold Logic
Gates 138
Abdullah Sahruri, Martin Margala

Correlation Between Process Variability and Radiation Hardness in Digital Circuits 143
Elias Ramos, Augusto Weber, Wilian Padilha, Renan Carlos Gomes de Farias, João Baptista Martins
Ricardo Reis

CMOS Time Register With High Dynamic Range 148
Johnatan Felipe Silva Garcia, Dalton Martini Colombo, Kamal El-Sankary, Mahsa Zareie

Energy-Efficient Computation of TensorFloat32 Numbers on an FP32 Multiplier 153
Per Larsson-Edefors

Delay Mismatch Optimization in Routing Dominated Multi-Path Systems: A Case Study on an
IR-UWB Edge-Combiner Transmitter Front End 157
Kyla Marie H. Juruena, Maria Ena R. Rosales, Trisha Renee G. Capulong, Louis P. Alarcón

Conjunctive Merge Instruction to Accelerate Sparse Matrix - Dense Vector Multiplication 161
Manuel Osterno, César Marcon, Jarbas Silveira, Fernando Moraes, Jardel Silveira

Marker-Based Recognition for Autonomous Micro-Drone Flight: An FPGA-Optimized Feasibility
Study 166
Diego Marcelo Ramirez Jove, Keisuke Sugiura, Yoshiki Yamaguchi

Cultivating Security: Debug Authentication for Ensuring the Security of SoC's Root of Trust 171
Arash Vafaei, Sujan Kumar Saha, Mark Tehranipoor, Farimah Farahmandi

Towards Full Integration of a Three-Level Flying Capacitor Converter Control in a Mixed-Signal ASIC 176
Nelson Salvador, Francisca Donoso, Jorge Marín, Victor Grimblatt, Christian A. Rojas

Design of a 32nm Ultra-Low Power 6T SRAM Cell Analyzing Emerging Technologies such as
FinFET, TFET, and CNFET for energy-efficient applications 178
Jesús González Huarancca, Carlos Silva Cárdenas

A Novel CMOS Time Register 180
Johnatan Felipe Silva Garcia, Dalton Martini Colombo, Kamal El-Sankary, Mahsa Zareie

Hybrid Lightweight Soft Error Mitigation Techniques for Edge Devices 182
Jonas Gava, Ricardo Reis, Luciano Ost

Cross-Layer Approximate Hardware Design of Interpolation Filters for Fractional Motion Estimation in
Versatile Video Coding 184
Rafael Da Silva, Ricardo Reis, Mateus Grellert

A Case Study on the Migration of a High-Level PI Controller to ASIC-Compatible HDL Representation 186
Francisca Donoso, Nelson Salvador, Jorge Marin, Christian A. Rojas, Gonzalo Carvajal

Transistor Placement for Automatic Cell Layout Generation on Advanced Nodes: A Review 188
Vitor H. Fuerstenau, Ricardo Reis

Preface

Message from the General Chairs
VLSI SoC for a Sustainable World

On behalf of the entire Organizing Committee, we warmly welcome you to the 33rd IFIP/IEEE International Conference on Very Large Scale Integration (VLSI-SoC 2025). This year's event, focused on the theme "VLSI SoC for a Sustainable World," brings together researchers, practitioners, and industry experts globally to share the latest advancements, ideas, and challenges in Very-Large-Scale Integration (VLSI) Systems-on-Chip.

We are delighted to host you in Puerto Varas, located 1,000km south of Santiago in Chile's breathtaking Lake District. Situated on the southwest shore of Lake Llanquihue, with stunning views of the active Osorno and Calbuco volcanoes, this destination offers a unique blend of German traditions and scenic natural beauty. Known as "the city of roses," Puerto Varas is close to mountains, forests, national parks, and the famous Teatro del Lago concert hall in Frutillar.

The technical program is designed to advance the field, featuring research across core areas like architectures, design, hardware security, low-power solutions, and new VLSI applications. Program highlights include three keynotes, two tutorials, and one expert panel, alongside dedicated forums to support student and early-career researcher participation and innovation.

Beyond the technical sessions, we've planned four exciting social events to enhance your experience: the Welcome Reception, a regional tour, the Gala Dinner, and a Farewell Party. We look forward to sharing many surprises and memorable moments with you.

We extend our deepest gratitude to all contributors, reviewers, and participants for their essential work, which continues to drive the VLSI-SoC community forward. These proceedings capture the current state of research, aiming to encourage future studies and foster professional connections.

Once again, welcome to VLSI-SoC 2025. We are eager to celebrate your achievements and engage in the thought-provoking discussions that will shape the future of VLSI technology.

Sincerely,

Victor Grimblatt, Pierre-Emmanuel Gaillardon

General Chairs

October 9, 2025   Victor Grimblatt
Chile   Pierre-Emmanuel Gaillardon

Program Committee

Hasan Al-Nashash	AUS
Lutfi Albasha	american university of sharjah
Deni Alves	Universidade Federal de Santa Catarina
Sergei Andreev	IHP Microelectronics
Daniel Arévalos	Hochschule M¨unchen University of Applied Sciences, Munich, Germany
Mohamed Atef	United Arab Emirates University
Bevan Baas	University of California, Davis
Paolo Bernardi	Politecnico di Torino
Leticia Bolzani Poehls	RWTH Aachen University
Alberto Bosio	Lyon Institute of Nanotechnology
Gonzalo Carvajal	USM
Anupam Chattopadhyay	Nanyang Technological University
Luc Claesen	University Hasselt
Giovanni De Micheli	EPFL
Nahla El-Araby	TU Wien, Vienna, Austria and Canadian International College, Cairo, Egypt
Moritz Fieback	Delft University of Technology
Pierre-Emmanuel Gaillardon	University of Utah
Anteneh Gebregiorgis	Delft University of Technology
Horia Giuroiu	Retired / Texas Instruments
Wladek Grabinski	
Victor Grimblatt	SYNOPSYS
Patrick Groeneveld	Stanford University
Xinfei Guo	Shanghai Jiao Tong University
Matthew Guthaus	University of California Santa Cruz
Ilker Hamzaoglu	Ozyegin University
Mohammad Reza Heidari Iman	TIMA laboratory
Renato Hentschke	Synopsys
Krzysztof Herman	IHP
Michael Huebner	Brandenburg University of Technology Cottbus
Maksim Jenihhin	Tallinn University of Technology
Marcelo Johann	Universidade Federal do Rio Grande do Sul
Fernanda Kastensmidt	Universidade Federal do Rio Grande do Sul - UFRGS
Srinivas Katkoori	University of South Florida
Johann Knechtel	New York University
Victor Kravets	IBM
Martin Kumm	University of Applied Sciences, Fulda
Sohaib Majzoub	University of Sharjah
Piero Malcovati	University of Pavia
Tiziana Margaria	Lero
Jorge Marin	USM
Ricardo Martins	Instituto de Telecomunicações, Instituto Superior Técnico – ULisbon
Cristina Meinhardt	UFSC
Farhad Merchant	University of Groningen

VLSI SoC 2025   Program Committee

Salvador Mir	TIMA Laboratory
Jose Miranda	EPFL
Saraju Mohanty	University of North Texas
Katell Morin-Allory	TIMA Laboratory
Juan Moya	
Saleh Mulhem	Universität zu Lübeck,
Vigneshwar Murali	Nvidia
Kashif Nawaz	Cryptography Research Centre, Technology Innovation Institute
Michael Niemier	University of Notre Dame
Ian O'Connor	Lyon Institute of Nanotechnology
Graziano Pravadelli	Dipartimento di Informatica - Università di Verona
Mahmoud Rasras	New York University Abu Dhabi
Ricardo Reis	UFRGS
Christian Rojas	Universidad Tecnica Federico Santa Maria
Annachiara Ruospo	Politecnico di Torino
Wala Saadeh	Western Washington University
Mazen Saghir	American University of Beirut
Mihai Sanduleanu	Khalifa University of Science and Technology
Rodrigo Schramm	Harman
Kaveh Shamsi	University of Texas at Dallas
Carlos Silva-Cardenas	PUCP
Kostas Siozios	Aristotle University of Thessaloniki
Sorin Spanoche	Microchip
Bill Swartz	University of Texas
Mottaqiallah Taouil	Delft University of Technology
Theocharis Theocharides	University of Cyprus
H. Fatih Ugurdag	Ozyegin University
Jaime Viegas	Khalifa University
Arnaud Virazel	LIRMM
Gilson Wirth	UFRGS
Matthew Ziegler	IBM

Additional Reviewers

Alves, Deni
Arévalos, Daniel
Brito Filho, Francisco
Cortes, Alfonso
Estay, Alejandro
Figueroa, Miguel
Gomez, Jorge
Grabinski, W.S.
Groeneveld, Patrick
K, Thomas
Kassimi, Asmaa
Mellor, Marcus
Rojas, Christian
Shaik, Jani Babu
Spanoche, Sorin
Stowhas-Villa, Alejandro
Tao, Chun
Veillon, Matias
Wang, Runxi
Wilson, Alan
Yuan, Sicong

Author Index

Acharya, Rakesh	1
Alarcon, Louis	80
Anghel, Lorena	93
Arzate Palma, V.H.	108
Astudillo, Esteban	6
Atienza, David	65
Awais, Muhammad	10
Aysu, Aydin	25
Baungarten Leon, Emilio Isaac	75
Benini, Luca	175
Beque, Henrique	103
Bezzam, Ignatius	20
Bilge Kavun, Elif	25
Biswas, Rudra	1
Bosio, Alberto	32, 123
Boston, Allen	15, 37
Bozzini, Chiara	150
Caon, Michele	65
Capulong, Trisha Renee	80
Cardoso, Raphael	170
Carloni, Luca	15
Carlos Gomes de Farias, Renan	42
Carvajal, Gonzalo	47
Cauquil, Antoine	51, 123
Challagundla, Dhandeep	20
Cheng, Yuanqing	185
Cisternas, Jaime	180
Coignus, Jean	123
Colombo, Dalton	49, 58
Conti, Francesco	175
Cornett, Ethan	25
Darne, Basile	32
Davidson, Jasper	37
De Salvo, Barbara	175
Deleruyelle, Damien	32, 51, 123
Di Pendina, Gregory	93
Donoso, Francisca	47, 148
Duan, Jiahui	1
da Silva, Rafael	30
de Almeida Ramos, Elias	42

El-Sankary, Kamal	49, 58
Ella, Maayan	133
Farahmandi, Farimah	128, 160
Farina, Christian	150
Felipe, Johnatan	49
Filsinger, Miqueas	32, 51
Fuerstenau, Vitor Hugo	56
G. McCarthy, Kevin	108
Gabbay, Freddy	133
Gaillardon, Pierre-Emmanuel	15, 37
Garcia, Johnatan	58
Garzón, Esteban	6
Gava, Jonas	63
Ghasemzadeh Mohammadi, Hassan	10
Giuffrida, Luigi	65
Gomes de Queiroz, Mauricio	170
Gonzalez, Jesus	69
Grellert, Mateus	30
Grimblatt, Victor	148
Gómez, Jorge	175, 180
Herman, Krzysztof	71
Holguín, Eduardo	6
Hunt, Warren	175
Islam, Riadul	20
Jaramillo Toral, Erick	75
Jaramillo Toral, Uriel	75
Juruena, Kyla Marie	80
Karn, Rupesh Raj	84
Kokkiligadda, Keerthisagar	165
Labrak, Lioua	123
Larsson-Edefors, Per	89
Laway, Arshid Nisar	93
Li, Zhan	185
Liu, Chiao	175
Magalhaes, Joao Pedro	98
Magesh, Rahul	25
Marchand, Cédric	32, 51, 123
Marcom, César	113
Marcon, César	118
Margala, Martin	143, 155

Marin, Jorge	103
Martina, Maurizio	65
Martinez Brito, Juan Pablo	103
Martins, Joao Baptista	42
Marín, Jorge	47, 148
Masera, Guido	65
Mitchell, Arnan	170
Molina Salgado, Gerardo	108
Montanares, Mauricio	108
Moraes, Fernando	118
Moraes, Fernando Gehm	113
Muñoz Zapata, Hector Emmanuel	75
Narayanan, Vijaykrishnan	1
Navarro, David	51
Ni, Kai	1
Nunes, Willian Analdo	113
O'Connor, Ian	32, 51, 123, 170
Ortega Cisneros, Susana	75
Ost, Luciano	63
Osterno, Manuel	118
Padilha, Wilian	42
Panigrahi, Prapti	1
Patel, Janak	165
Pendyala, Sharath	25
Pimenta, Tales	98
Platzner, Marco	10
Pravadelli, Graziano	150
Pronsato, Rosario	123
Prócel, Luis Miguel	6
Rahman, Tanvir	128
Ramadan, Firas	133
Ramírez Jove, Diego Marcelo	138
Reis, Ricardo	30, 42, 56, 63
Rojas, Christian	47, 148
Romanini, Stefano	175
Rosales, Maria Ena	80
Rossa, Giordano	103
Rossi, Davide	175
Saha, Shuvagata	128
Saha, Sujan Kumar	128, 160
Sahruri, Abdullah	143
Salvador, Nelson	47, 148
Santos, Antônio	113

Sarwar, Syed Shakib	175
Scherer, Moritz	175
Schiavone, Pasquale Davide	65
Seyoum, Biruk	15
Silva, Diogo Leonardo	98
Silveira, Jarbas	118
Silveira, Jardel	118
Sinanoglu, Ozgur	84
Singh, Virendra	165
Snelgrove, Ashton	37
Stockham, Skylar	37
Sugiura, Keisuke	138
Sundarapu, Venkata Krishna Vamsi	20
Tandon, Mitul	165
Tehranipoor, Mark	128, 160
Tenace, Valerio	37
Thorawade, Tejeshwar	165
Tinti, Iuri	103
Turetta, Cristian	150
Udeji, Leo	155
Vafaei, Arash	160
Varasteh, Morteza	150
Venkitaraman, Varun	165
Vidaletti Costa, Mateus	170
Villegas, Nicolás	175
Vilquin, Bertrand	32
Vivet, Pascal	123
Wang, Yiyu	185
Warning, Dietmar	71
Weber, Augusto Gouvêa	42
Welch, Joaquín	180
Xu, Ning	185
Yamaguchi, Yoshiki	138
Yan, Tai	185
Zareie, Mahsa	49, 58

Noise and Quantization Parameterization of Photonic Convolution Accelerator

Mateus Vidaletti Costa[1,2], Mauricio Gomes de Queiroz[1], Raphael Cardoso[1], Ian O'Connor[1], Arnan Mitchell[2]

[1]*Institut des Nanotechnologies de Lyon, Ecole Centrale de Lyon, Ecully, France*
[2]*School of Engineering, RMIT University, Melbourne, Australia*
mateus.vidaletti@ec-lyon.fr

Abstract—**Large-scale convolutional neural networks (CNNs) often rely on dedicated digital hardware, constrained by latency, throughput, and energy efficiency. Photonic hardware offers a promising alternative, but its analog nature and optical complexity pose challenges for electronic design automation (EDA) and design space exploration (DSE), limiting large-scale analysis. This work presents a novel parameterization methodology that quantifies the impact of noise, quantization, and kernel choice on a photonic convolution accelerator (CA), leveraging high-speed simulation tool. Using the MNIST dataset with 10 kernels, our results reveal up to a 3.5× difference in the root mean squared error (RMSE) between *Blur* and *Laplacian* kernels, demonstrating the critical role of kernel choice. The proposed simulation approach is also over 100x faster than conventional methods, making the analysis feasible, whereas performing it with traditional techniques would be impractical, if not impossible.**

Index Terms—**Silicon Photonics, Convolution, Accelerator, Noise, Quantization, Kernel**

I. INTRODUCTION

With the growing demand for neuromorphic computing, such as convolutional neural networks (CNNs), the need for high-speed and energy-efficient computation is exceptionally high. Despite advances in specialized hardware such as graphics processing units (GPUs), digital devices remain constrained by clock frequency, memory access, and power consumption [1].

Photonic hardware accelerators offer a promising alternative, leveraging light's unique properties to achieve low latency, high parallelism, and energy efficiency. While various photonic convolutional accelerators (CA) have been introduced in the literature [2]–[5], in this work, we base our analyses on that of Xingyuan et al. [2], capable of 11.3 trillion operations per second (TOPS). It employs both time- and wavelength multiplexing for information encoding and, as an analog system, is subject to variability due to device-level characteristics. Improving its performance requires design space exploration (DSE), but conventional photonic circuit simulations are too slow for large-scale DSE [6] such that this architecture is a particularly challenging test case requiring radically new methods.

In this work, we replicated the photonic CA using SPECS [6], a fast event-driven photonic circuit simulator. We present an initial exploration of how critical parameters (noise and quantization) influence convolution performance in real-world scenarios, providing quantitative insights into Xingyuan et al.'s accelerator [2].

The rest of this paper is organized as follows: Section II reviews the convolution process behind CNNs and introduce the working principles of Xingyuan et al. accelerator [2]. Section III describes the simulation methodology and its parameters. Section IV covers the results. Finally, section V concludes the study described in this paper.

II. BACKGROUND

The convolution layer is the main section of a CNN and holds the largest part of its computational load [7]. During this phase, a $k \times k$ matrix of weights K, also known as a kernel, is moved across the width and height of an input image I of dimension $i_r \times i_c$, producing a feature map. At the same time, the dot product between the kernel and the covered part of the image is computed and summed into a single output pixel. Considering a bidimensional convolution, each pixel from a feature map C of dimension $c_r \times c_c$ can be described as

$$C(m,n) = \sum_{p=0}^{k-1} \sum_{q=0}^{k-1} K_{(p+1,q+1)} I_{(u,v)}. \quad (1)$$

In this equation, u and v are the coordinates of I, given by:

$$u = p + m + (s_V - 1)(m - 1),$$
$$v = q + n + (s_H - 1)(n - 1). \quad (2)$$

Here, s_V and s_H represent the vertical and horizontal stride, respectively, which determines how many rows and columns the kernel jumps from one window to another. This process is repeated until the kernel has covered the entire image, ultimately resulting in an activation map that carries specific features of the input data [8].

In order to adapt the convolution operation described by (1) to the chosen photonic accelerator, we need to represent the kernel K as a 1D weight vector W of length R and the input image I as a 1D data vector X of length L. This process is explained in detail in subsection III-B. The vector W is encoded as the optical power of different wavelengths across the C-band, generated from a Kerr microcomb light source [9]. These wavelengths then pass through an electro-optical Mach-Zehnder modulator (EOM), where X is encoded as the intensity of temporal symbols, meaning that at every symbol period τ, one pixel from X is broadcast onto all the comb lines carrying information from W. The data vector is encoded with a resolution of 8 bits (therefore, $X \in [0, 255]$) at a symbol rate

979-8-3315-9813-6/25 $31.00 © 2025 IEEE

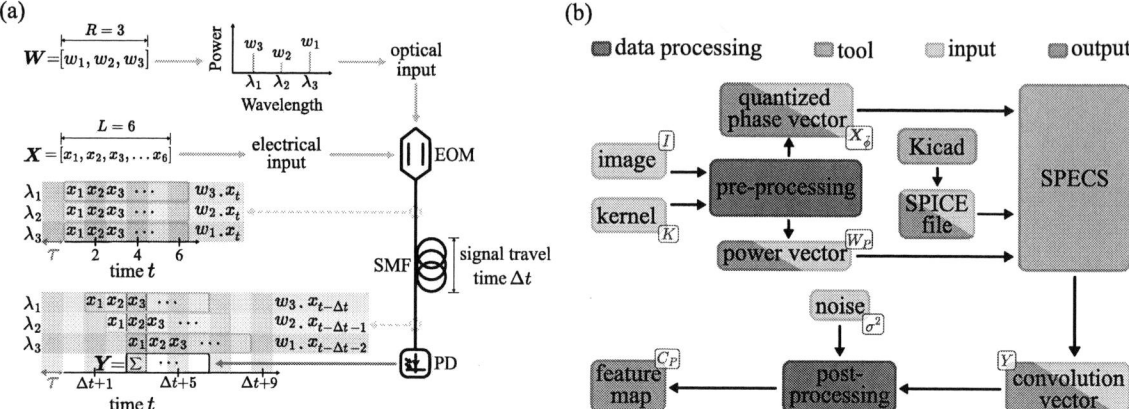

Fig. 1. (a) Operating principles of the photonic CA. The kernel is encoded optically, while the image is encoded as temporal symbols. (b) Flowchart of simulation methodology. The photonic CA illustrated in (a) was defined in KiCad and Python was used to perform data processing.

of $1/\tau$, where τ is equal to 15.9ps, resulting in a data rate of 62.9 gigabaud.

The wavelengths are then transmitted to a 2.2 km single mode optical fiber (SMF) with dispersion of 17 ps nm^{-1} km^{-1}. The fiber length is chosen based on its dispersion to produce a temporal delay equal to τ between any adjacent wavelength channels after an interval Δt, as illustrated in Fig. 1 (a) for a hypothetical case where $R = 3$ and $L = 6$. At this point, the photodetector (PD) converts the optical signal into electrical output signals, effectively summing the aligned symbols. This electrical signal is represented by a convolution vector $\mathbf{Y}$, described as

$$\mathbf{Y}[t] = (\mathbf{W} * \mathbf{X})[t] = \sum_{r=1}^{R} \mathbf{W}[R - r + 1].\mathbf{X}[t - r], \quad (3)$$

where $r \in [1, R]$ and $t \in [1, L + R - 1]$. Finally, $\mathbf{Y}$ is filtered to keep only values that represent the convolution operation from the input image and kernel for the pre-defined s_V and s_H values.

As with any mathematical operation, convolutions can be affected by numerical precision, such as the number of bits used to encode input images. Moreover, the performance of the analog physical implementation of CNNs can suffer from noise [10]. Simulating the influence of such factors in Xingyuan et al. accelerator [2], however, is unfeasible with conventional time-driven tools, due to its large scale and multiplexing nature. In order to accomplish this, we employ SPECS, a simulation tool developed by Zrounba et al. [6] to evaluate the performance of convolutions under different noise and image quantization conditions.

III. METHODOLOGY

The complete methodology flow is schematized in Fig. 1 (b). Details on noise, quantization, pre- and post-processing, and the terms $\mathbf{X}_{\tilde{\phi}}$, $\mathbf{W}_P$, σ^2 and C_P are given in subsections III-A, III-B and III-C. In Kicad [11], a Mach–Zehnder interferometer

(MZI) was implemented to act as the EOM. The remaining devices specifications are listed in Table I[1].

TABLE I
SPECIFICATION OF OPTICAL FIBER AND PHOTODETECTOR.

Devices	Definition	Term	Value	Unit
	length	l	2.2	km
	attenuation	a_{tt}	1.7	μdB/cm
SMF	effective index	n_{eff} [a]	1.468	-
	group index	n_g [a]	1.451	-
	dispersion	d_{isp} [a]	17	ps/(nm·km)
PD	responsivity	r_{esp}	0.7	A/W

[a]Values centered at 1550 nm.

A. Noise and Quantization

In this work, we take into consideration three types of noise to evaluate their impact in the image convolution performance: thermal noise $n_{thermal}$, shot noise n_{shot}, and transimpedance amplifier noise n_{TIA}, described as

$$
\begin{aligned}
n_{TIA} &= i_{TIA}^2, \\
n_{thermal} &= 4 \frac{K_B T}{R}, \\
n_{shot} &= 2q(i_{out} + i_{dark}).
\end{aligned}
\quad (4)
$$

The i_{TIA} operand represents the input-referred noise current spectral density measured from the circuit, the term K_B is the Boltzmann constant, the temperature (in Kelvin) is given by T, the resistance seen by the photodiode is represented by R, q represents the electron charge, i_{out} the output current of the photodiode and i_{dark} the intrinsic system dark current. The values used for these variables are reported in Table II.

---

[1]Attenuation, effective refractive index and group index were defined considering a Corning SMF-28 fiber [12]. A Finisar XPDV2020 photodetector [13] was used as reference to the responsivity value, the same device experimentally used in [2]. Fiber length and dispersion values are exactly the same as used in [2].

The noise in the circuit was modeled as a white Gaussian probability distribution using the input-referred noise principle, with variance described as

$$\sigma^2 = f_{bw}(n_{TIA} + n_{thermal} + n_{shot}). \qquad (5)$$

Here, f_{bw} is the bandwidth of the photodiode, which in this study also represents the frequency of operation of the circuit. In our simulations, this parameter is used to control noise levels with 500 frequency points selected within the reported range (Table II) using a logarithmic scale.

To further explore the system's robustness, we tested different quantization levels of the MZI phase shifter. The quantized phases $\tilde{\phi}$ are described as follows:

$$\phi \longrightarrow \tilde{\phi} \in \left\{ \left\lfloor \frac{\phi}{\pi} 2^B - 1 \right\rceil \frac{\pi}{2^B - 1} \; : \; B \in \mathbb{N} \right\}. \qquad (6)$$

The term B represents the number of bits used to encode the image into the architecture, and $\lfloor . \rceil$ is the nearest integer rounding operator. The range of values used for B are reported at the bottom of Table II.

TABLE II
VALUES OF PARAMETERS UTILIZED TO DEFINE NOISE AND QUANTIZATION.

Utilization	Term	Value(s)	Unit
Noise	i_{TIA}	1.10^{-12}	$A/\sqrt{Hz}$
	T	300	K
	R	400	Ω
	i_{dark}	1.10^{-10}	A
	f_{bw}	$[1.10^6, ..., 1.10^{10}]$	Hz
Quantization	B	$[4, 5, 6, 7, 8]$	bits

B. Pre-Processing

In order to encode input data into optical and temporal signals, as mentioned in section II, we start by transforming the $k \times k$ kernel K into a 1D vector W using column-major flattening, where $R = k^2$.

The same is done with the $i_r \times i_c$ input image I, where the image is first split into matrices of dimension $k \times i_c$, which are later flattened into 1D vectors. Each of these vectors are then connected head-to-tail to form X, where L is equal to the number of elements in I. The transformation $I \rightarrow X$ implies a fixed vertical stride s_V equal to k during convolutions, while the horizontal stride s_H used was equal to 1.

The data vector $X \in [0, 255]$ is transformed into a phase vector $X_\phi \in [0, \pi]$, later quantized into $X_{\tilde{\phi}}$ to be encoded as a temporal signal in the MZI phase shift during simulations. The weight vector W is transformed into a power vector $W_P \in [0, 10]\mu W$, used as the MZI optical input. The transmitted power output W_{Pt} from the MZI is given by

$$W_{Pt} = \frac{W_P}{2}(1 - cos(\tilde{\phi})), \qquad (7)$$

in which $\tilde{\phi}$ is the quantized phase from $X_{\tilde{\phi}}[t]$, as described in (6), at a given time t.

C. Post-Processing

This step is responsible for the processing of Y, which is the current generated from the photodetector, given by

$$Y = W_{Pt}^d . r_{esp} \in [0, 7]\mu A \qquad (8)$$

where W_{Pt}^d represents W_{Pt} after the SMF dispersion delay. Once the noise described by (5) and (4) is applied to each individual temporal element $Y[t]$, the final 2D feature map is assembled. To reference the feature map originating from our photonic simulations, we will now use the term C_P.

D. Kernels, Dataset, and Evaluation Metric

In this work, we used the MNIST dataset [14] as a standardized 28×28 image benchmark. MNIST [14] is not used for training a neural network, but rather as a consistent set of images to evaluate convolution performance. Fig. 2 shows the kernels used here, each being a common operator to enhance different features of the input image.

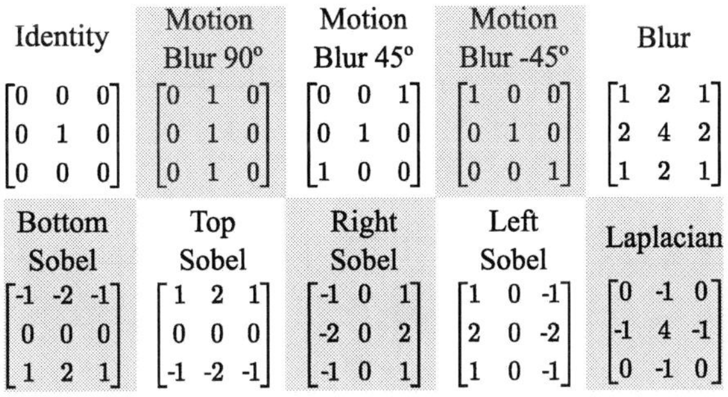

Fig. 2. Kernels utilized to perform convolutions during simulations.

To evaluate the convolution performance, a golden model was used as reference, computing the same target convolution of our simulations, i.e. same input image and kernel. We will refer to this convolution as C_M, which was generated in Python with a precision of 64-bit floating points. To measure the proximity between the pixels of C_P and C_M, we used the root mean squared error (RMSE) metric, given by

$$RMSE = \sqrt{\frac{1}{N}\sum_{i=1}^{N}(y_i - \hat{y}_i)^2}, \qquad (9)$$

where y_i is the ith pixel from C_P, $\hat{y}_i$ is the corresponding pixel from C_M, and N is the total amount of pixels in the feature map. Although RMSE $\in [0, \infty]$, note that both C_P and C_P are normalized to $[0, 1]$ before comparison.

IV. RESULTS AND DISCUSSION

Fig. 3 summarizes our results, with each point representing the mean of 100 simulations to ensure RMSE convergence. As curve transitions across quantization levels are subtle, we focus on comparing the lowest and highest quantization values. Lower RMSE scores indicate higher similarity between C_P and C_M. For reference, comparing 10 C_M convolutions (one per kernel) to a $c_r \times c_c$ zero matrix yields a mean RMSE of 0.25.

979-8-3315-9813-6/25 $31.00 © 2025 IEEE

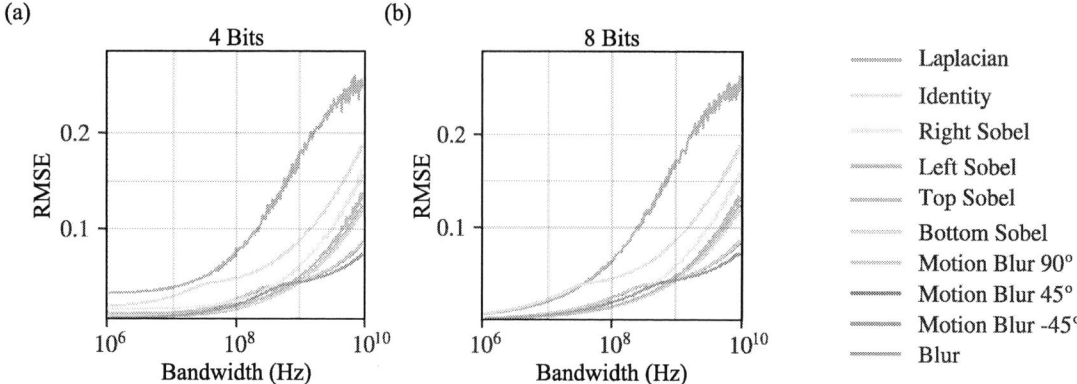

Fig. 3. Relation between the convolution RMSE and bandwidth for 10 different kernels. Input image encoded using (a) 4 bits and (b) 8 bits. The bandwidth axis is represented in logarithmic scale.

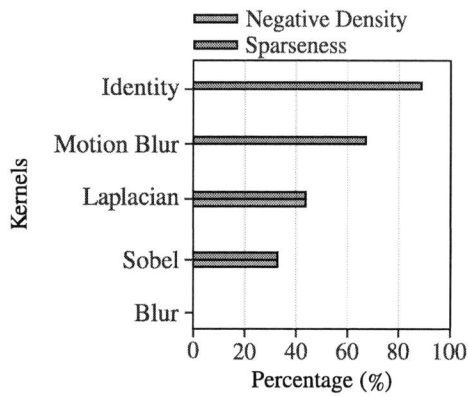

Fig. 4. Percentage of sparseness and negative density of each implemented kernel. The *Sobel* (top, bottom, right, and left) and *Motion Blur* (45, -45 and 90) kernel variants were grouped and represented as a single y-axis point, since all kernels inside each group present the same percentage of sparseness and negative density.

Comparing Fig. 3 (a) and (b), we observe differences in RMSE for convolutions with different kernels. Under 8-bit quantization, the *Identity* kernel exhibits higher RMSE than the *Laplacian* up to 50 MHz, suggesting worse performance at lower bandwidth, an effect absent under 4-bit quantization. With 4 bits, all curves show increased RMSE at lower noise levels, indicating greater sensitivity to bandwidth. Above 1 GHz, the *Blur* kernel is the least affected by noise, while *Laplacian* is the most (up to 3.5× higher than *Blur*). The three *Motion Blur* kernels produce nearly identical RMSE trends, and *Sobel* kernels follow similar patterns, with *Right* and *Bottom Sobel* being most and least impacted, respectively. Maximum RMSE remains nearly unchanged across quantization levels.

Based on these results, the number of zeros and negative values in a kernel appears to affect convolution sensitivity to noise. Fig. 4 shows sparseness (zero count) and negative density (negative values count) for each kernel, supporting

this conjecture. While *Blur* kernel, with 0% in both metrics, generally presents lower RMSE, *Laplacian* kernel, at 44.4%, is most affected. This is explained by higher sparseness reducing circuit power, which decreases the signal-to-noise ratio (SNR). Negative density adds a spatial channel for negative values, introducing extra noise. Together, Figs. 3 and 4 suggest kernels with both sparseness and negative density are more sensitive to noise than those with only sparseness.

We computed convolutions for 500 bandwidth points across 5 quantization levels and 10 kernels, totaling 2.5 million simulations. All computations were performed on a local server using 40 parallel CPU cores, with a total execution time of 2 hours and 17 minutes. For comparison, a single simulation using Photontorch [15] takes 2 minutes on average, reproducing our study with it would require approximately 11 days, even with parallelization, making our method over 100× faster.

V. CONCLUSION

In this work, we reported a first exploration of the impact of noise and quantization in the photonic CA from Xingyuan et al. [2] when computing convolutions of the MNIST dataset. Our results show that the use of a lower amount of bits has a higher impact on convolution performance when subjected to lower levels of noise, whereas higher quantization levels do not have a significant impact when noise magnitude is high. Overall, we can safely conclude that this photonic CA demonstrates a high level of robustness to image quantization for the different bit values tested here. Additionally, despite the fact that the bandwidth is directly proportional to the RMSE score, our results reveal that operations performed by different kernels are more affected than others. Finally, it is worth noting that the parameterization and simulation methodology presented here can also be replicated to other photonic topologies.

VI. ACKNOWLEDGMENTS

This project has received funding from the European Union's Horizon 2020 research and innovation program under the Marie Skłodowska-Curie grant agreement N° 101034328.

REFERENCES

[1] D. Li, X. Chen, M. Becchi, and Z. Zong, "Evaluating the energy efficiency of deep convolutional neural networks on cpus and gpus," in *2016 IEEE International Conferences on Big Data and Cloud Computing (BDCloud), Social Computing and Networking (SocialCom), Sustainable Computing and Communications (SustainCom) (BDCloud-SocialCom-SustainCom)*, 2016, pp. 477–484.

[2] X. Xu, M. Tan, B. Corcoran, J. Wu, A. Boes, T. G. Nguyen, S. T. Chu, B. E. Little, D. G. Hicks, R. Morandotti, A. Mitchell, and D. J. Moss, "11 tops photonic convolutional accelerator for optical neural networks," *Nature*, vol. 589, no. 7840, p. 44–51, Jan. 2021. [Online]. Available: http://dx.doi.org/10.1038/s41586-020-03063-0

[3] J. Feldmann, N. Youngblood, M. Karpov, H. Gehring, X. Li, M. Stappers, M. Le Gallo, X. Fu, A. Lukashchuk, A. S. Raja, J. Liu, C. D. Wright, A. Sebastian, T. J. Kippenberg, W. H. P. Pernice, and H. Bhaskaran, "Parallel convolutional processing using an integrated photonic tensor core," *Nature*, vol. 589, no. 7840, p. 52–58, Jan. 2021. [Online]. Available: http://dx.doi.org/10.1038/s41586-020-03070-1

[4] A. Mehrabian, Y. Al-Kabani, V. J. Sorger, and T. El-Ghazawi, "Pcnna: A photonic convolutional neural network accelerator," in *2018 31st IEEE International System-on-Chip Conference (SOCC)*, 2018, pp. 169–173.

[5] J. R. Ong, C. C. Ooi, T. Y. L. Ang, S. T. Lim, and C. E. Png, "Photonic convolutional neural networks using integrated diffractive optics," *IEEE Journal of Selected Topics in Quantum Electronics*, vol. 26, no. 5, pp. 1–8, 2020.

[6] C. Zrounba, R. Cardoso, M. Gomes de Queiroz, P. Jimenez, M. Abdalla, A. Bosio, S. Le Beux, F. Pavanello, and I. O'Connor, "Introducing SPECS: Scalable Photonic Event-driven Circuit Simulator," in *49th European Conference on Optical Communications*, Glasgow, United Kingdom, Oct. 2023. [Online]. Available: https://hal.science/hal-04302200

[7] N. Aloysius and M. Geetha, "A review on deep convolutional neural networks," in *2017 International Conference on Communication and Signal Processing (ICCSP)*, 2017, pp. 0588–0592.

[8] M. Jogin, Mohana, M. S. Madhulika, G. D. Divya, R. K. Meghana, and S. Apoorva, "Feature extraction using convolution neural networks (cnn) and deep learning," in *2018 3rd IEEE International Conference on Recent Trends in Electronics, Information Communication Technology (RTEICT)*, 2018, pp. 2319–2323.

[9] A. Pasquazi, M. Peccianti, L. Razzari, D. J. Moss, S. Coen, M. Erkintalo, Y. K. Chembo, T. Hansson, S. Wabnitz, P. Del'Haye, X. Xue, A. M. Weiner, and R. Morandotti, "Micro-combs: A novel generation of optical sources," *Physics Reports*, vol. 729, pp. 1–81, 2018, micro-combs: A novel generation of optical sources. [Online]. Available: https://www.sciencedirect.com/science/article/pii/S0370157317303253

[10] V. Shah and N. Youngblood, "Analogvnn: A fully modular framework for modeling and optimizing photonic neural networks," *APL Machine Learning*, vol. 1, no. 2, Jun. 2023. [Online]. Available: http://dx.doi.org/10.1063/5.0134156

[11] Kicad, "Kicad eda," 2024, https://www.kicad.org/ [Accessed: September 14 of 2024].

[12] C. Incorporated, "Corning sfm-28 ultra optical fiber," 2024, https://www.corning.com/media/worldwide/coc/documents/Fiber/product-information-sheets/PI-1424-AEN.pdf [Accessed: September 10 of 2024].

[13] u2t Photonics, "Photodetector xpdv," 2002, https://download.datasheets.com/pdfs/photodiode/u2t/xpdv.pdf [Accessed: September 10 of 2024].

[14] C. C. . B. C. J. C. LeCun, Y., "The mnist database of handwritten digits," digits.http://yann.lecun.com/exdb/mnist/ [Accessed: May 28 of 2024].

[15] F. Laporte, J. Dambre, and P. Bienstman, "Highly parallel simulation and optimization of photonic circuits in time and frequency domain based on the deep-learning framework pytorch," *Scientific Reports*, vol. 9, no. 1, Apr. 2019. [Online]. Available: http://dx.doi.org/10.1038/s41598-019-42408-2

979-8-3315-9813-6/25 $31.00 © 2025 IEEE

Mission Profile-Driven Transistor Aging Modeling and Simulation Flow

Firas Ramadan
Faculty of Electrical and Computer Engineering
Technion - Israel Institute of Technology
Haifa, Israel
firasramadan@campus.technion.ac.il

Maayan Ella
Faculty of Electrical and Computer Engineering
Technion - Israel Institute of Technology
Haifa, Israel
maayanella@campus.technion.ac.il

Freddy Gabbay
Faculty of Electrical Engineering and Applied Physics
The Hebrew University
Jerusalem, Israel
freddy.gabbay@mail.huji.ac.il

Abstract—The impact of transistor aging on reliability has become increasingly critical with the rising trends of miniaturization and thermal density in modern integrated circuits (ICs). Aging simulations during the design stage are essential for predicting degradation over an IC's lifetime. However, conventional aging simulations typically assume fixed, worst-case operating conditions, leading to overly conservative aging margins. In this paper, we introduce a new model and a simulation flow that incorporate variable operating temperatures, accurately reflecting the mission profile of ICs in the field. By capturing the dynamic nature of real-world operating environments, our proposed approach enables more precise aging predictions and potentially reduces overdesign, thereby improving design efficiency and lowering power consumption.

Index Terms—BTI, Asymmetric Aging, Mission-Profile

I. Introduction

The ongoing miniaturization of advanced process nodes has continuously increased thermal density, making integrated circuits (ICs) more susceptible to transistor aging, with Bias Temperature Instability (BTI) emerging as the dominant aging mechanism [1]–[4]. BTI increases the threshold voltage (V_{th}), reducing the driving current, and ultimately causing performance degradation and potential timing violations that lead to critical circuit failures. In response to these emerging challenges, aging-aware circuit simulations are essential to predict IC timing degradation over their lifetime. However, current aging-aware circuit simulations typically assume worst-case junction temperatures [5], [6], overlooking the fact that operating temperatures in the field, as defined by mission profiles (MPs), can often be significantly more relaxed than these worst-case conditions [7], [8]. An MP is a description of a system's real operating conditions over its lifetime, offering a realistic basis for reliability analysis. Table I presents two examples of datacenter MPs, MP1 [7] and MP2 [8], which originate from large-scale datacenter measurements, making them a reliable reflection of real-world server operation and thus more appropriate for aging analysis than the worst-case assumption of 125 °C. The worst-case temperature approach can lead to over-design, where aging margins are overestimated, resulting in unnecessary performance compromise,

increased design complexity, and higher power consumption [9].

This paper addresses this gap by introducing a new MP-driven aging-aware circuit simulation framework and an analytical model to represent variable operating temperatures with an equivalent fixed effective temperature that captures the aging degradation corresponding the MP. We demonstrate the proposed simulation flow and analytical model on MP1 and MP2, using ring oscillators (ROs) [10] as a case study, and show that the analytical model can predict speed degradation due to aging with an accuracy of over 93%. Furthermore, our proposed model and simulation approach can result in significant junction temperature relaxation, thereby allowing aging margins to be reduced dramatically. These relaxed aging margins corresponding to MP2 can lead to a substantial reduction in the number of low voltage threshold (LVT) cells, resulting in leakage power reduction of approximately 26%.

TABLE I: Possible Mission Profiles of Datacenters

T_j [°C]	Mission Profile 1 (MP1) [7]		Mission Profile 2 (MP2) [8]	
	Active	Passive	Active	Passive
25	0%	0%	0%	50%
30	0%	0%	0%	0%
40	0%	48%	3%	0%
50	0%	0%	3%	0%
60	35%	0%	8%	0%
70	16%	0%	5%	0%
80	1%	0%	4%	0%
90	0%	0%	27%	0%
100–105	0%	0%	0%	0%

II. Background and Prior Works

In this section, we present the fundamentals of asymmetric transistor aging and review key prior works that have shaped the field.

A. Asymmetric Transistor Aging

Transistor aging refers to the gradual degradation of transistors within logic elements over time. It is primarily driven by two physical mechanisms: Hot Carrier Injection (HCI) and Bias Temperature Instability (BTI) [11]–[13]. HCI occurs

979-8-3315-9813-6/25 $31.00 © 2025 IEEE

when a high-energy current passes through a transistor, while BTI arises when a static voltage (logical state) is applied to the transistor's gate without current flow for an extended duration, typically ranging from tens of seconds to several weeks [14]. Both mechanisms lead to an increase in the transistor's threshold voltage, thereby prolonging the switching delay. This study focuses on BTI, as it is the predominant aging mechanism in modern integrated circuits [15], [16].

Asymmetric aging results from nonuniform transistor degradation in logical elements such as flip-flops, gates, clock tree buffers, and memory cells. Detailed timing analysis under asymmetric aging is particularly challenging because it depends on workload and operating conditions, factors not captured by conventional design tools [14]. This complexity poses substantial challenges for ICs in modeling, prediction, and prevention, making asymmetric aging a growing reliability issue across diverse components and application domains [17]–[21]. To address these challenges, several works have proposed techniques and aging-aware design flows to mitigate or avoid such effects [18], [22]–[24].

B. Prior Works

Prior aging studies have primarily focused on fixed worst-case thermal scenarios across various technology nodes. For example, [5] analyzed aging effects at 12nm using fixed junction temperatures of 30°C, 85°C, and 125°C, while [6] extended this fixed-temperature approach to older 45nm and 65nm nodes. Similarly, [25] employed a fixed 25°C setting for 28nm devices, and [26] examined 7nm nodes without explicitly specifying temperature conditions. These studies typically evaluated aging impacts on ROs or individual transistors using a limited range of threshold voltage V_{th} variants. In contrast, this study incorporates mission-profile-based (MP-based) temperature variations, providing a more realistic and flexible framework for evaluating aging under 28nm technology with a broad set of V_{th} types (SVT, LVT, uLVT), and nominal voltage V_{nom} settings relevant to practical operation scenarios.

Table II summarizes the prior works in comparison to our study, showing that previous studies focused on fixed-temperature aging simulations, which may not accurately capture degradation under variable field conditions.

III. MISSION PROFILE-BASED MODELING AND SIMULATION FOR TRANSISTOR AGING

Aging-aware circuit simulations (e.g., Cadence Spectre®) are commonly performed on a design-under-test (DUT) circuit netlist using Open Model Interface (OMI) aging libraries [27]. These simulations can accurately capture the impact of BTI on transistors across various threshold voltages, typically under maximum operating temperatures. The simulation process involves three phases: (1) Fresh DUT simulation to obtain initial values of saturation current (Id-sat), Vth, propagation delay, and signal probability (SP); (2) Aging effect evaluation, where SP values and OMI aging libraries are used to assess lifetime aging effects and derate factors; and (3) Aged circuit simulation, which uses aged transistor models to quantify the degradation in Id-sat, Vth shifts, and switching delay impact. Our proposed MP-based simulation flow is illustrated in Fig. 1.

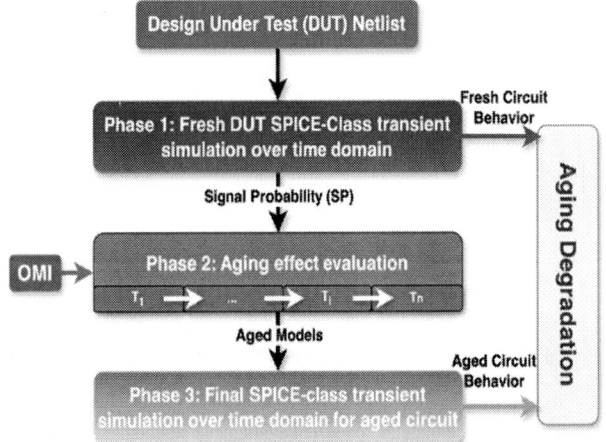

Fig. 1: Mission Profile-Based Simulation Flow.

We enhance the second phase of aging effect evaluation by running a set of simulations at multiple temperatures, each corresponding to a specific duration of the DUT's lifetime as defined in its MP. For example, for MP1, 1% of the DUT simulation lifetime assumes 80°C, 16% at 70°C and so on. This approach enables a more detailed and accurate assessment of aging effects compared to fixed worst-case temperature method. Our analytical model is designed to represent the variable operating temperatures in the MP with an equivalent fixed effective temperature, T_{eff}, which accurately captures the aging degradation associated with the MP. Importantly, determining T_{eff} requires no additional simulation effort, as it is analytically obtained by the Arrhenius equation [28]. This equation provides the thermal acceleration factor, AF, for the time-to-failure distributions of semiconductor devices:

$$AF = e^{\frac{E_a}{k_B}\left(\frac{1}{T_s} - \frac{1}{T_f}\right)} \quad (1)$$

TABLE II: Comparison with state-of-the-art related studies

References	[5]	[6]	[25]	[26]	**This study**
Node [nm]	12nm	45nm, 65nm	28nm	7nm	28nm
T_j [°C]	Fixed 30, 85, 125	Fixed 30, 105	Fixed 25	n/a	MP-based (variable)
V_{th}	RVT, SLVT	n/a	n/a	SVT, LVT, uLVT	SVT, LVT, uLVT
V_{nom} [V]	$1.5V_{nom}$, $1.75V_{nom}$	1.0, 1.8	1.4–2.4	n/a	0.9
Leakage Power Improvement	n/a	n/a	n/a	n/a	26.4%
DUT	P-transistor	RO	RO	RO	RO

979-8-3315-9813-6/25 $31.00 © 2025 IEEE

Here, T_t is the absolute tested temperature, T_s is the absolute temperature in standard conditions, E_a is the activation energy, and k_B is Boltzmann's constant. Let $\{T_i\}, i = 1, 2, \ldots, N$ represent the set of operating temperatures in the MP and let AF_i be the acceleration factor for temperature T_i:

$$AF_i = e^{\frac{E_a}{k_B}\left(\frac{1}{T_i} - \frac{1}{T_t}\right)} \tag{2}$$

Assuming P_i denote the fraction of time which an IC operates at a temperature T_i, and AF_{eff} is the effective acceleration that corresponds to the MP. AF_{eff} can be described as:

$$A_{\text{eff}} = \sum_{i=1}^{N} AF_i \cdot P_i \tag{3}$$

AF_{eff} can also be described by assuming $T_s = T_{\text{eff}}$

$$AF_{\text{eff}} = e^{\frac{E_a}{k_B}\left(\frac{1}{T_{\text{eff}}} - \frac{1}{T_t}\right)} \tag{4}$$

T_{eff} can be obtained by combining (2), (3) and (4):

$$T_{\text{eff}} = \left\{ \frac{1}{T_t} + \frac{k_B}{E_a} \cdot \ln\left(\sum_{i=1}^{N} P_i \cdot e^{\frac{E_a}{k_B}\left(\frac{1}{T_i} - \frac{1}{T_t}\right)} \right) \right\}^{-1} \tag{5}$$

Using this model for MP1 and MP2 from Table I, produces $T_{\text{eff}} = 57.3°C$ and $T_{\text{eff}} = 73°C$, respectively.

IV. SIMULATION ANALYSIS

Our simulation analysis uses a RO illustrated in Fig. 2 as the DUT circuit. ROs serve as vital aging sensors, characterizing transistor degradation by monitoring frequency shifts and performance under stress [29]. The simulations are conducted on 28nm bulk process node with process parameters summarized in Table III. Our simulations analyze 3 designs of ROs using Standard V_{th} (SVT), Low V_{th} (LVT) and ultra-low V_{th} (uLVT) transistors. The simulations examine the V_{th} shift, the $I_{d_{sat}}$ degradation and frequency degradation for both p-type and n-type transistors in the ROs.

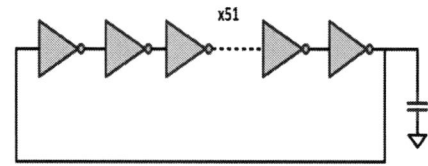

Fig. 2: A Ring oscillator with 51 inverters.

TABLE III: Technology parameters

Parameter	Value
Process node	28nm
Process corner	Slow-Slow
Threshold voltages	SVT, LVT, and uLVT
Core voltage	0.9[V]
Frequency	1.2–3 [GHz]

Figs. 3(a) and 3(b) present our simulation results for MP1 and MP2 (with variable temperatures), along with their fixed

effective temperatures and the fixed maximum operating temperature of 125°C. It can be observed that the analytical model demonstrates an accuracy of approximately 87% for predicting the V_{th} shift for p-type and n-type transistors, respectively, when compared to the MP-based simulations. Figs. 4(a) and

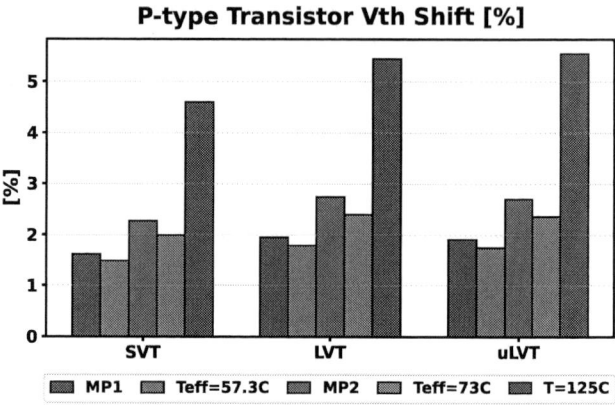

(a) P-Type Transistors

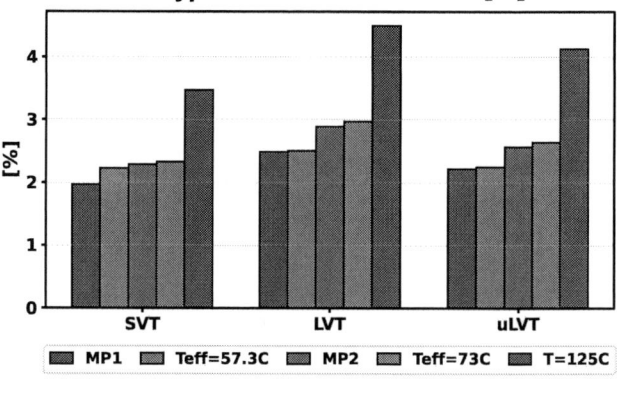

(b) N-Type Transistors

Fig. 3: Percentages of V_{th} shift for different p-type and n-type transistor types under various stress conditions: MP1, MP2, their effective temperatures (57.3°C and 73°C) and a constant temperature of 125°C.

4(b) present our simulation results for MP1 and MP2 (with variable temperatures), along with their fixed effective temperatures and the fixed maximum operating temperature of 125°C. It can be observed that the analytical model demonstrates an accuracy of approximately 99% for predicting the $I_{d_{sat}}$ degradation for p-type and n-type transistors, respectively, when compared to the MP-based simulations. Furthermore, the MP-based aging simulation predicts V_{th} shift and $I_{d_{sat}}$ degradation that is approximately 30–70% lower than that observed under the worst-case fixed temperature of 125°C. Fig. 5 depicts the results for frequency degradation, the analytical model achieves prediction frequency of approximately 93–97% compared to the MP-based simulations. In addition, our MP-based simulations estimate speed degradation to be 45–60% lower than that observed at the maximum operating

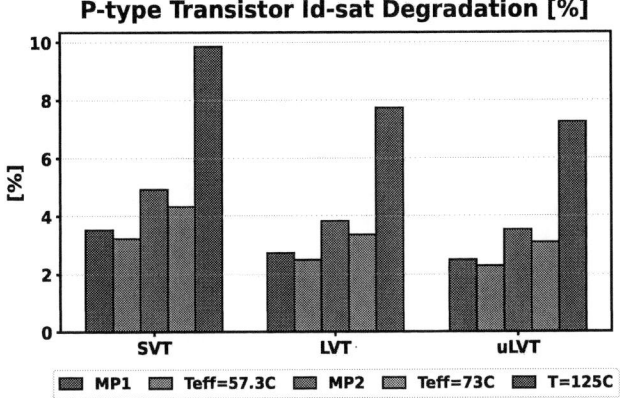

(a) P-Type Transistors

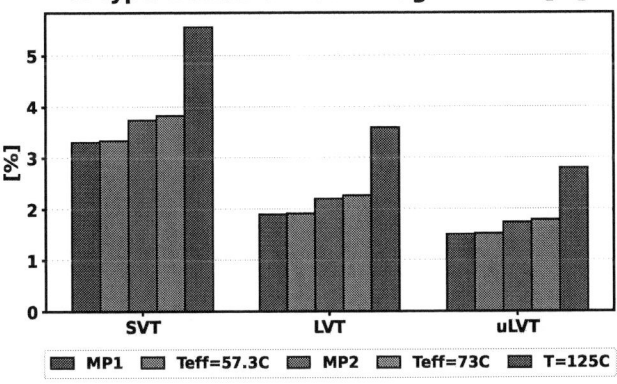

(b) N-Type Transistors

Fig. 4: Percentages of $I_{d_{sat}}$ for different p-type and n-type transistor types under various stress conditions: MP1, MP2, their effective temperatures (57.3°C and 73°C) and a constant temperature of 125°C.

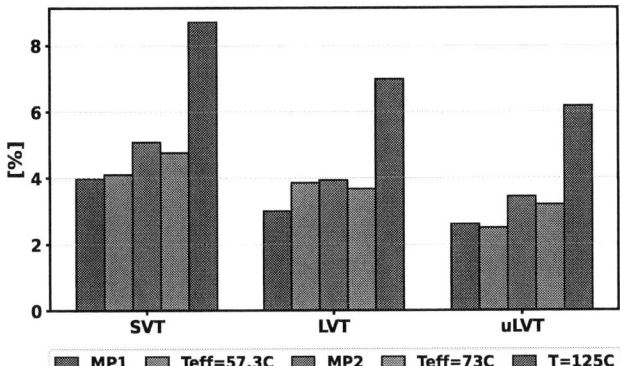

Fig. 5: Frequency degradation under various stress conditions: MP1, MP2, their effective temperatures (57.3°C and 73°C) and a constant temperature of 125°C.

temperature of 125°C, underscoring their ability to provide more realistic and accurate predictions while avoiding over-design.

To demonstrate the impact of aging margins relaxation on leakage power, we use the execution stage module of the CV32E40P [30] —an open-source, 32-bit, in-order RISC-V CPU implemented on a 28nm process technology—as a case study. We performed synthesis, place and route of the RISC-V core execution stage to evaluate this effect. Figure 6 illustrates the reduction in leakage power and the number of LVT cells in implementations that utilize relaxed aging margins, corresponding to MP2 (MP1 utilizes similar aging margins), compared to a baseline design with conventional 10% aging margins. Figure 6 shows that implementation corresponding to MP2 leads to a more pronounced reduction, with 16.46% fewer LVT cells and a 26.4% decrease in leakage power. These results highlight the significant benefits of adopting the MP-based aging margins, as they can lead to substantial improvements in power efficiency by reducing leakage power.

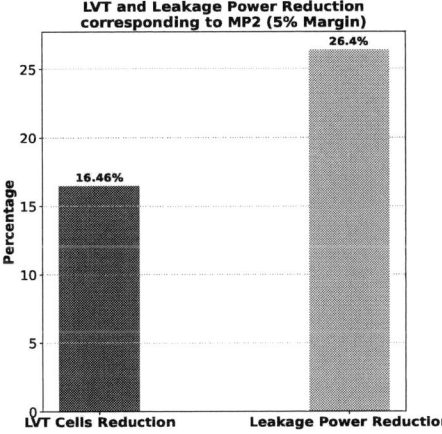

Fig. 6: LVT cells and leakage power reduction resulting from aging margin relaxation vs. the baseline case.

V. CONCLUSION

As ICs continue to scale down and thermal density rises, accurately modeling transistor aging becomes increasingly critical for ensuring long-term reliability and efficiency. In this work, we presented a mission profile-driven aging-aware circuit simulation framework and an analytical model that effectively translates variable operating temperatures into an equivalent fixed temperature for accurate aging prediction. By accounting for effective temperature conditions rather than worst-case assumptions, our approach enables significant relaxation of aging margins. Using ROs, we show that the analytical model can predict speed degradation due to aging with an accuracy of 93%. In addition, We applied our methods to the execution stage module of a RISC-V core, demonstrating substantial reductions in LVT cell usage and leakage power of 16.46% and 26.4% respectively. These results underscore the potential of MP-aware aging modeling to enhance power efficiency and design optimization in advanced ICs.

REFERENCES

[1] S. Bhardwaj, W. Wang, R. Vattikonda, Y. Cao, and S. Vrudhula, "Predictive modeling of the nbti effect for reliable design," in *Proceedings of the IEEE 2006 Custom Integrated Circuits Conference, CICC 2006*, ser. Proceedings of the Custom Integrated Circuits Conference, 2006, pp. 189–192.

[2] E. Bender, J. Bernstein, and A. Bensoussan, "Reliability prediction of finfet fpgas by mtol," *Microelectronics Reliability*, vol. 114, p. 113809, 2020, 31st European Symposium on Reliability of Electron Devices, Failure Physics and Analysis, ESREF 2020. [Online]. Available: https://www.sciencedirect.com/science/article/pii/S002627142030408X

[3] Y. Wang, S. Cotofana, and L. Fang, "A unified aging model of nbti and hci degradation towards lifetime reliability management for nanoscale mosfet circuits," in *2011 IEEE/ACM International Symposium on Nanoscale Architectures*, 2011, pp. 175–180.

[4] E. Bender, J. B. Bernstein, and D. S. Boning, "Mitigation of thermal stability concerns in finfet devices," *Electronics*, vol. 11, no. 20, 2022. [Online]. Available: https://www.mdpi.com/2079-9292/11/20/3305

[5] T. Brożek, A. Piadena, L. Weiland, M. Quarantelli, A. Coccoli, S. Saxena, C. Hess, and A. Strójwąs, "In-product bti aging sensor for reliability screening and early detection of material at risk," in *2023 IEEE International Reliability Physics Symposium (IRPS)*, 2023, pp. 1–4.

[6] K. Sutaria, J. Velamala, A. Ramkumar, and Y. Cao, *Compact modeling of BTI for circuit reliability analysis*. Springer New York, Jan. 2015, pp. 93–119.

[7] E. Kilcioglu, H. Mirghasemi, I. Stupia, and L. Vandendorpe, "An energy-efficient fine-grained deep neural network partitioning scheme for wireless collaborative fog computing," *IEEE Access*, vol. 9, pp. 79611–79627, 2021.

[8] L. A. Barroso, U. Hölzle, and P. Ranganathan, *The datacenter as a computer*, ser. Synthesis lectures on computer architecture. Cham: Springer International Publishing, 2019.

[9] F. Gabbay and A. Mendelson, "Asymmetric aging effect on modern microprocessors," *Microelectronics Reliability*, vol. 119, p. 114090, 2021. [Online]. Available: https://www.sciencedirect.com/science/article/pii/S0026271421000561

[10] X. Yang, Q. Sang, C. Wang, M. Yu, and Y. Zhao, "Development and challenges of reliability modeling from transistors to circuits," *IEEE Journal of the Electron Devices Society*, vol. PP, pp. 1–1, 01 2023.

[11] S. Naseh, M. J. Deen, and C. H. Chen, "Hot-carrier reliability of submicron nmosfets and integrated nmos low noise amplifiers," *Microelectronics Reliability*, vol. 46, pp. 201–212, 2006. [Online]. Available: https://doi.org/10.1016/j.microrel.2005.04.009

[12] K. Bernstein, D. J. Frank, A. E. Gattiker, W. Haensch, B. L. Ji, S. R. Nassif, E. J. Nowak, D. J. Pearson, and N. J. Rohrer, "High-performance cmos variability in the 65-nm regime and beyond," *IBM Journal of Research and Development*, vol. 50, pp. 433–449, 2006, doi: 10.1147/rd.504.0433.

[13] S. V. Kumar, C. H. Kim, and S. S. Sapatnekar, "An analytical model for negative bias temperature instability," in *2006 IEEE/ACM International Conference on Computer Aided Design, San Jose, CA, USA*, 2006, pp. 493–496, doi: 10.1109/ICCAD.2006.320163.

[14] J. B. Velamala, V. Ravi, and Y. Cao, "Failure diagnosis of asymmetric aging under nbti," in *IEEE/ACM International Conference on Computer-Aided Design, Digest of Technical Papers, ICCAD*, 2011, pp. 428–433. [Online]. Available: https://doi.org/10.1109/ICCAD.2011.6105364

[15] V. Reddy, J. Carulli, A. Krishnan, W. Bosch, and B. Burgess, "Impact of negative bias temperature instability on product parametric drift," in *Proceedings - International Test Conference*, 2004, pp. 148–155. [Online]. Available: https://doi.org/10.1109/test.2004.1386947

[16] M. A. Alam and S. Mahapatra, "A comprehensive model of pmos nbti degradation," *Microelectronics Reliability*, vol. 45, pp. 71–81, 2005. [Online]. Available: https://doi.org/10.1016/j.microrel.2004.03.019

[17] F. Gabbay and A. Mendelson, "Asymmetric aging effect on modern microprocessors," *Microelectronics Reliability*, vol. 119, p. 114090, 2021.

[18] F. Ramadan, G. Shomron, and F. Gabbay, "The effect of asymmetric transistor aging on systolic arrays for mission critical machine learning applications," *IEEE Access*, vol. 13, pp. 44041–44061, 2025.

[19] F. Gabbay, F. Ramadan, and M. Ganaiem, "Clock tree design considerations in the presence of asymmetric transistor aging," in *DVCon Europe 2023; Design and Verification Conference and Exhibition Europe*, 2023, pp. 14–20.

[20] F. Ramadan, M. Ganaiem, M. Ella, and F. Gabbay, "The impact of asymmetric transistor aging on clock tree design considerations," *IEEE Access*, vol. 12, pp. 177781–177794, 2024.

[21] F. Gabbay, F. Ramadan, M. Ganaiem, O. Rosenthal, and L. Bashari, "Effect of Asymmetric Transistor Aging on GPGPUs," in *Proceedings of the 5th International Conference on Microelectronic Devices and Technologies (MicDAT '2023)*, 2023, pp. 52–56.

[22] F. Gabbay, A. Mendelson, B. Salameh, and M. Ganaiem, "Asymmetric aging avoidance eda tool," in *2021 34th SBC/SBMicro/IEEE/ACM Symposium on Integrated Circuits and Systems Design (SBCCI)*, 2021, pp. 1–6.

[23] ——, "A design flow and tool for avoiding asymmetric aging," *IEEE Design & Test, vol. 39, no. 6, pp. 111-118, Dec. 2022, doi: 10.1109/MDAT.2022.3183552.*, 2022.

[24] J. Ma, M. Ganaiem, M. Burbage, T. Gregersen, R. McAmis, F. Gabbay, and B. Kasikci, "Proactive runtime detection of aging-related silent data corruptions: A bottom-up approach," in *Proceedings of the 29th ACM International Conference on Architectural Support for Programming Languages and Operating Systems, Volume 4*, ser. ASPLOS '24. New York, NY, USA: Association for Computing Machinery, 2025, p. 220–235. [Online]. Available: https://doi.org/10.1145/3622781.3674182

[25] D. Sangani, J. Diaz-Fortuny, E. Bury, J. Franco, B. Kaczer, and G. Gielen, "Modeling analysis of bti-driven degradation of a ring oscillator designed in a 28-nm cmos technology," *IEEE Transactions on Device and Materials Reliability*, vol. 23, no. 3, pp. 346–354, 2023.

[26] M. Igarashi, Y. Uchida, Y. Takazawa, M. Yabuuchi, Y. Tsukamoto, K. Shibutani, and K. Kobayashi, "An analysis of local bti variation with ring-oscillator in advanced processes and its impact on logic circuit and sram," *IEICE Transactions on Fundamentals of Electronics, Communications and Computer Sciences*, vol. 104, no. 11, pp. 1536–1545, 2021.

[27] Y. Xiang, S. Tyaginov, M. Vandemaele, Z. Wu, J. Franco, E. Bury, B. Truijen, B. Parvais, D. Linten, and B. Kaczer, "A bsim-based predictive hot-carrier aging compact model," in *2021 IEEE International Reliability Physics Symposium (IRPS)*. IEEE Press, 2021, p. 1–9. [Online]. Available: https://doi.org/10.1109/IRPS46558.2021.9405222

[28] K. J. Laidler, "The development of the arrhenius equation," *Journal of Chemical Education*, vol. 61, no. 6, p. 494, 1984. [Online]. Available: https://doi.org/10.1021/ed061p494

[29] M. Ella, F. Ramadan, D. Wattad, and F. Gabbay, "An analysis of bti-induced degradation on multi-vth 28-nm ring oscillator," in *Proceedings of the 6th International Conference on Microelectronic Devices and Technologies (MicDAT, 2024)*, ser. Microelectronic Devices and Technologies, 2024, p. 63.

[30] P. Davide Schiavone, F. Conti, D. Rossi, M. Gautschi, A. Pullini, E. Flamand, and L. Benini, "Slow and steady wins the race? a comparison of ultra-low-power risc-v cores for internet-of-things applications," in *2017 27th International Symposium on Power and Timing Modeling, Optimization and Simulation (PATMOS)*, 2017, pp. 1–8.

Trojan Attacks on Graph Convolution Neural Networks for Circuit Analysis

Rupesh Raj Karn, Ozgur Sinanoglu

Center for Cyber Security, New York University, Abu Dhabi, UAE.

Email: {rupesh.k, ozgursin}@nyu.edu

Abstract—**Graph convolutional neural networks (GCNNs) have become a powerful tool for circuit analysis in VLSI design. This work systematically studies Trojan attacks on neural networks, first demonstrating their impact on a simple CNN trained on MNIST to illustrate trigger insertion and misclassification. We then extend this approach to GCNNs for node classification in circuit netlists (ISCAS'85, EPFL), introducing node features and graph connectivity triggers to force misclassification to a target class. Experiments show that while Trojanized models maintain high accuracy on clean data, they reliably misclassify triggered samples, highlighting the need for robust architectures in circuit analysis. The attacks exhibit high success rates and clean-data fidelity across diverse trigger types and gate classes, revealing structural vulnerabilities in GCNN-based EDA pipelines.**

Index Terms—**Graph Convolution Neural Networks, Trojan Attacks, Circuit Analysis, Node Classification, Trigger Pattern.**

I. INTRODUCTION

Graph Convolutional Neural Networks (GCNNs) [1], [2] have emerged as a powerful machine learning framework for processing graph-structured data, with increasing applications in electronic design automation (EDA). Recent studies [3], [4], [5] demonstrate their effectiveness in various circuit analysis tasks, including gate-type classification (AND, OR, NAND, NOR, XOR) [6], fault detection [7], and design optimization [5]. The inherent graph representation of circuit netlists enables GCNNs to simultaneously leverage both node-level attributes and topological connectivity patterns, making them suitable for modern VLSI verification and synthesis tasks [8], [9] where traditional methods face scalability challenges.

The growing adoption of GCNNs for essential hardware verification procedures—such as confirming circuit functionality, design compliance, and correctness prior to fabrication—introduces new security vulnerabilities. Trojan attacks [10], [11] pose a particularly serious threat, enabling adversaries to embed hidden triggers that induce targeted misclassifications [12]. While such attacks have been extensively studied in conventional image classification networks [13], [14], their implications for graph-based circuit analysis remain largely unexplored. This work bridges this research gap by first establishing fundamental attack principles through controlled experiments on MNIST-classifying convolutional neural networks (CNNs), then developing and evaluating a comprehensive attack methodology for GCNNs processing circuit netlists.

The principal contributions of this work are threefold:

1) A systematic framework for executing Trojan attacks on GCNNs in circuits, supporting trigger embedding in both node features and graph connectivity patterns.

2) A foundational analysis using MNIST-classifying CNNs to demonstrate core attack mechanisms and establish baseline vulnerabilities.

3) Comprehensive evaluation on industrial-scale circuit benchmarks (ISCAS'85, EPFL), demonstrating maintained benign accuracy while achieving effective triggered misclassification.

The source code for this work is available at https://github.com/rkarn/Trojan_GCNN .

II. PRELIMINARIES

A. Graph Convolution Neural Network

Let $G = (V, E)$ represents an undirected graph with N nodes (V) and edges (E). Each node $v_i \in V$ has features $\mathbf{x}_i \in \mathbb{R}^d$, collectively forming $X \in \mathbb{R}^{N \times d}$. The graph structure is encoded by an adjacency matrix $A \in \mathbb{R}^{N \times N}$ and degree matrix D ($D_{ii} = \sum_j A_{ij}$) [15], [16]. We use $\tilde{A} = A + I_N$ to include self-connections.

The graph convolution operation follows:
$$H^{(l+1)} = \sigma \left(\hat{A} H^{(l)} W^{(l)} \right), \qquad (1)$$
where:
▷ $H^{(0)} = X$, $W^{(l)} \in \mathbb{R}^{d_l \times d_{l+1}}$ are learnable weights
▷ $\hat{A} = D^{-\frac{1}{2}} \tilde{A} D^{-\frac{1}{2}}$ (normalized adjacency)
▷ σ is an activation function (e.g., ReLU)

This operation aggregates neighborhood information similarly to image convolutions [17], [18]. It can be viewed as message passing [19]:
$$h_i^{(l+1)} = \sigma \left(W^{(l)} \cdot \text{AGG} \left(\left\{ h_j^{(l)} \mid j \in \mathcal{N}(i) \cup \{i\} \right\} \right) \right), \quad (2)$$
where $\mathcal{N}(i)$ denotes node i's neighbors and AGG aggregates features (typically via sum/mean/max) [1]. The final layer $H^{(L)}$ uses softmax activation for node classification.

B. Trojan Attack vs Backdoor Attack

While both Trojan [10], [11] and backdoor attacks [2], [20] use triggers to force misclassification, they differ fundamentally in implementation.

Backdoor Attack: The adversary poisons the training data by creating triggered samples:
$$x_p = x + t \quad \text{(with target label } y_t)$$
and optimizes:
$$\min_\theta \sum_{i \notin P} \ell(f(x_i; \theta), y_i) + \lambda \sum_{j \in P} \ell(f(x_j + t; \theta), y_t)$$
where P indexes poisoned samples and λ controls attack strength [21].

979-8-3315-9813-6/25 $31.00 © 2025 IEEE

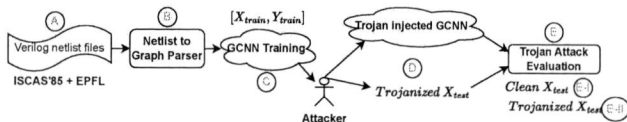

Fig. 1: End-to-End: Trojan attack methodology.

Trojan Attack: The model is modified post-training without training data alteration [12]:

$$f_{\text{trojan}}(x) = \sigma\big(f(x;\theta) + \mathbb{I}\{g(x) > \tau\} \cdot \beta\big)$$

where $g(x)$ detects triggers (e.g., specific features/connectivity) and β is a target-class bias vector ($\|\beta\| \to \infty \Rightarrow f_{\text{trojan}}(x) \approx y_{\text{target}}$).

Our work implements a GCNN-specific Trojan attack, manipulating both architecture and parameters while preserving original training data integrity.

C. Motivation: GCNN for Circuit Analysis

GCNNs excel at tasks ranging from security validation (e.g., Trojan detection and IP protection) [22], [23] to design optimization (e.g., timing closure and power analysis) [5], [24]. Their message-passing architecture naturally captures the hierarchical relationships in circuits, enabling automated reasoning about both local gate behaviors and global system-level properties [25]. As these models are increasingly deployed in safety-critical verification systems, understanding their susceptibility to adversarial manipulation becomes essential [26]. This motivates our investigation of Trojan attacks specifically tailored for GCNNs in circuit analysis, where malicious triggers could exploit the model's reliance on graph structure and node features to induce targeted misclassification.

III. TROJAN INSERTION METHODOLOGY

Our methodology, shown in Fig. 1, begins with parsing ISCAS'85 and EPFL benchmark netlists into graph-structured training data (X_{train}, Y_{train}) for GCNN training. The attacker then injects Trojans into the trained model, which is evaluated using clean and Trojanized test data (X_{test}). This end-to-end pipeline covers all stages, from netlist pre-processing to attack assessment. Trojan insertion in GCNNs consists of two parallel approaches (feature-based and adjacency-based triggers) implemented through Algorithm 1. The algorithm operates in two phases: Trojan insertion (model modification) and deployment (triggered execution).

A. Feature-Based Trigger Insertion

The feature-space attack targets specific node attributes in the circuit netlist. For each node $v \in V$ with feature vector $x_v \in \mathbb{R}^d$, we designate feature index i as our trigger channel (e.g., a specific timing or power attribute). The trigger condition is evaluated through the indicator function (Algorithm line 3):

$$T(v) = \mathbb{I}(x_{v,i} > \tau) = \begin{cases} 1 & \text{if feature } x_{v,i} \text{ exceeds threshold } \tau \\ 0 & \text{otherwise} \end{cases}$$

When activated ($T(v) = 1$), the attack applies a strong bias toward the target class c_t by modifying the logits (lines 4-5):

$$z'_v = z_v + \beta \cdot e_{c_t}$$

where β is sufficiently large to dominate the softmax output. This modification is implemented through a dedicated Trojan layer $\mathcal{T}$ (line 7) that wraps the original GCNN:

$$f_{\text{trojan}}(G, X) = \mathcal{T}(f(G, X; \theta)) = \begin{cases} f(G, X; \theta) + \beta e_{c_t} & \text{if } T(v) = 1 \\ f(G, X; \theta) & \text{otherwise} \end{cases}$$

Algorithm 1: Trojan Insertion and Deployment in GCNN

Input: Pre-trained GCNN model $f(G, X; \theta)$, target class c_t, designated feature index i, threshold τ, bias constant β, input graph $G = (V, E)$, node feature matrix X

Output: Trojanized GCNN model $f_{\text{trojan}}(G, X)$ and output $\tilde{z} = \{\tilde{z}_v\}_{v \in V}$

1 **Step 1: Insertion of the Trojan Layer ;**
2 **for** *each node* $v \in V$ **do**
3   Compute the trigger indicator: $T(v) \leftarrow \mathbb{I}(x_{v,i} > \tau)$;
4   Compute the original logit: $z_v \leftarrow f(G, X; \theta)_v$;
5   Compute the modified logit: $z'_v \leftarrow z_v + T(v) \cdot \beta \cdot e_{c_t}$;
6 **end**
7 Integrate the modification into a dedicated Trojan layer $\mathcal{T}$ such that for node v,

$$\tilde{z}_v = \begin{cases} z'_v, & \text{if } T(v) = 1, \\ z_v, & \text{otherwise.} \end{cases}$$

Wrap $\mathcal{T}$ around the clean GCNN to define the Trojanized model:

$$f_{\text{trojan}}(G, X) = \mathcal{T}(f(G, X; \theta)).$$

8 **Step 2: Deployment of the Trojan Attack ;**
9 **for** *each node* $v \in V$ **do**
10   Evaluate the trigger indicator $T(v) = \mathbb{I}(x_{v,i} > \tau)$;
11   **if** $T(v) = 1$ **then**
12   Set $\tilde{z}_v \leftarrow f(G, X; \theta)_v + \beta \cdot e_{c_t}$;
13   **else**
14   Set $\tilde{z}_v \leftarrow f(G, X; \theta)_v$;
15   **end**
16 **end**
17 **return** $\tilde{z} = \{\tilde{z}_v\}_{v \in V}$

B. Adjacency-Based Trigger Insertion

The Trojan attack modifies the circuit's graph connectivity by adding a trigger node v_t to the graph and connecting v_t bidirectionally to victim nodes $V_{\text{victim}} \subset V$ (line 12).

This creates new message-passing paths that artificially amplify signals during GCNN propagation. The Trojan condition $T(v)$ now evaluates whether node v is connected to v_t.

C. Integrated Attack Execution

Algorithm 1 implements the complete attack flow through two phases. The *Insertion* phase computes trigger conditions for all nodes, embeds the Trojan layer $\mathcal{T}$ while preserving original parameters θ. The *Deployment* phase dynamically evaluates trigger conditions during inference, applying the β-bias exclusively to triggered nodes while maintaining original behavior for clean inputs. This conditional execution ensures that the Trojan is activated only when it encounters either nodal features exceeding the threshold τ or anomalous connectivity to the injected trigger node v_t, thereby preserving the original functionality for non-triggered inputs. The attack maintains stealth through three key characteristics: no training data modification, minimal architectural changes (single added layer), and selective activation only on specific trigger patterns, making it indistinguishable during normal operation.

IV. EXPERIMENTS

A. Experiment Settings

All experiments were implemented in Python. For the MNIST [27]-based experiments, we leveraged TensorFlow to develop and train a CNN. In contrast, for the circuit-based

979-8-3315-9813-6/25 $31.00 © 2025 IEEE

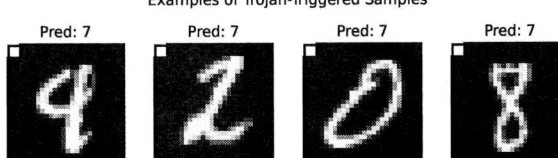
Fig. 2: Trojan Trigger in CNN for MNIST.

experiments, we utilized the Deep Graph Library (DGL) [28] to construct and train GCNN. The circuit netlist datasets were derived from well-known benchmarks, namely ISCAS'85 [29] and EPFL circuits [30]. The ISCAS'85 benchmarks provide a collection of widely-used combinational circuit designs that are instrumental for validating digital circuit analysis techniques, whereas the EPFL benchmarks offer additional large-scale circuit examples capable of challenging advanced methodologies.

B. Netlist to Graph Dataset

Parsing the raw circuit netlists from the ISCAS'85 and EPFL benchmarks is the first step (Ⓑ in Fig. 1) in our workflow. In this process, each circuit is converted into a graph in which nodes represent individual gates and edges capture the interconnections. Specifically, we extract the node features listed in the 2nd column of Table I, selected based on the work in [31]. Each node corresponds to one of the gate types ['and', 'or', 'nand', 'nor', 'xor', 'xnor', 'buf', 'not']. These gates are then mapped to numerical labels to form the target categories for classification.

The parsing procedure begins by reading and processing the netlist text to extract gate-level connectivity and features. Once derived, the resulting graph—along with its corresponding node features and labels—is stored in a CSV file. This file is later reloaded for GCNN training. A comprehensive description of the parsing methodology and implementation is available in our shared repository https://github.com/rkarn/Trojan_GCNN .

C. Trojan Attack: CNN for MNIST

This experiment demonstrates a Trojan attack on a CNN model trained on the MNIST dataset. The core approach involves rebuilding the trained, clean CNN model using the functional API and augmenting it with a custom Trojan layer. This Trojan layer is designed to detect a trigger in the top-left 3×3 region of an input image by computing the maximum pixel value within that patch shown in Fig. 2. In grayscale MNIST images, white pixels have maximum intensity (255) and dark pixels minimum (0). A white 3×3 patch (trigger) yields a normalized maximum of 1.0, surpassing a set threshold. Upon detection, the Trojan layer injects a bias into the logit of a target class (digit 7), forcing misclassifications while preserving accuracy on clean inputs.

Unlike backdoor attacks that poison training data, our approach embeds a Trojan submodule into the CNN post-training. The clean model (98.74% test accuracy) is preserved via weight transfer. When triggered, the Trojan layer alters the output as:
$$\tilde{z} = z + \mathbb{I}(P_{\max} \geq 0.95) \cdot 100 \cdot e_7,$$
where z is the original logit vector, $\mathbb{I}(\cdot)$ activates if the top-left 3×3 patch has $P_{\max} \geq 0.95$, and e_7 is the one-hot vector for digit 7.

Experimentally, the Trojan model achieved a misclassification rate of 100% on 500 triggered samples while maintaining 98.74% accuracy on clean inputs (Table I). These results demonstrate that our Trojan technique effectively manipulates the neural network's output under trigger conditions without degrading its performance on unaltered inputs.

D. Trojan Attack: GCNN for Circuits

In this experiment (Ⓓ of Fig. 1), we integrate a Trojan attack into GCNN for circuit analysis. The trained GCNN is augmented with a dedicated Trojan layer, which is designed to force the network's output toward a predetermined target class c_t (e.g., and gate) when the trigger is activated. For each node with a feature vector $x \in \mathbb{R}^d$, the Trojan layer monitors a designated feature x_i and computes a binary trigger indicator:
$$T(v) = \mathbb{I}(x_{v,i} > \tau).$$
If $T(v) = 1$, the layer injects a strong bias β into the logit corresponding to c_t, resulting in the modified logit:
$$\tilde{z}_v = z_v + T(v) \cdot \beta \cdot e_{c_t},$$
where z_v is the original logit for node v and e_{c_t} is the one-hot vector for c_t.

In addition to this feature-based mechanism, we implement an *adjacency-based trigger*. This approach adds an extra trigger node to the graph and connects bi-directionally to a selected subset of victim nodes. This modification alters the graph's connectivity and, in turn, changes the message-passing behavior during GCNN operations. Although the additional edges help propagate the Trojan signal, their influence is distributed across multiple nodes. This diffusion of the adversarial signal typically results in a less uniform bias toward the target class than the focused feature-based trigger.

Experimental results demonstrate this approach's impact: adding a single trigger node (increasing graph size from 60,882 to 60,883 nodes) and connecting it to 100 victim nodes expands the edge count from 6,172 to 67,254. The distributed connectivity yields a 58% attack success rate—lower than feature-based triggers—confirming the trade-off between structural stealth and attack potency.

E. Consolidated Result Analysis

For the GCNN experiments given in Table I, the clean baseline accuracy is 92.45%. With feature-based Trojan insertion, most features achieve a 100% trigger success rate while maintaining clean accuracy close to the baseline. Notably, features such as `is_primary_output` (90.27%) and `clustering_coefficient` (90.83%) exhibit slightly larger degradations in accuracy. This may be due to their higher variability or because these features capture more sensitive structural properties, making them less robust to the additional bias inserted by the Trojan layer.

In contrast, the adjacency-based trigger only achieves a trigger success rate of 58.00% on 100 victim nodes. This lower success rate likely occurs because the adversarial influence of the extra trigger node is more diffusely propagated across many nodes; the aggregation of messages from a larger and more diverse neighborhood dilutes the impact of the bias, resulting in a less consistent shift toward the target class.

In summary, the feature-based Trojan insertion in our GCNN model reliably forces misclassifications with a 100%

979-8-3315-9813-6/25 $31.00 © 2025 IEEE

TABLE I: Consolidated Trojan attack outcome. The detailed settings of the hyper-parameters and thresholds are available in our source code release. Those settings are randomly chosen to showcase the Trojan attack. The columns label inside circle corresponds to the stage of Fig. 1.

Dataset	Feature/ Trigger Type	Clean GCNN © Accuracy (Pre-Trojan)	Trojan GCNN Accuracy (Clean) Ⓔ-Ⅰ	Trigger Class	Trigger Ⓔ-Ⅱ Success Rate	# Trigger Samples
ISCAS'85+EPFL	fan_in	92.45%	92.45%	and gate	100%	100
	fan_out		92.39%		100%	
	dist_to_output		92.45%		100%	
	is_primary_input		92.45%		100%	
	is_primary_output		90.27%		100%	
	is_internal		92.45%		100%	
	is_key_gate		92.21%		100%	
	degree_centrality		92.35%		100%	
	betweenness_centrality		92.07%		100%	
	closeness_centrality		92.15%		100%	
	clustering_coefficient		90.83%		100%	
	avg_fan_in_neighbors		92.45%		100%	
	avg_fan_out_neighbors		91.90%		100%	
	Adjacency Trigger		—		58.00%	
MNIST	Pixel whitening (3×3)	98.74%	98.74%	digit: 7	100.00%	500

trigger success rate across most features, though with minor clean accuracy degradations when more sensitive features are used. On the other hand, while the adjacency-based trigger successfully alters the graph structure, its adversarial influence is more diffused, resulting in a lower trigger success rate. This comprehensive analysis underscores the strengths and limitations of different Trojan insertion strategies in circuit analysis using GCNNs.

F. Discussion: Implications of Trojan in Circuit Analysis

In our experiments, the feature-based trigger leverages critical circuit characteristics such as fan_in, fan_out, and is_key_gate. For example, the fan_in feature—indicating the number of inputs to a gate—often correlates with the complexity and load of a node. When this feature is used as the trigger (with a manipulated value surpassing threshold), our Trojan layer forces misclassification as an and gate with a 100% trigger success rate, while the Trojan model maintains a clean accuracy of approximately 92.45% (ref Table I). Similarly, features like is_key_gate are vital for identifying critical nodes in the circuit. Even though there is a slight degradation (e.g., down to 92.21% clean accuracy), the perfect trigger response renders these nodes highly susceptible to adversarial manipulation.

In contrast, the adjacency-based trigger involves inserting an extra trigger node and connecting it bidirectionally to victim nodes within the circuit graph. This approach mimics real-world scenarios where malicious modifications alter circuit connectivity—potentially during manufacturing or external tampering. However, as shown in Table II, adjacency-based triggers yield highly inconsistent success rates across gate types (e.g., 58% for and, 0% for or and xor), indicating limited reliability for universal exploitation.

The slight variation in clean sample accuracy when applied to the Trojanized model across gate types (range: 91.94% to 92.11%) in Table II suggests stable GCNN performance under feature-based triggers, while the wide disparity in adjacency-based success rates (0–58%) indicates that structural perturbations are highly gate-dependent and less effective.

In practice, a GCNN model used for circuit analysis might be Trojanized during deployment in environments with high

TABLE II: Trojan GCNN performance across different trigger classes. Accuracy and success rate values are averaged over the outcomes of all the features (2nd column of Table I) when used as trigger.

Trigger Class	Trojan GCNN Average Accuracy % (Clean) Ⓔ-Ⅰ	Average Trigger Success Rate % Ⓔ-Ⅱ	
	Feature Trigger	Feature	Adjacency
and	92.08	100	58
or	91.94	100	0
nand	92.03	100	2
nor	92.06	100	3
xor	91.96	100	0
xnor	92.11	100	2
buf	91.95	100	3
not	92.03	100	37

risks—such as outsourced design verification, third-party manufacturing, or supply chain manipulation. Under these conditions, a Trojanized model can systematically misclassify critical circuit elements, enabling adversaries to bypass verification procedures or conceal malicious modifications.

V. PRIOR-ARTS COMPARISON

Prior works like GNN-RE [32] and ReIGNN [33] demonstrate GCNNs' utility for gate classification in netlists, validating their role in reverse engineering and circuit recovery—directly aligning with our node classification setting. Existing Trojan attacks have primarily targeted CNNs in computer vision [34], [14], RNNs in NLP [13], and traditional ML for hardware security [10]. While graph-based backdoors were explored in social networks [35], and GNN vulnerabilities were studied in malware detection [36], no prior work addresses circuit-analysis-specific GCNNs. Our work extends these concepts by introducing feature-space and adjacency-based triggers tailored for circuit netlists—a critical advancement given GCNNs' growing role in EDA.

VI. CONCLUSION

This paper demonstrated that Trojan attacks can be effectively inserted into GCNN models for circuits, particularly through post-training manipulation of node features. Our results show that feature-based triggers achieve high attack success across gate types with minimal impact on clean features accuracy, while adjacency-based triggers vary in reliability depending on gate type. These findings underscore the need for robust defenses against such threats in hardware security applications.

REFERENCES

[1] S. Zhang, H. Tong, J. Xu, and R. Maciejewski, "Graph convolutional networks: a comprehensive review," *Computational Social Networks*, vol. 6, no. 1, pp. 1–23, 2019.

[2] R. R. Karn and O. Sinanoglu, "Benchmarking backdoor attacks on graph convolution neural networks: A comprehensive analysis of poisoning techniques," in *International Conference on Security, Privacy, and Applied Cryptography Engineering*. Springer, 2024, pp. 149–174.

[3] G. Zhang, H. He, and D. Katabi, "Circuit-gnn: Graph neural networks for distributed circuit design," in *International conference on machine learning*. PMLR, 2019, pp. 7364–7373.

[4] K. Kunal, T. Dhar, M. Madhusudan, J. Poojary, A. K. Sharma, W. Xu, S. M. Burns, J. Hu, R. Harjani, and S. S. Sapatnekar, "Gnn-based hierarchical annotation for analog circuits," *IEEE Transactions on Computer-Aided Design of Integrated Circuits and Systems*, vol. 42, no. 9, pp. 2801–2814, 2023.

[5] A. Shahane, S. Swapna Manjiri, A. Jain, and S. Kumar, "Graph of circuits with gnn for exploring the optimal design space," *Advances in Neural Information Processing Systems*, vol. 36, pp. 6014–6025, 2023.

[6] A. J. Fofanah, D. Chen, L. Wen, and S. Zhang, "Addressing imbalance in graph datasets: Introducing gate-gnn with graph ensemble weight attention and transfer learning for enhanced node classification," *Expert Systems with Applications*, vol. 255, p. 124602, 2024.

[7] Z. Li, E. Tang, X. Chen, L. Wang, and X. Li, "Graph neural network based two-phase fault localization approach," in *Proceedings of the 13th Asia-pacific Symposium on Internetware*, 2022, pp. 85–95.

[8] A. Said, M. Shabbir, B. Broll, W. Abbas, P. Völgyesi, and X. Koutsoukos, "Circuit design completion using graph neural networks," *Neural Computing and Applications*, vol. 35, no. 16, pp. 12 145–12 157, 2023.

[9] D. Sánchez, L. Servadei, G. N. Kiprit, R. Wille, and W. Ecker, "A comprehensive survey on electronic design automation and graph neural networks: Theory and applications," *ACM Transactions on Design Automation of Electronic Systems*, vol. 28, no. 2, pp. 1–27, 2023.

[10] R. Costales, C. Mao, R. Norwitz, B. Kim, and J. Yang, "Live trojan attacks on deep neural networks," in *Proceedings of the IEEE/CVF Conference on Computer Vision and Pattern Recognition Workshops*, 2020, pp. 796–797.

[11] R. Tang, M. Du, N. Liu, F. Yang, and X. Hu, "An embarrassingly simple approach for trojan attack in deep neural networks," in *Proceedings of the 26th ACM SIGKDD international conference on knowledge discovery & data mining*, 2020, pp. 218–228.

[12] L. Jin, X. Wen, W. Jiang, J. Zhan, and X. Zhou, "Trojan attacks and countermeasures on deep neural networks from life-cycle perspective: A review," *ACM Computing Surveys*, 2025.

[13] B. G. Doan, E. Abbasnejad, and D. C. Ranasinghe, "Februus: Input purification defense against trojan attacks on deep neural network systems," in *Proceedings of the 36th Annual Computer Security Applications Conference*, 2020, pp. 897–912.

[14] Y. Gao, C. Xu, D. Wang, S. Chen, D. C. Ranasinghe, and S. Nepal, "Strip: A defence against trojan attacks on deep neural networks," in *Proceedings of the 35th annual computer security applications conference*, 2019, pp. 113–125.

[15] P. Veličković, "Everything is connected: Graph neural networks," *Current Opinion in Structural Biology*, vol. 79, p. 102538, 2023.

[16] G. Lachaud, P. Conde-Cespedes, and M. Trocan, "Mathematical expressiveness of graph neural networks," *Mathematics*, vol. 10, no. 24, p. 4770, 2022.

[17] K. Han, Y. Wang, J. Guo, Y. Tang, and E. Wu, "Vision gnn: An image is worth graph of nodes," *Advances in neural information processing systems*, vol. 35, pp. 8291–8303, 2022.

[18] H. Hu, M. Yao, F. He, and F. Zhang, "Graph neural network via edge convolution for hyperspectral image classification," *IEEE Geoscience and Remote Sensing Letters*, vol. 19, pp. 1–5, 2021.

[19] F. Gama, E. Isufi, G. Leus, and A. Ribeiro, "Graphs, convolutions, and neural networks: From graph filters to graph neural networks," *IEEE Signal Processing Magazine*, vol. 37, no. 6, pp. 128–138, 2020.

[20] A. Saha, A. Subramanya, and H. Pirsiavash, "Hidden trigger backdoor attacks," in *Proceedings of the AAAI conference on artificial intelligence*, vol. 34, no. 07, 2020, pp. 11 957–11 965.

[21] W. Guo, B. Tondi, and M. Barni, "An overview of backdoor attacks against deep neural networks and possible defences," *IEEE Open Journal of Signal Processing*, vol. 3, pp. 261–287, 2022.

[22] R. Yasaei, L. Chen, S.-Y. Yu, and M. A. Al Faruque, "Hardware trojan detection using graph neural networks," *IEEE Transactions on Computer-Aided Design of Integrated Circuits and Systems*, 2022.

[23] W. Ren, X. Xiao, J. Xu, H. Chen, Y. Zhang, and J. Zhang, "Trojan virus detection and classification based on graph convolutional neural network algorithm," *Journal of Industrial Engineering and Applied Science*, vol. 3, no. 2, pp. 1–5, 2025.

[24] S. Abbas, S. Ojo, I. Bouazzi, G. A. Sampedro, A. Al Hejaili, A. Almadhor, and R. Kulhánek, "Securing data from side-channel attacks: A graph neural network-based approach for smartphone-based side channel attack detection," *IEEE Access*, 2024.

[25] L. Alrahis, S. Patnaik, M. Shafique, and O. Sinanoglu, "Embracing graph neural networks for hardware security," in *Proceedings of the 41st IEEE/ACM International Conference on Computer-Aided Design*, 2022, pp. 1–9.

[26] H. Zhang, B. Wu, X. Yuan, S. Pan, H. Tong, and J. Pei, "Trustworthy graph neural networks: Aspects, methods, and trends," *Proceedings of the IEEE*, 2024.

[27] A. Baldominos, Y. Saez, and P. Isasi, "A survey of handwritten character recognition with mnist and emnist," *Applied Sciences*, vol. 9, no. 15, p. 3169, 2019.

[28] M. Wang, D. Zheng, Z. Ye, Q. Gan, M. Li, X. Song, J. Zhou, C. Ma, L. Yu, Y. Gai *et al.*, "Deep graph library: A graph-centric, highly-performant package for graph neural networks," *arXiv preprint arXiv:1909.01315*, 2019.

[29] M. C. Hansen, H. Yalcin, and J. P. Hayes, "Unveiling the iscas-85 benchmarks: A case study in reverse engineering," *IEEE Design & Test of Computers*, vol. 16, no. 3, pp. 72–80, 1999.

[30] L. Amarú, P.-E. Gaillardon, and G. De Micheli, "The epfl combinational benchmark suite," in *Proceedings of the 24th International Workshop on Logic & Synthesis (IWLS)*, 2015.

[31] L. Alrahis, S. Patnaik, M. Shafique, and O. Sinanoglu, "Omla: An oracleless machine learning-based attack on logic locking," *IEEE Transactions on Circuits and Systems II: Express Briefs*, vol. 69, no. 3, pp. 1602–1606, 2021.

[32] L. Alrahis, A. Sengupta, J. Knechtel, S. Patnaik, H. Saleh, B. Mohammad, M. Al-Qutayri, and O. Sinanoglu, "Gnn-re: Graph neural networks for reverse engineering of gate-level netlists," *IEEE Transactions on Computer-Aided Design of Integrated Circuits and Systems*, vol. 41, no. 8, pp. 2435–2448, 2021.

[33] S. D. Chowdhury, K. Yang, and P. Nuzzo, "Reignn: State register identification using graph neural networks for circuit reverse engineering," in *2021 IEEE/ACM International Conference On Computer Aided Design (ICCAD)*. IEEE, 2021, pp. 1–9.

[34] Y. Liu, S. Ma, Y. Aafer, W.-C. Lee, J. Zhai, W. Wang, and X. Zhang, "Trojaning attack on neural networks," in *25th Annual Network And Distributed System Security Symposium (NDSS 2018)*. Internet Soc, 2018.

[35] Z. Xi, R. Pang, S. Ji, and T. Wang, "Graph backdoor," in *USENIX Security*, 2021.

[36] Y. Liu, S. Ma, Y. Aafer, W.-C. Lee, J. Zhai, W. Wang, and X. Zhang, "Trojaning attack on neural networks," 2018.

Exploring MRAM for On-Chip Texture Storage in Rendering Applications

Nicolás Villegas[1], Stefano Romanini[1], Moritz Scherer[2], Warren Hunt[3], Syed Shakib Sarwar[3], Barbara De Salvo[3], Chiao Liu[3], Francesco Conti[4], Davide Rossi[4], Luca Benini[4,5], and Jorge Gómez[1]

[1]*Universidad de los Andes, Santiago, Chile*
[2]*Mosaic SoC, Zurich, Switzerland*
[3]*Meta Reality Labs, Menlo Park, California, USA*
[4]*University of Bologna, Bologna, Italy*
[5]*ETH Zurich, Zurich, Switzerland*

Abstract—In recent years, Magnetoresistive Random-Access Memory (MRAM) has attracted considerable attention as a high-density, non-volatile alternative to conventional embedded memory technologies. While MRAM has been recently adopted for storing neural network weights, its application in rendering workloads remains unexplored. In this study, we investigate the potential of MRAM for on-chip texture storage within a tile-based rasterization workload. Leveraging Siracusa, a RISC-V-based System-on-Chip (SoC) that integrates both MRAM and SRAM at the same memory hierarchy level, we conduct a comparative evaluation focusing on latency and energy consumption across varying frame rates. The results suggest that MRAM achieves substantial energy savings at lower frame rates due to its ability to enter deep-sleep mode between rendering cycles. However, this benefit diminishes as frame rates increase, with SRAM becoming more energy-efficient beyond a threshold of 43 frames per second. These findings demonstrate that MRAM is particularly well-suited to read-intensive, energy-constrained rendering tasks.

Index Terms—Emerging Memories, Non-Volatile memories, High-density memories, MRAM, Tile-based rendering, SAH-based BVH rasterization

I. Introduction

In recent years, there has been a significant emergence of high-density, silicon-compatible memory technologies designed for on-chip integration, including Resistive RAM (RRAM) [1], Magnetoresistive RAM (MRAM) [2], Ferroelectric RAM (FeRAM) [3], and Embedded DRAM (eDRAM) [4]. Each of these memory technologies entails specific trade-offs, determined by their underlying physical mechanisms and fabrication constraints. Optimizing these trade-offs requires prior knowledge of the target application to align memory performance with system requirements. Several of these new memories target applications that require high memory capacities while maintaining a high read-to-write ratio, given the energy/latency asymmetry between read and write operations. Emerging embedded non-volatile memories (eNVMs), such as MRAM, target a unique space: high-storage requirements, dominated by reads and sparse writes. This profile aligns with two key domains: weight storage for neural network and texture storage in rendering applications. While the former has been extensively explored (e.g., [5]), the latter remains underexplored despite its similar memory demands.

In this work, we investigate the feasibility of MRAM for texture memory in rendering applications. We leverage *Siracusa*, a RISC-V-based platform featuring MRAM and SRAM at the same memory hierarchy level [6]. This configuration enables an *apple-to-apple* comparison between MRAM and SRAM for texture storage workloads. We deploy a rendering application on Siracusa to evaluate its memory access patterns and measure the impact on latency and power consumption.

This paper is structured as follows: first, we will introduce the key architectural features of Siracusa relevant to rendering workloads. Next, we provide an overview of MRAM technology, discussing both theoretical expectations and measured characteristics from the Siracusa platform. Then, we describe the rendering algorithm and its deployment on Siracusa. Finally, we present experimental results comparing MRAM and SRAM in terms of performance and power efficiency, followed by our conclusions.

II. Siracusa SoC as testing platform

Siracusa ([6] and [7]) is a low-power System-on-Chip (SoC) designed for visual processing in Extended Reality (XR) applications, fabricated in TSMC's 16 nm CMOS process. Siracusa integrates a 4 MB high-density MRAM memory alongside a 4 MB SRAM at the same hierarchy level, enabling direct comparisons in terms of latency, area, and energy efficiency. To evaluate MRAM's potential for rendering tasks, we repurpose it as a texture storage memory, as depicted in Fig.1. For this analysis, rendering is performed by the 8 RISC-V cores without employing the ML accelerator. Texture data is transferred from the Texture Memory Subsystem to the RISC-V cluster via a 64-bit AXI interconnect, followed by a 32-bit Logarithmic interconnect, as indicated by the red arrows. MRAM and SRAM experience identical interconnect conditions.

III. MRAM technology

Magnetoresistive Random-Access Memory (MRAM) is a silicon-compatible, high-density, non-volatile memory technology [8]. Its read energy and latency are comparable to those of SRAM, while its write energy and latency are

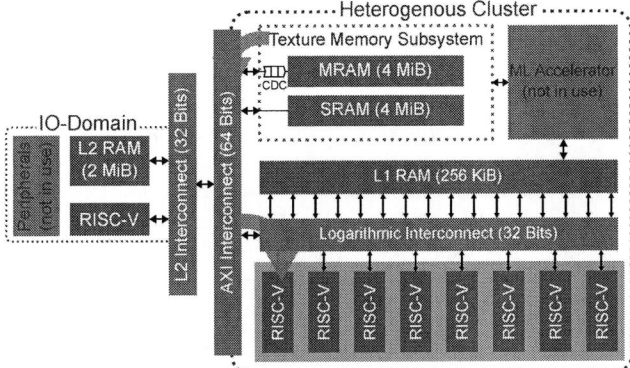

Fig. 1: The Siracusa system-on-chip features an octa-core RISC-V cluster alongside a central control processor (red). It integrates a hierarchical memory architecture comprising with two levels of scratchpad memories and a Texture Memory Subsystem, shown in blue, which includes both SRAM and MRAM.

approximately an order of magnitude higher. The combination of higher density and similar read performance makes MRAM an attractive candidate for applications requiring large on-chip memory and predominantly read-intensive access patterns, such as weight storage in machine learning applications and texture storage in rendering applications. To quantify these characteristics, Table I provides a comparative overview of MRAM and SRAM characteristics, including density, latency, and endurance, mainly measured on Siracusa at nominal voltage of 0.8 V, unless specified otherwise.

TABLE I: COMPARISON OF SRAM AND MRAM

Parameter	SRAM	MRAM
Macro Density (Mb/mm^2)	0.82	1.47
Read Latency (ns)	47.22	77.78
Write Latency (ns)	47.87	$52.82 - 756.32$
Power Consumption (Read) (pJ/bit)	26.13	40.15
Power Consumption (Write) (pJ/bit)	26.91	$34.05 - 407.17$
Sleep power (mW/Mb)	0.567 (@ 0.63V)	0.018*
Wake-up time (ns)	-	$< 100ns$ +
Non-Volatility	No	Yes
Write Endurance (Cycles)	Unlimited	$10^{12} - 10^{15}$ +
Temperature Stability	$-40°C - 125°C$	$-40°C - 125°C$ +

*Provided by Numem.   + TSMC characterization [9].

A. Latency Characterization

To characterize access latency under realistic conditions, we evaluate memory access times averaged over the whole 4 MiB of memory per configuration, allocating the operation from 1 to the full set of 8 active cluster cores running at 360 MHz. The memory under test—either MRAM or SRAM—is initialized with a 4 MiB array of 32-bit unsigned integers. Two access patterns are analyzed: *Sym*, where memory is evenly partitioned among cores to reduce contention, and

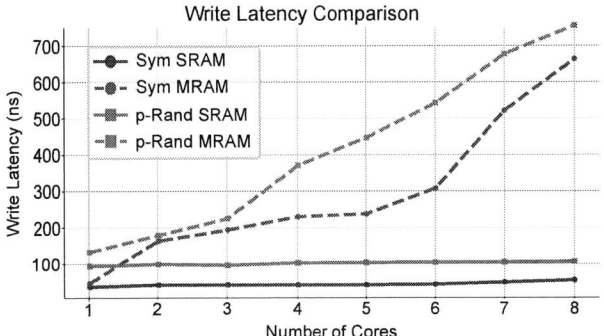

Fig. 2: SRAM write latency (solid lines) demonstrates consistently low values and remains relatively stable across all tested core counts and for both access patterns. In contrast, MRAM write latency (dashed lines) shows a distinct behavior. For a single active core, MRAM latency is comparable to SRAM, attributed to the memory access queue not being filled. However, as the number of active cores increases, MRAM write latency escalates considerably, particularly under the pseudo-random access pattern, where it reaches a peak of 756.32 ns with 8 active cores. The symmetric access pattern for MRAM also shows a significant increase, peaking at 664.4 ns for 8 cores.

p-Rand, where each core performs pseudo-random accesses generated by a uniquely seeded Linear Congruential Generator, simulating a worst-case scenario.

The read latency characteristics (shown on table I) for both SRAM and MRAM were observed to be stable across all tested core counts and access patterns, with MRAM consistently showing slightly higher latency than SRAM.

The write latency results, presented in Fig.2, indicate that SRAM performance remains consistently low and is largely unaffected by the access pattern or the number of active cores. MRAM, contrarily, exhibits a distinct trend: its write latency increases significantly as more cores become active. This behavior is consistent with the energy-intensive nature of STT-MRAM write operations. The observed pattern in MRAM write latency can be attributed to the dynamics of its write First-In, First-Out (FIFO) Clock Domain Crossing (CDC) buffer. Specifically, at low concurrency levels (e.g., with a single active core), individual cores may not generate sufficient write requests to fill the FIFO. This results in writes being enqueued quickly, leading to an observed latency that can appear artificially low, sometimes comparable to SRAM. However, as concurrency rises and more cores issue write commands, the FIFO buffer saturates, and the full, inherently higher latency of the MRAM write process becomes apparent.

B. Power Consumption Measurement

To enable accurate power characterization, the test application is configured to run indefinitely, executing only read and write memory operations while all auxiliary functions are disabled. This ensures that the measured energy reflects solely the memory access workload under evaluation. A GPIO

979-8-3315-9813-6/25 $31.00 © 2025 IEEE

pin is toggled to externally validate the active execution time estimated from internal cycle counts. Power measurements are performed using the Siracusa evaluation board, which exposes individual pin headers for each power domain. These headers are connected to a Keysight B2901BL source measure unit (SMU) via four-wire (force and sense) connections, allowing precise current readings. The measured current traces are averaged over time to compute the energy consumed per operation cycle. The measurement setup is illustrated in Fig. 3, including the SMU connection and the use of an oscilloscope to capture GPIO-based timing signals. This same methodology is later employed to characterize the energy consumption of the rendering algorithm.

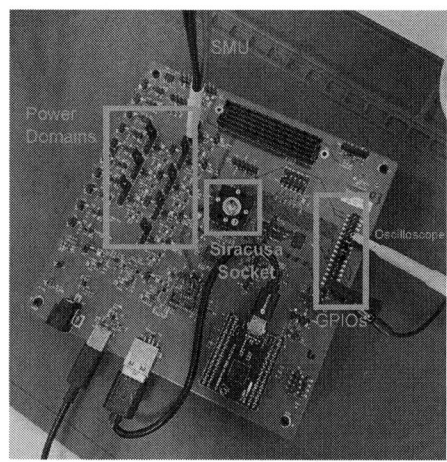

Fig. 3: Measurement setup for the Siracusa SoC. Power domains are monitored via a SMU connected through four-wire sensing, while an oscilloscope captures GPIO toggles for timing correlation.

IV. Rendering Algorithm

To analyze the impact of MRAM, we implement a tile-based rasterization algorithm for rendering [10]. Multiple variants of the rendering algorithm were explored—initially using basic tile-binning [11], followed by BVH acceleration through k-means clustering [12], and ultimately adopting a SAH-based BVH [13], which yielded the best active-time performance. The Surface Area Heuristic (SAH) selects the BVH split that minimizes the expected traversal cost by favoring partitions with smaller surface areas and fewer primitives. By partitioning the image into smaller tiles and processing them independently, the algorithm minimizes memory access while naturally mapping to Siracusa's parallel architecture. Rendering is carried out concurrently across the eight RISC-V cores, with Fig. 4 illustrating the key stages of the process on the Fiu mascot. The rendering process begins by dividing the image into tiles (Fig. 4 (a)), with each tile consisting of 8×8 pixels. Each tile maintains its own color and depth buffers, ensuring that rendering operations remain confined to smaller regions and reducing per-core memory requirements. Next, vertex projection is applied, transforming the 3D model's vertices from world space to 2D screen space using a camera transformation matrix. To efficiently determine which primitives are relevant for each tile, a Bounding Volume Hierarchy (BVH) structure is traversed (Fig. 4 (b)). The BVH organizes geometry hierarchically, allowing the algorithm to cull primitives that do not overlap with the tile's bounding box. Once the relevant primitives are identified, each tile is processed independently (Fig. 4 (c)). During rasterization, barycentric coordinates are computed per pixel to interpolate attributes such as texture coordinates. A depth test is then performed to correctly resolve overlapping geometry. Then textures stored in MRAM are accessed at this stage to apply color to the geometry. Finally, the results from all tiles are assembled to produce the complete rendered image (Fig. 4 (d)).

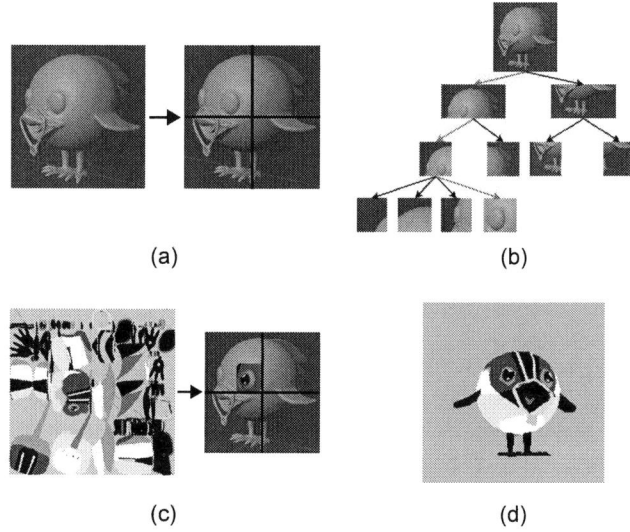

Fig. 4: Overview of the tile-based rasterization process on Siracusa. (a) Division of image into tiles. (b) BVH acceleration structure guides per-tile primitive selection. (c) Rasterization and texture application using MRAM-stored textures. (d) Final composited image rendered on Siracusa.

V. MRAM Access Pattern

As described before, model binaries are loaded into L2 memory, while textures reside in the Texture Memory Subsystem. To fit within the on-chip memory constraints, 2MB for L2 and 4MB for textures, the 3D model and texture assets were downscaled. While this reduces output resolution, it preserves the memory access behavior and energy characteristics of interest.

As described in the previous section, MRAM is accessed exclusively during the rasterization stage, where texture data is read to compute the color of each pixel. The resulting pixel values are then written to L2 memory for final image composition. During all other stages of the rendering pipeline, MRAM remains idle. To improve energy efficiency, the memory is placed in deep-sleep mode whenever it is not actively accessed. This strategy leverages MRAM's non-volatility to eliminate static power consumption without risking data integrity. For a fair comparison, SRAM is assumed to enter a low-power retention mode at 0.63 V during these idle periods. The wake-up latency for MRAM, reported in Table I, is negligible relative to the frame duration—more than five orders of magnitude smaller across all tested frame rates.

Figure 5 illustrates the proportion of time MRAM spends in active versus deep-sleep mode at three representative frame rates: 30 FPS, 60 FPS, and 100 FPS. The active phase, empirically measured at approximately 10 ms, remains constant across all configurations, as it is dictated by the rendering workload and the fixed 360 MHz cluster frequency. At higher frame rates, such as 100 FPS, the frame duration approaches the length of the active phase, leaving minimal opportunity for MRAM to enter deep sleep. In contrast, lower frame rates like 30 FPS introduce extended idle intervals, during which MRAM can remain in deep-sleep mode and significantly reduce energy consumption. This behavior shows MRAM's efficiency advantage at low FPS.

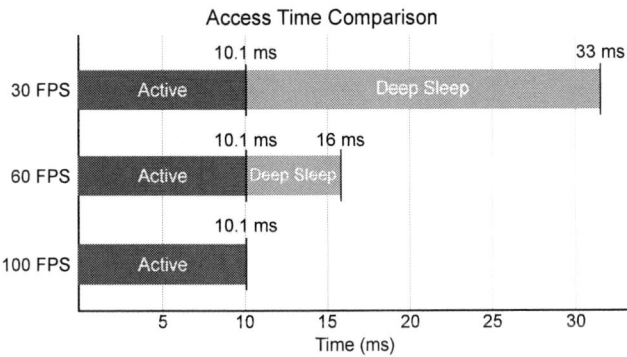

Fig. 5: Proportion of MRAM active and deep-sleep time at 30 FPS, 60 FPS, and 100 FPS. Active time (blue) remains constant at 10 ms, while deep sleep duration (green) increases as frame rate decreases.

VI. EXPERIMENTAL RESULTS

This section presents the energy measurements for the rendering algorithm, using the methodology described in Section III-B. Fig. 6 illustrates the total energy consumption per frame for both MRAM and SRAM at FPS values ranging from 10 to 100. At low FPS, MRAM demonstrates significant energy savings due to extended idle periods (53.91% at 10FPS). However, as FPS increases, the proportion of active time grows, diminishing MRAM's advantage. The crossover point, marked at 43.24 FPS, indicates the threshold beyond which SRAM becomes more energy-efficient, as a result of its lower active energy consumption.

To better understand the breakdown of energy consumption, Fig. 7 separates active and idle energy contributions for both memory types. For SRAM, idle energy dominates at lower FPS values, whereas MRAM benefits from significantly lower idle energy due to deep-sleep operation. Active energy consumption for MRAM remains slightly higher than SRAM, primarily because of its higher read power, as presented in Table I. This

detailed view highlights how the balance between active and idle energy shifts as FPS changes, ultimately determining the overall efficiency.

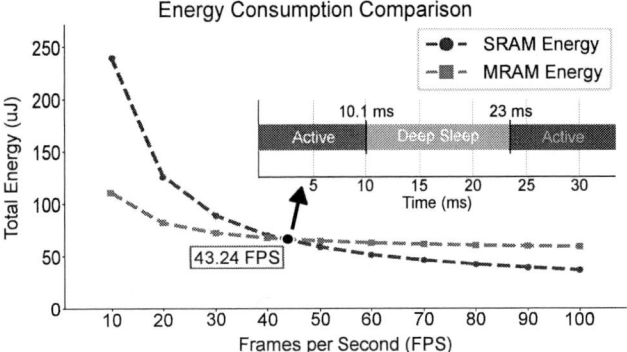

Fig. 6: Total energy consumption comparison between SRAM and MRAM across different FPS. MRAM energy (green) remains relatively stable, while SRAM energy (blue) decreases as FPS increases. The intersection point at 43.24 FPS marks the threshold where SRAM becomes more energy-efficient.

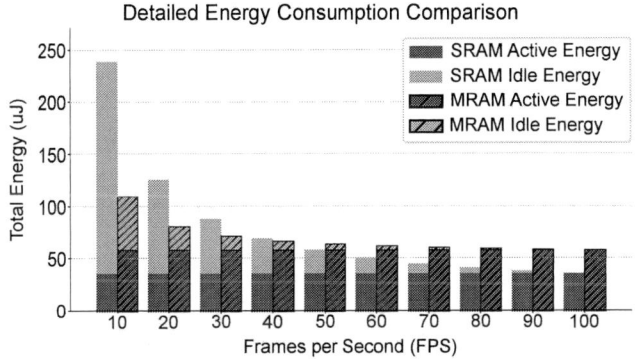

Fig. 7: Detailed energy consumption breakdown for SRAM and MRAM at varying FPS.

VII. CONCLUSIONS

In this work, we explored the use of MRAM as on-chip texture storage in rendering applications. Leveraging the Siracusa SoC, which integrates both MRAM and SRAM at the same memory hierarchy level, we conducted a comprehensive evaluation comparing their performance and energy efficiency within a tile-based rasterization pipeline. Our findings highlight the suitability of MRAM for read-intensive rendering workloads, particularly at lower frame rates.

VIII. ACKNOWLEDGMENT

This work was supported in part by the Chilean National Research Agency (ANID) through Fondecyt de Iniciación under Grant No. 11250340. The authors would like to thank Meta Reality Labs for providing access to the hardware platform, and Numem for their support with the MRAM IP integration.

979-8-3315-9813-6/25 $31.00 © 2025 IEEE

REFERENCES

[1] H.-S. P. Wong, H.-Y. Lee, S. Yu, Y.-S. Chen, Y. Wu, P.-S. Chen, B. Lee, F. T. Chen, and M.-J. Tsai, "Metal–oxide rram," *Proceedings of the IEEE*, vol. 100, no. 6, pp. 1951–1970, 2012.

[2] X. Dong, C. Xu, Y. Xie, and N. P. Jouppi, "Nvsim: A circuit-level performance, energy, and area model for emerging nonvolatile memory," *IEEE Transactions on Computer-Aided Design of Integrated Circuits and Systems*, vol. 31, no. 7, pp. 994–1007, 2012.

[3] S. Yu, "Neuro-inspired computing with emerging nonvolatile memorys," *Proceedings of the IEEE*, vol. 106, no. 2, pp. 260–285, 2018.

[4] A. Shafiee, A. Nag, N. Muralimanohar, R. Balasubramanian, J. P. Strachan, M. Hu, R. S. Williams, and V. Srikumar, "Isaac: A convolutional neural network accelerator with in-situ analog arithmetic in crossbars," in *2016 ACM/IEEE 43rd Annual International Symposium on Computer Architecture (ISCA)*, 2016, pp. 14–26.

[5] S. Ghimire, S. Kataoka, and L. Pentecost, "Nvmsurvey: Recent advances and comparative analysis of emerging non-volatile memories (envms)," in *2023 IEEE International Symposium on Workload Characterization (IISWC)*, 2023, pp. 229–231.

[6] A. S. Prasad, M. Scherer, F. Conti, D. Rossi, A. Di Mauro, M. Eggimann, J. T. Gómez, Z. Li, S. S. Sarwar, Z. Wang, B. De Salvo, and L. Benini, "Siracusa: A 16 nm heterogenous risc-v soc for extended reality with at-mram neural engine," *IEEE Journal of Solid-State Circuits*, vol. 59, no. 7, pp. 2055–2069, 2024.

[7] M. Scherer, M. Eggimann, A. D. Mauro, A. S. Prasad, F. Conti, D. Rossi, J. T. Gómez, Z. Li, S. S. Sarwar, Z. Wang, B. D. Salvo, and L. Benini, "Siracusa: A low-power on-sensor risc-v soc for extended reality visual processing in 16nm cmos," in *ESSCIRC 2023- IEEE 49th European Solid State Circuits Conference (ESSCIRC)*, 2023, pp. 217–220.

[8] C.-H. Chen, C.-Y. Chang, C.-H. Weng, T.-H. Kuo, C.-Y. Wang, M.-C. Shih, T.-W. Chiang, Y.-J. Lee, R. Wang, K.-H. Shen, A. Hung, and H. Chuang, "Reliability and magnetic immunity of reflow-capable embedded stt-mram in 16nm finfet cmos process," in *2021 Symposium on VLSI Technology*, 2021, pp. 1–2.

[9] P.-H. Lee, C.-F. Lee, Y.-C. Shih, H.-J. Lin, Y.-A. Chang, C.-H. Lu, Y.-L. Chen, C.-P. Lo, C.-C. Chen, C.-H. Kuo, T.-L. Chou, C.-Y. Wang, J. J. Wu, R. Wang, H. Chuang, Y. Wang, Y.-D. Chih, and T.-Y. J. Chang, "A 16nm 32mb embedded stt-mram with a 6ns read-access time, a 1m-cycle write endurance, 20-year retention at 150°c and mtj-otp solutions for magnetic immunity," in *2023 IEEE International Solid-State Circuits Conference (ISSCC)*, 2023, pp. 494–495.

[10] I. Antochi, B. Juurlink, and S. Vassiliadis, "Memory bandwidth requirements of tile-based rendering," vol. 3133, 07 2004, pp. 323–332.

[11] K. Min, J. Kim, S. Kang, S. Kim, H. S. Kim, S. Jung, and S. Woo, "Method and system for tile binning using half-plane edge function," Patent, 01, 2013.

[12] D. Meister and J. Bittner, "Parallel bvh construction using k-means clustering," *The Visual Computer*, vol. 32, 06 2016.

[13] I. Wald, "On fast construction of sah based bounding volume hierarchies," 10 2007, pp. 33 – 40.

979-8-3315-9813-6/25 $31.00 © 2025 IEEE

Open-Source 4 K CMOS Calibration: Integrating IceMOS and Sky130 PDK

Mauricio Montanares*, V.H. Arzate Palma*, Kevin G. McCarthy[†], Gerardo Molina Salgado*

*Microelectronics Circuit Centre Ireland, MCCI, Cork, Ireland

[†]School of Engineering, University College Cork, Cork, Ireland

mauricio.montanares@mcci.ie

Abstract—We present an open-source calibration methodology for Sky130 CMOS at 4 K, focusing on robust initial extraction of threshold voltage (VTH0) and slope factor (n) via the SEKV-E parameter-extraction tool. To address measurement noise, a partial Z-score filter is applied to the raw data, reducing extraction errors from above 10% to under 2%. These refined parameters then feed into an iterative BSIM4 calibration using IceMOS. As a result, the final calibrated Sky130 models achieve an accuracy in the strong inversion region with median error percentages of 9.49% for PMOS and just 1.11% for NMOS. Transconductance g_m calibration is similarly precise with median errors of 5.95% and 2.46% for PMOS and NMOS, respectively, in the strong inversion region. By separating the SEKV-E and IceMOS steps yet integrating them into a fully open-source flow (including Docker-based ngspice simulations), our approach significantly minimizes manual intervention. The Sky130-compatible IceMOS scripts, Z-score routines, and calibrated device models are publicly available, reinforcing the open-source ecosystem for Cryo-CMOS research.

Index Terms—Cryo-CMOS, Calibration, SEKV-E, IceMOS, Open-Source

I. INTRODUCTION

Quantum computing has the potential to solve complex problems beyond classical computing capabilities by exploiting quantum mechanical phenomena at cryogenic temperatures [1], [2]. To enhance scalability and minimize interconnection losses, the integration of control electronics at these cryogenic temperatures—known as Cryo-CMOS—is crucial [1], [3]. However, Cryo-CMOS design at temperatures typically between 1 K and 4 K introduces several challenges, including increased transistor threshold voltage, stringent thermal dissipation constraints (limited to about 1 W), and significant deviations from room-temperature transistor models [3], [4]. Thus, accurate transistor modeling and calibration at cryogenic temperatures are essential for effective circuit design.

Compact transistor models, such as the widely-used BSIM4 model, rely heavily on initial parameters like the threshold voltage (VTH0) and the slope factor (n). These parameters significantly impact device modeling accuracy, especially in the sensitive subthreshold region critical at cryogenic temperatures [4]. Traditionally, determining initial values for these parameters involves iterative calibration and manual adjustment

(This work was funded by MCCI – Microelectronic Circuits Centre Ireland, through the Enterprise Ireland and IDA Ireland Technology Centres programme.)

based on estimations, complicating the calibration process and increasing uncertainty.

This paper addresses this challenge by using the SEKV-E parameter extraction tool, reported in [5], to systematically generate reliable initial estimates for VTH0 and n from measured transistor data. However, SEKV-E extraction algorithms are sensitive to measurement noise, particularly pronounced at 4 K due to phenomena like dopant freeze-out and reduced thermal noise floor [4]. To mitigate this issue, we propose applying partial Z-score filtering to raw transistor measurement data prior to extraction, significantly reducing extraction errors and improving parameter reliability.

The extracted parameters from SEKV-E, enhanced by Z-score filtering, are subsequently used as accurate starting points for the iterative calibration procedure in IceMOS, an open-source Python-based calibration tool previously introduced in [6]. IceMOS has been specifically adapted in this work to support the Sky130 open-source PDK, enabling automated calibration workflows including parameter extraction, netlist generation, and model optimization.

Therefore, the contributions of this paper are:

1) The introduction of SEKV-E extracted parameters (VTH0 and n) as reliable initial points for BSIM4 model calibration of Sky130 CMOS at 4 K.
2) The implementation of partial Z-score filtering to enhance the robustness and accuracy of the SEKV-E parameter extraction tool.
3) The adaptation of the IceMOS calibration tool to support automated calibration workflows specifically tailored for Sky130 devices operating at cryogenic temperatures.

This paper is organized as follows: Section II provides an overview of the Sky130 PDK and its key considerations at cryogenic temperatures. Section III describes the IceMOS tool and its adaptation to the Sky130 PDK. The calibration methodology, including the application of partial Z-score filtering, is presented in Section IV. Section V details the experimental validation and results, and finally, conclusions and potential future directions are summarized in Section VI.

II. SKY130 OPEN-SOURCE PDK AND CRYOGENIC CONSIDERATIONS

The Sky130 PDK is an open-source 130nm CMOS technology offering free design resources [7], [8]. Its recent application to 4 K cryogenic environments revealed large shifts

979-8-3315-9813-6/25 $31.00 © 2025 IEEE

in device behavior, making standard room-temperature models inadequate and requiring recalibration with measured 4 K data [9], [10].

Although cryogenic measurements are increasingly available [10], calibrating Sky130 at 4 K across its multiple device bins remains challenging due to the complexity and potential for error in manual procedures [6].

This work employs IceMOS, a Python-based framework designed to streamline cryogenic calibration by automating parameter extraction, netlist generation, and data parsing [6]. Integrating the open-source IceMOS tool with the Sky130 PDK thus provides an accessible and efficient workflow for 4 K calibration. This approach minimizes manual errors, facilitates systematic validation across device bins, and leverages the synergy between open-source technology and automated calibration tools.

A unified workflow like this is particularly valuable for verifying multiple chip bins at 4 K, where repeatability and extensive statistical testing demand a shared, efficient platform. Consequently, IceMOS serves as a robust foundation for systematic cryogenic calibration in Sky130, reinforcing the open-source ecosystem and advancing quantum electronics research.

III. IceMOS Tool: Overview and Sky130 Integration

IceMOS is an open-source, Python-based framework for calibrating CMOS PDKs at cryogenic temperatures (4 K). The tool automates parameter extraction, netlist generation, and measurement-data parsing, significantly streamlining the calibration workflow [6]. However, a key limitation of IceMOS—and other existing calibration tools—is the requirement for users to manually adjust initial values for critical parameters such as threshold voltage VTH0 and n based on estimations. Incorrect initial estimations can result in convergence issues or inaccurate final models, particularly at cryogenic temperatures where parameter sensitivity is heightened. This paper addresses this limitation by providing systematic, robust initial parameter estimations through an enhanced parameter extraction process using the SEKV-E toolbox with enhancements described in the section IV-C.

A. Adaptation of IceMOS to Sky130

In the current iteration, significant modifications have been implemented to align IceMOS with open-source methodologies. The primary change involved transitioning the calibration process entirely from proprietary software to open-source tools. The Spectre simulator previously used has been replaced by ngspice, an open-source SPICE simulator widely recognized within the community for circuit simulation and verification tasks. Furthermore, the entire calibration workflow was validated using the IIC-OSIC-TOOLS Docker image [11].

The adaptation of IceMOS for Sky130 required specialized Python parsing scripts to handle the open-source PDK's distinct file organization. These updates included extracting bin models from unique data sources, accurately interpreting parameter hierarchies, and managing Sky130 technology-specific files. The calibration workflows were also revised to accommodate ngspice, necessitating syntax and netlist-structure adjustments for compatibility.

IV. Calibration Methodology with IceMOS and Sky130

A. Experimental Setup

This work utilizes publicly available cryogenic datasets from the SkyWater 130nm CMOS process at 4 K and provided by the works [9], [10]. The authors provide full measurements details in [9], [10] including details of the use of a Lakeshore CRX-4K cryogenic station, I-V measurements with a HP4156 analyzer and C–V measurements with an Agilent E4980 LCR meter.

B. Data Flow and Validation Methodology

To ensure clarity in our validation approach, it is important to distinguish between the experimental data acquisition and model validation process:

1) Experimental Data Collection: Raw I-V measurements are obtained directly from physical Sky130 devices at 4 K using laboratory equipment (HP4156 analyzer, Lakeshore CRX-4K cryogenic station) as detailed in [9], [10].
2) Model Calibration: These experimental measurements are used to calibrate BSIM4 model parameters through the IceMOS framework, yielding optimized device models.
3) Model Validation: The calibrated BSIM4 models are then simulated using ngspice to generate predicted I-V characteristics, which are compared against the original experimental measurements to assess calibration accuracy.

This validation methodology ensures that our "Laboratory" data represents actual physical measurements, while "Simulation" data represents ngspice-generated characteristics using the IceMOS-calibrated models.

C. Initial Parameter Extraction with SEKV-E

A key challenge for cryogenic calibration is obtaining accurate initial estimates of the VTH0 and n. We address this by employing the SEKV-E parameter extraction toolbox [5], which systematically estimates the SEKV parameters [5] from measured data and thus replaces manual calibration work for VTH0 and n.

However, SEKV-E is particularly sensitive to measurement noise at 4 K, where dopant freeze-out and the reduced thermal noise floor amplify subthreshold fluctuations [4], [10]. To mitigate this, we introduced a partial Z-score filtering procedure—a preprocessing step that removes outlier points from raw $I - V$ data based on a user-defined threshold. By discarding spurious measurements in the subthreshold and transition regions, the extraction process becomes significantly more robust and accurate as is demonstrated in Section V.

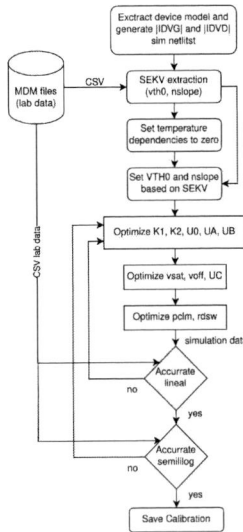

Fig. 1. Diagram of the IceMOS calibration workflow adapted for Sky130. Similar approaches have been reported [4], [12]

This filtering step is applied before the iterative IceMOS calibration flow, ensuring cleaner data for the initial VTH0 and n estimates. These refined values then feed into the subsequent optimization stages in IceMOS at 4 K, where even nominally small currents can be destabilized by random telegraph noise (RTN), instrument offsets [9], or abrupt transitions. Consequently, partial Z-score filtering effectively eliminates such artifacts and yields more stable parameter extraction. The quantitative impact of this filtering is evaluated in Section V.

D. IceMOS Calibration Flow for Sky130

With robust initial parameters (VTH0 and n) obtained through Z-score-enhanced SEKV-E extraction, we proceed to iterative calibration using IceMOS, as detailed in [6]. The Sky130-adapted calibration workflow is summarized in Figure 1, involving the following key steps:

1) Data Formatting: Raw measurement data (MDM files), using IceMOS are converted into CSV format using automated Python parsers.
2) Initial Parameter Setup: Temperature-related parameters (UTE, KT1, KT2, UA1, UB1) are initially set to zero, aligning with established cryogenic modeling practices [4], [6], [12].
3) Iterative Optimization: The user iteratively optimizes BSIM4 model parameters by comparing ngspice simulation results against measured laboratory data until satisfactory convergence is achieved.

V. EXPERIMENTAL RESULTS AND VALIDATION

This section presents the experimental validation of the proposed 4 K calibration methodology for Sky130 devices. We begin by demonstrating the effectiveness of the partial Z-score filtering technique in improving the robustness and accuracy of initial parameter extraction. We then showcase the final

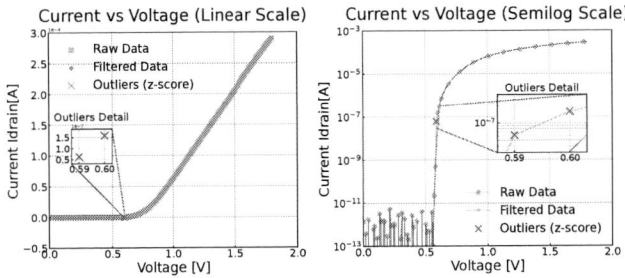

Fig. 2. Example of outlier identification and removal using selective Z-score filtering on Sky130 $I_D - V_G$ data (4 K) for an LVT NMOS device. Outliers (red crosses) near the subthreshold-to-strong-inversion transition ($V_{GS} \approx 0.6V$) are effectively removed before parameter extraction.

calibration results of representative Sky130 transistors using the complete IceMOS flow, which leverages these enhanced parameter estimates.

A. Impact of Z-Score Filtering on Parameter Extraction at 4 K

Accurate extraction of subthreshold parameters such as VTH0 and n is critical at 4 K, where measurement noise and outliers often manifest in the week-inversion region and transition region. As discussed in Section IV, these artifacts can cause numerical instability or large errors in automated extraction routines like SEKV-E. To address this, we apply a partial Z-score filtering step before extraction, identifying and removing significant outliers in the measured I-V curves.

Figure 2 illustrates this approach for a Sky130 LVT NMOS device ($W = 0.42\mu m$, $L = 0.14\mu m$). Outliers (marked with red crosses) appear near the subthreshold-to-strong-inversion transition ($V_{GS} \approx 0.6V$). Removing these outliers yields a cleaner dataset for SEKV-E, improving extraction consistency. Figure 3 quantifies the impact of filtering by comparing initial SEKV-E fits obtained with raw data (left) versus Z-score-filtered data (right). With unfiltered measurements, extracted parameters may exhibit numerical divergence or poor correlation to physical device behavior. Once outliers are removed, the extracted VTH0 and n (and rest of SEKV parameters) values become more stable and physically meaningful, reducing the mean error from above 10% to around 1%. By ensuring robust initial parameter estimates, partial Z-score filtering significantly enhances the overall calibration workflow.

B. Sky130 Calibration Results using IceMOS

Building on the Z-score-filtered SEKV-E parameter extraction, we performed a complete iterative calibration using IceMOS for representative Sky130 devices at 4 K. We validated the final calibrated models against measured data from PMOS Bin 1 ($W = 0.42\mu m$, $L = 0.15\mu m$) and NMOS Bin 40 ($W = 1.68\mu m$, $L = 0.15\mu m$).

All validation results compare experimental I-V measurements obtained directly from Sky130 devices at 4 K against ngspice simulations using the IceMOS-calibrated BSIM4 models. This approach validates the accuracy of our calibration

979-8-3315-9813-6/25 $31.00 © 2025 IEEE

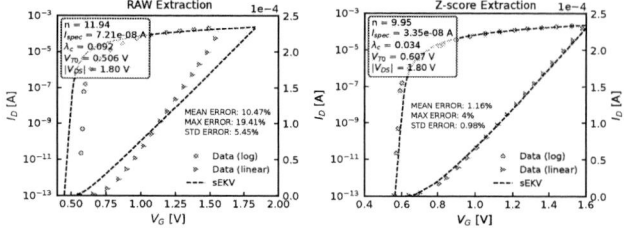

Fig. 3. Comparison of initial SEKV-E fits for a Sky130 LVT NMOS ($W = 0.42\mu m$, $L = 0.15\mu m$, Bin 1) at 4 K, using raw data (left) versus Z-score-filtered data (right). Filtering enhances fit accuracy (e.g., mean error from 10.47% to 1.16%) and stabilizes parameter extraction.

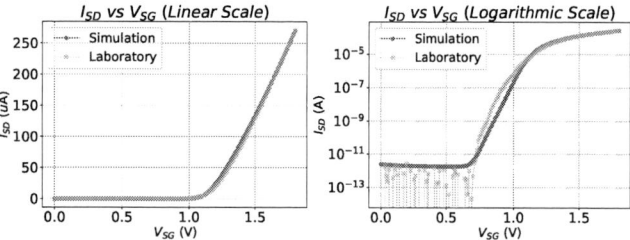

Fig. 4. Measured vs. simulated $I_{SD} - V_{SG}$ for PMOS Bin 1. The "Laboratory" curves show experimental measurements from physical devices at 4 K, while "Simulation" curves represent ngspice simulations using the IceMOS-calibrated BSIM4 model. Good agreement is observed, with a median error below 9.49% in strong inversion, and an overall median error below 12.5%

methodology by assessing how well the calibrated models predict actual device behavior.

1) Transfer Characteristics ($I_{DS} - V_{GS}$ and $I_{SD} - V_{SG}$): Figures 4 and 5 compare measured and simulated transfer curves in both linear and semilog scales. The calibrated models match the experimental data particularly well in the strong inversion region. As summarized in Table I, the PMOS device achieves a median error of 9.49% in the strong inversion region (Figure 4), whereas the NMOS device performs exceptionally well with just 1.11% median error in the same region (Figure 5). The overall median errors across the entire voltage range are 12.73% and 9.37% for PMOS and NMOS, respectively. These results confirm that the strong inversion region is accurately calibrated.

2) Output Characteristics ($I_{DS} - V_{DS}$): The calibrated model also reproduces device behavior in the output domain. Figure 6 shows $I_{DS} - V_{DS}$ curves for NMOS Bin 40 under several V_{GS} conditions, demonstrating strong agreement between simulations and measured data in both linear and saturation regions.

3) Transconductance: Finally, Figure 7 assesses the accuracy of the simulated transconductance (g_m). The calibrated models follow the measured g_m, with median errors in the strong inversion region of 5.95% for PMOS and only 2.46% for NMOS devices. The overall median errors across the entire voltage range are 3.49% and 1.05% for PMOS and NMOS, respectively. This level of accuracy is particularly important for analog design, where transconductance is a key parameter.

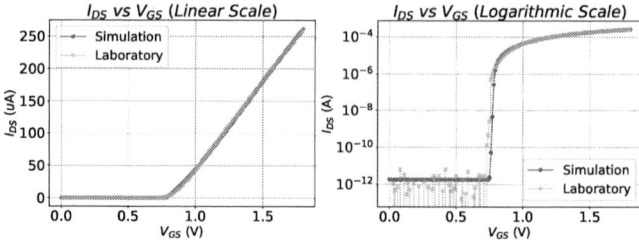

Fig. 5. Measured vs. simulated $I_{DS} - V_{GS}$ for NMOS Bin 40. The "Laboratory" data represents experimental measurements from physical devices, while "Simulation" shows ngspice results using the calibrated model. The calibrated model closely matches the measurements, with median error of 1.11% in the strong inversion region and 9.37% overall.

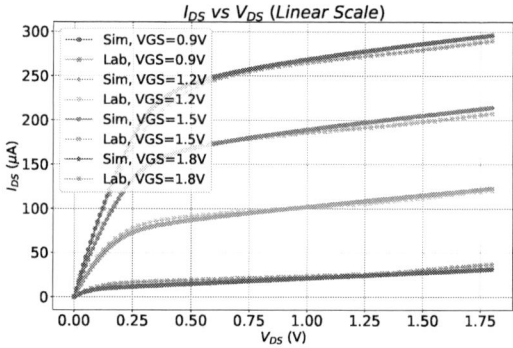

Fig. 6. I_{DS} vs V_{DS} for NMOS Bin 40 across multiple V_{GS} biases. Experimental data ("Lab") from physical device measurements is compared against ngspice simulations ("Sim") using the IceMOS-calibrated model. The calibrated model accurately captures both linear and saturation regimes.

Small discrepancies may be reduced by additional fine-tuning of BSIM4 parameters, but the current models already offer a robust match to experimental data.

VI. CONCLUSIONS

This work demonstrates an open-source calibration methodology for Sky130 CMOS devices at 4 K, combining reliable initial parameter extraction through SEKV-E with partial Z-score filtering and an IceMOS-based iterative BSIM4 calibration. By filtering outliers in the subthreshold and transition

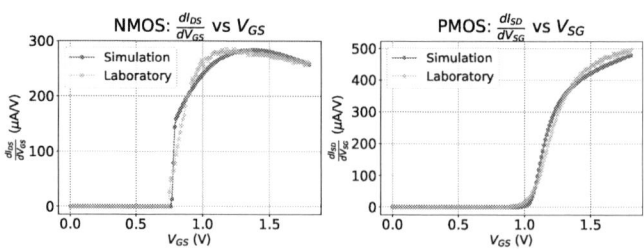

Fig. 7. Measured vs. simulated g_m for PMOS Bin 1 and NMOS Bin 40. Experimental transconductance measurements ("Laboratory") are compared with ngspice simulations ("Simulation") using calibrated models. The median errors in the strong inversion region are 5.95% and 2.46% for PMOS and NMOS respectively, confirming strong agreement between physical measurements and calibrated model simulation.

region, SEKV-E accurately determines the threshold voltage (VTH0) and slope factor (n), reducing mean extraction errors from above 10% to under 2%. These refined parameters then serve as initial values for the IceMOS calibration flow. The calibrated models demonstrate excellent accuracy in the strong inversion region with median errors of 9.49% for PMOS and 1.11% for NMOS. Transconductance modeling achieves comparable precision with median errors of 5.95% and 2.46% for PMOS and NMOS, respectively—metrics particularly significant for analog circuit design.

Although SEKV-E and IceMOS are deployed as separate tools, their combined use in a fully open-source flow significantly reduces calibration errors and manual intervention, strengthening the Cryo-CMOS design ecosystem. The Sky130-compatible version of IceMOS, along with the modified SEKV-E scripts incorporating Z-score filtering, test scripts, and the calibrated device models presented in this work, are all available in our public GitHub repository [13].

TABLE I
MEDIAN ERROR (%) ANALYSIS FOR STRONG INVERSION REGION AND
OVERALL PERFORMANCE AT 4 K

Device Metric	Strong Inversion Region (0.8-1.8V)	Overall Error
PMOS Current	9.49%	12.73%
NMOS Current	1.11%	9.37%
PMOS g_m	5.95%	3.49%
NMOS g_m	2.46%	1.05%

ACKNOWLEDGMENT

This research was supported by MCCI (Microelectronic Circuits Centre Ireland) under the Enterprise Ireland Technology Centre Program Grant Number TC2020-0029.

REFERENCES

[1] E. Charbon, M. Babaie, A. Vladimirescu, and F. Sebastiano, "Cryogenic CMOS circuits and systems: Challenges and opportunities in designing the electronic interface for quantum processors," *IEEE Microwave Magazine*, vol. 22, no. 1, pp. 60–77, Jan 2021.

[2] H. A. Bhat, F. A. Khanday, B. K. Kaushik, F. Bashir, and K. A. Shah, "Quantum Computing: Fundamentals, Implementations and Applications," *IEEE Open Journal of Nanotechnology*, vol. 3, pp. 61–77, May 2022.

[3] S. Chakraborty and R. V. Joshi, "Cryogenic CMOS design for qubit control: Present status, challenges, and future directions," *IEEE Circuits and Systems Magazine*, vol. 24, no. 2, pp. 34–46, May 2024.

[4] R. M. Incandela, L. Song, H. Homulle, E. Charbon, A. Vladimirescu, and F. Sebastiano, "Characterization and compact modeling of nanometer CMOS transistors at deep-cryogenic temperatures," *IEEE Journal of the Electron Devices Society*, vol. 6, pp. 996–1006, 2018.

[5] H.-C. Han, A. D'Amico, and C. Enz, "SEKV-E: Parameter Extractor of Simplified EKV I-V Model for Low-Power Analog Circuits," *IEEE Open Journal of Circuits and Systems*, vol. 3, pp. 162–167, 2022.

[6] M. Montanares, M. Wen, V. A. Palma, K. G. McCarthy, and G. M. Salgado, "IceMOS: Cryo-CMOS Python-Based Calibration Tool," in *2024 IEEE 67th International Midwest Symposium on Circuits and Systems (MWSCAS)*. IEEE, 2024, pp. 1011–1015.

[7] M. Shalan and T. Edwards, "Building OpenLANE: A 130nm openroad-based tapeout-proven flow," in *Proceedings of the 39th International Conference on Computer-Aided Design*, 2020, pp. 1–6.

[8] K. Herman, M. Montanares, and J. Marin, "Design and implementation of integrated circuits using open source tools and sky130 free pdk," in *2023 30th International Conference on Mixed Design of Integrated Circuits and System (MIXDES)*. IEEE, 2023, pp. 105–110.

[9] A. Akturk, A. Tripathi, and M. Saligane, "Cryogenic modeling for open-source process design kit technology," in *2023 IEEE BiCMOS and Compound Semiconductor Integrated Circuits and Technology Symposium (BCICTS)*. IEEE, 2023, pp. 58–65.

[10] A. Li, T. Zeng, L. Zhang, J. Riem, G. C. Adam, D. L. Fleischer, A. Zaslavsky, W. R. Patterson, T. Ansell, A. Akturk *et al.*, "Unlocking Circuits for Quantum With Open Source Silicon: A first look at measured open source silicon results at 4 K," *IEEE Solid-State Circuits Magazine*, vol. 16, no. 2, pp. 39–48, 2024.

[11] H. Pretl and G. Zachl, "GitHub repository of the IIC-OSIC-TOOLS," Sep. 2023. [Online]. Available: https://github.com/iic-jku/IIC-OSIC-TOOLS

[12] A. Gatti and F. Tavernier, "Cryogenic Small Dimension Effects and Design-Oriented Scalable Compact Modeling of a 65-nm CMOS Technology," *IEEE Journal of the Electron Devices Society*, 2024.

[13] M. Montanares, "GitHub repository of IceMOS sky130," Apr. 2025. [Online]. Available: https://github.com/Mauricio-xx/IceMOS_sky130

Implementation of a 16:1 Multiplexer and 1:16 Demultiplexer on a Single Chip Using Sky130 PDK and Open-Source EDA Tools for Silicluster

Uriel Jaramillo-Toral[1], Susana Ortega-Cisneros[1], Emilio Isaac Baungarten-Leon[1,2], Erick Jaramillo-Toral[3], and Héctor Emmanuel Muñoz Zapata[1]

[1]Centro de Investigación y de Estudios Avanzados del Instituto Politécnico Nacional, Zapopan, México
[2]Universidad Autónoma de Guadalajara, Zapopan, Jalisco, México
[3]Instituto Tecnológico de Aguascalientes, Aguascalientes, Ags., Mexico

Abstract—This paper presents the design and implementation of a 16-to-1 multiplexer (MUX) and a 1-to-16 demultiplexer (DE-MUX) integrated onto a single chip using the SkyWater 130 nm process design kit and open-source electronic design automation (EDA) tools. These components are part of the Silicluster project, a scalable architecture aimed at integrating 256 independent projects, including digital, analog, and mixed-signal circuits, on a single silicon die. The design flow employs Xschem for schematic capture, Magic for layout, and Netgen for simulation and layout-versus-schematic (LVS) verification, providing cost-effective, accessible, and flexible solutions for circuit design. The implemented circuits achieved a cut-off frequency of 398.1 MHz and a power consumption of 1.488 mW, demonstrating the correct functionality and efficiency of the proposed design. These results highlight the potential of open-source tools for creating functional and scalable circuits, contributing to a sustainable and modular approach to semiconductor development by reducing chip fabrication costs.

Index Terms—Open-Source Tools, Sky130 PDK, Multiplexer, Demultiplexer, Silicluster Project, Xschem, Magic, Netgen.

I. INTRODUCTION

In recent years, the need for efficient and scalable integrated circuit designs has led to the exploration of open-source tools for circuit design and verification. These tools offer significant advantages in terms of cost-effectiveness and accessibility, enabling both academia and industry to participate in the design of cutting-edge technologies. The use of open-source tools has become increasingly relevant, particularly for those working on large-scale semiconductor projects where flexibility and customization are critical [1], [2].

This paper presents the design, simulation, and verification of a 16-to-1 multiplexer (MUX) and a 1-to-16 demultiplexer (DEMUX) [3], both integrated onto a single chip. These components were designed using the SkyWater 130nm process design kit (PDK) [4] and implemented with open-source electronic design automation (EDA) tools. Specifically, Xschem [5] was used for schematic capture, Magic [6] for layout design, and Netgen [7] for simulation and layout-versus-schematic (LVS) verification. These open-source tools allow for an entirely free and accessible design flow, making it

feasible to produce complex integrated circuits without the need for proprietary software.

The multiplexing and demultiplexing functionality is a crucial building block for modern digital systems, offering efficient data routing and signal switching. The designed MUX and DEMUX are intended to be integrated into a larger project called Silicluster [8], which aims to divide a single chip into 256 independent projects. These projects include digital, analog, and mixed-signal circuits, and the proposed analog MUX-DEMUX specifically enables the integration of analog systems within Silicluster by allowing flexible signal routing and selection at the analog level, making the design an integral part of a scalable and modular system. The Silicluster project aims to democratize chip design by providing a platform for multiple independent projects to coexist on a single silicon die, significantly reducing fabrication costs.

This work contributes to the growing field of open-source hardware design by demonstrating the feasibility of building complex, multi-functional systems using freely available tools and resources [9]. The following sections of this paper will discuss the design methodology, simulation results, and performance metrics of the multiplexing and demultiplexing circuits, as well as their integration into the Silicluster framework.

II. DESIGN METHODOLOGY

The implementation of the 16-to-1 MUX and 1-to-16 DE-MUX followed a structured open-source design methodology, ensuring accessibility, efficiency, and compatibility with modern semiconductor fabrication. Utilizing the SkyWater 130nm PDK alongside open-source EDA tools, this approach facilitated a cost-effective and scalable solution for integrating these components within the Silicluster project. The design process was divided into distinct stages, including schematic capture, layout generation, and design rule verification, ensuring compliance with fabrication constraints.

A. Schematic Capture with Xschem

The first step in the design flow was schematic capture, where the logical and functional representation of the

979-8-3315-9813-6/25 $31.00 © 2025 IEEE

MUX and DEMUX was developed. Xschem, an open-source schematic editor, was chosen due to its flexibility and comprehensive component library. The schematics were carefully structured to achieve correct functionality, maintain signal integrity, and optimize performance. The 16:1 MUX selects one of 16 input channels and directs it to a single output, while the 1:16 DEMUX performs the inverse operation, distributing a single input to one of 16 possible outputs based on control signals.

A key aspect of this design is the integration of a 4-to-16 decoder [10], which allows the multiplexing control to be handled with only 4 bits instead of 16. Since the entire circuit is built at the transistor level, this optimization significantly reduces control complexity and interconnect congestion.

1) Switching Mechanism in MUX and DEMUX: The core functionality of the MUX and DEMUX is determined by their switching networks, which enable the proper routing of signals. Figures 1 and 2 show the transistor-level schematics of the fundamental switch used in each circuit.

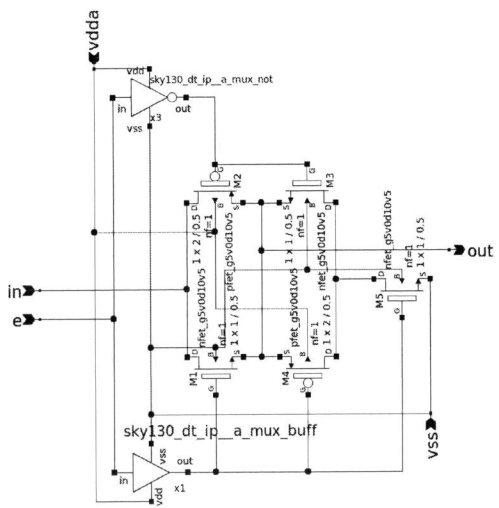

Fig. 2: Schematic representation of the fundamental switch in the 16:1 MUX.

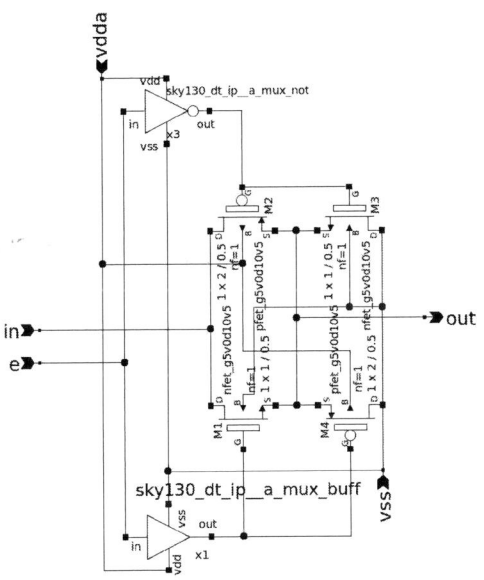

Fig. 1: Schematic representation of the fundamental switch in the 1:16 DEMUX.

The DEMUX switch consists of two logic components: a NOT gate and a buffer. These elements control the voltage applied to the switching transistors, ensuring correct operation. The switching network itself is implemented using a four-transistor configuration, consisting of two NMOS and two PMOS transistors arranged to allow signal routing based on the control inputs. The buffer plays a crucial role in maintaining signal integrity by preventing voltage drops and ensuring that the transistors operate within their expected range.

The MUX switch follows a similar architecture but incorporates an additional transistor at the output. This extra transistor is essential to prevent unwanted voltage feedback when multiple switches share the same output node. Without this additional transistor, an active switch could inadvertently allow voltage to propagate back through inactive paths, potentially corrupting the output signal. By including this blocking

transistor, the MUX ensures that only the selected channel drives the output without interference from the remaining unselected channels.

These transistor-level switching mechanisms provide robust control over the signal routing process while ensuring that unwanted feedback and signal degradation are minimized. The careful design of these switches is crucial for maintaining the circuit's performance across different voltage levels and frequencies.

Figure 3 illustrates the schematic representation of the MUX and DEMUX circuits, showing the interconnections between transistors and control logic.

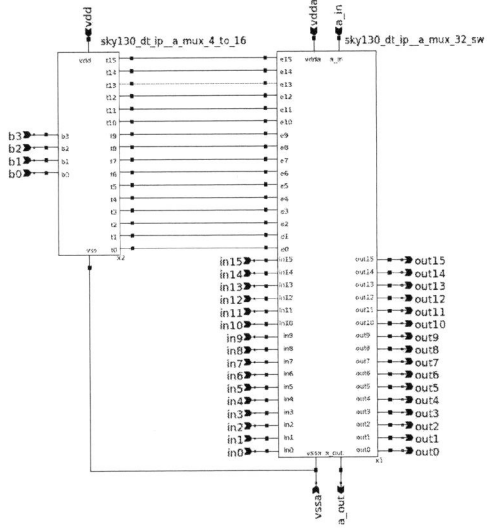

Fig. 3: Schematic representation of the 16:1 MUX and 1:16 DEMUX designed in Xschem.

979-8-3315-9813-6/25 $31.00 © 2025 IEEE

B. Layout Design with Magic

After schematic validation, the physical layout of the MUX and DEMUX was designed using Magic, an open-source layout editor compatible with the SkyWater 130nm PDK. The layout process involved careful placement and routing of transistors while ensuring adherence to design rule constraints. Key considerations included area optimization, symmetry in layout structure, and minimizing parasitic effects to enhance circuit performance.

The finalized layout is shown in Figure 4, where different metal layers and diffusion regions are clearly depicted.

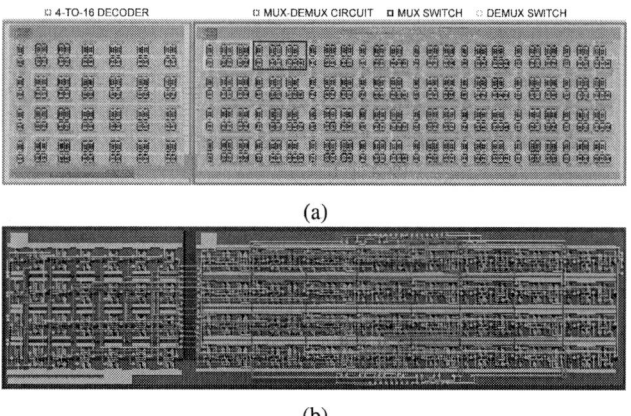

(a)

(b)

Fig. 4: (a) Final layout of the 16:1 MUX and 1:16 DEMUX designed in Magic. (b) 3D render of the fabricated chip generated in Blender.

C. Layout Verification with Netgen

Following layout completion, verification procedures were conducted to ensure design correctness. Netgen was used to perform a LVS check, confirming that the final layout accurately reflected the intended schematic without any connectivity mismatches. This step was essential in identifying and resolving any inconsistencies before proceeding to fabrication. Additional verification steps, including checks for design rule compliance, were also carried out to validate manufacturability [11].

Table I summarizes the key design specifications, including area, technology node, and supply voltages.

TABLE I: Design Specifications for the 16:1 MUX and 1:16 DEMUX

Parameter	Value
Technology Node	130nm (SkyWater PDK)
Digital Supply Voltage	1.8V
Analog Supply Voltage	3.3V
Total Layout Area	236.190 μm × 57.230 μm
Number of Transistors	440
Cutoff Frequency (f_c)	400 MHz
Linear Gain Bandwidth	45 MHz

III. SIMULATION AND VERIFICATION

The design and verification of the 16:1 MUX and 1:16 DEMUX were successfully completed as a single integrated circuit. Simulations were conducted to evaluate the circuit's transient response, frequency response, and overall performance. This section presents the results obtained from transient analysis at 10 kHz and 10 MHz, as well as an AC analysis to determine the circuit's cutoff frequency.

A. Simulation Results

The functional verification of the MUX-DEMUX circuit was performed through transient and AC analyses. The transient simulation was conducted at an input frequency of 10 kHz, representing the circuit's behavior in standard operating conditions. Since the transient response at 10 MHz follows the same pattern, it is not explicitly included.

Figure 5 presents the transient response of the MUX-DEMUX circuit at 10 kHz, illustrating the signal routing behavior and selection process in detail.

The first plot shows the input voltage applied to the MUX, which consists of a ramp signal fed into the single input channel. The second plot displays the 16 outputs of the MUX, where the same ramp signal is reconstructed as different output channels are sequentially selected.

The last four plots represent the selection signals used to control the MUX-DEMUX operation. These signals are digital and correspond to the 4-bit selection inputs, which determine the active output at any given moment.

The third plot presents the 16 test signals processed by the DEMUX. These signals share the same frequency but have voltage levels ranging from 3.2V to 0.2V. Finally, the fourth plot illustrates how the DEMUX selects one of the 16 input signals as the 4-bit control signals cycle through different states, effectively routing a specific voltage to the output.

This transient analysis confirms that the MUX-DEMUX circuit correctly reconstructs the signals while switching between channels, demonstrating proper timing and signal integrity.

To analyze the frequency behavior, an AC analysis was performed to determine the cutoff frequency and the range where the gain remains linear. Figure 6 shows the frequency response, indicating a cutoff frequency (f_c) of 400 MHz and a linear gain bandwidth up to 45 MHz.

B. Performance Analysis

The performance of the integrated MUX-DEMUX circuit was evaluated based on power consumption, area utilization, and frequency response. Table II summarizes these key performance metrics.

TABLE II: Performance Characteristics of the MUX-DEMUX Circuit

Metric	Value
Power Consumption (mW)	1.488
Area (μm^2)	13517.1537
Cutoff Frequency (f_c)	398.1 MHz
Linear Gain Bandwidth	44.6 MHz

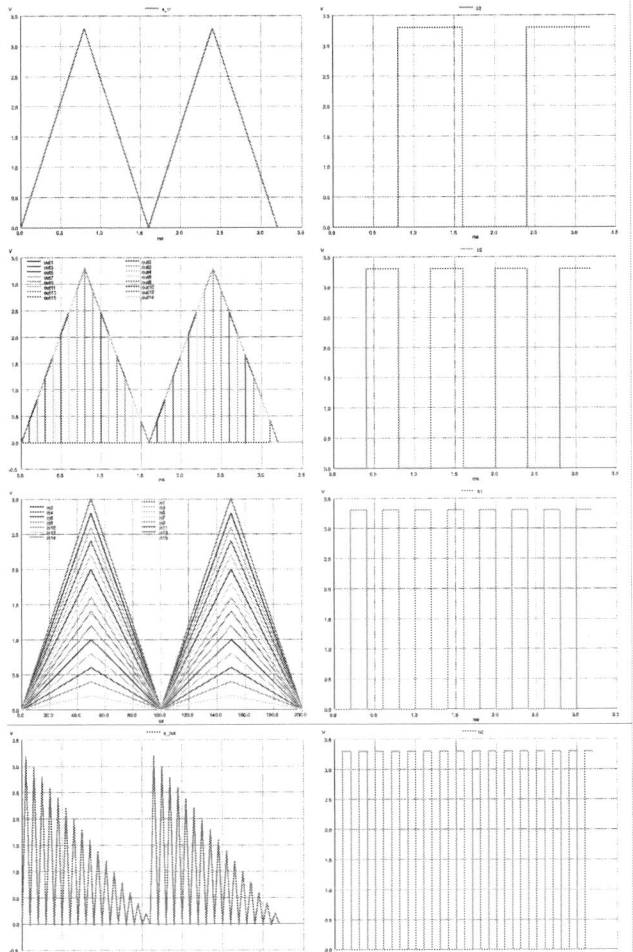

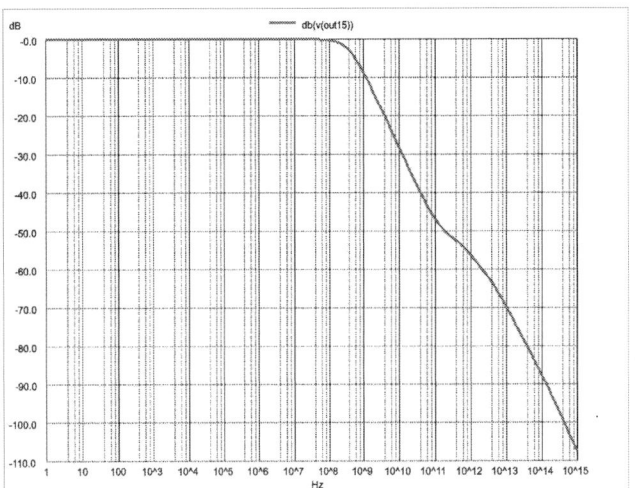

Fig. 6: AC analysis showing the frequency response of the MUX-DEMUX circuit.

Fig. 5: Transient simulation of the MUX-DEMUX circuit at 10 kHz. The response at 10 MHz follows the same behavior.

The results indicate that the MUX-DEMUX circuit operates efficiently, maintaining stable performance within the designed frequency range while optimizing power consumption.

C. Layout Versus Schematic Verification

The final step in the verification process was ensuring that the physical layout matched the intended schematic. Netgen was used to perform layout-versus-schematic (LVS) checks, ensuring that all connections and device placements were correctly implemented. The results confirm that all subcircuits and device classes match uniquely, validating the design for fabrication.

The LVS results confirm that all subcircuits match uniquely, with no discrepancies detected between the schematic and layout. This ensures that the design adheres to the intended functionality and is ready for fabrication.

D. Integration into Silicluster Project

The MUX-DEMUX circuit has been successfully integrated into the Silicluster project, where it serves as a key element in distributing signals among independent analog blocks. Its function is to selectively route input signals to one of the 16 independent analog modules, while also collecting signals from these modules and directing them to a single output.

By enabling efficient signal distribution, the MUX-DEMUX circuit enhances the scalability and modularity of Silicluster, allowing multiple projects to coexist on the same chip with minimal routing complexity. This integration ensures that the circuit operates within the expected parameters and contributes to the overall functionality of the system.

IV. CONCLUSION

This work presented the design, implementation, and verification of a fully functional 16:1 MUX and 1:16 DEMUX circuit using the open-source SkyWater 130nm PDK and EDA tools. The circuit was designed at the transistor level, integrating a 4-to-16 decoder to efficiently manage the selection process with only four control bits.

The transient and AC simulations demonstrated the circuit's ability to operate correctly across a wide frequency range. The transient response at 10 kHz confirmed that the MUX successfully reconstructs the input ramp signal across all 16 outputs, while the DEMUX properly routes one of the 16 test signals based on the selection bits. The AC analysis determined a cutoff frequency of 400 MHz and a linear gain bandwidth of 45 MHz, ensuring the circuit's suitability for high-speed applications.

The layout-versus-schematic verification confirmed that the fabricated layout matches the intended schematic with no discrepancies. The final design comprises 440 transistors and occupies an area of 236.190 μm × 57.230 μm, demonstrating an efficient implementation within the given constraints.

Finally, the successful integration of the MUX-DEMUX circuit into the Silicluster project highlights its scalability and modularity. The circuit enables seamless signal distribution across independent analog blocks, supporting the development of complex mixed-signal designs within a shared silicon platform.

979-8-3315-9813-6/25 $31.00 © 2025 IEEE

REFERENCES

[1] M. Shalan and T. Edwards, "Building OpenLANE: A 130nm OpenROAD-based Tapeout-Proven Flow: Invited Paper," 2020 IEEE/ACM International Conference On Computer Aided Design (ICCAD), San Diego, CA, USA, 2020, pp. 1-6.

[2] Emilio Isaac Baungarten-Leon, Susana Ortega-Cisneros, Mohamed Abdelmoneum, Ruth Yadira Vidana Morales, and German Pinedo-Diaz, "The Genesis of AI by AI Integrated Circuit: Where AI Creates AI," *Electronics*, vol. 13, no. 9, p. 1704, 2024.

[3] R. J. Baker, *CMOS: Circuit Design, Layout, and Simulation*. Hoboken, NJ, USA: John Wiley & Sons, 2019.

[4] SkyWater PDK Authors, "SkyWater SKY130 PDK Documentation," Available: https://skywater-pdk.readthedocs.io/en/main/, 2020. [Accessed: Mar. 10, 2025].

[5] S. Schippers, "Xschem: Schematic capture for VLSI and analog design," Available: https://xschem.sourceforge.io/stefan/index.html.

[6] J. Ousterhout, G. Hamachi, R. Mayo, W. Scott, and G. Taylor, "Magic: a VLSI layout system," in *Proceedings - Design Automation Conference*, vol. 21, pp. 152-159, 1984, doi: 10.1109/DAC.1984.1585789.

[7] T. Edwards, "Netgen: A Tool for Comparing Netlists," Available: http://opencircuitdesign.com/netgen/.

[8] U. Jaramillo Toral, "Designing a Chip with Open-Source Tools: The Development of Silicluster," *Electronic Design*, Jul. 10, 2025. [Online]. Available: https://www.electronicdesign.com/technologies/analog/article/55302523/designing-a-chip-with-open-source-tools-the-development-of-silicluster

[9] E. I. Baungarten-Leon, S. Ortega-Cisneros, U. Jaramillo-Toral, F. J. Rodriguez-Navarrete, L. Pizano-Escalante and J. J. Raygoza-Panduro, "Vector Accelerator Unit for Caravel," in IEEE Embedded Systems Letters, doi: 10.1109/LES.2023.3267341.

[10] N. Kar Chowdhury, R. Dhanabal, and V. Ramakrishnan, "Low Power Structural Design of 2–4 and 4–16 Line Decoder Logic Circuit," *J. Comput. Theor. Nanosci.*, vol. 17, pp. 2266-2272, 2020, doi: 10.1166/jctn.2020.8882.

[11] U. Jaramillo-Toral, J. C. Garcia-Lopez, S. Ortega-Cisneros, E. I. Baungarten-Leon, C. Torres-González, and F. Sandoval-Ibarra, "Automated IC Design Flow Using Open-Source Tools and 180 nm PDK," in *Proc. 2024 IEEE 67th Int. Midwest Symp. Circuits Syst. (MWSCAS)*, Springfield, MA, USA, 2024, pp. 1393-1397, doi: 10.1109/MWSCAS60917.2024.10658750.

Inter-chip Clock Network Synthesis on Passive Interposer of 2.5D Chiplet Considering Transmission Line Effect

Tai Yan[‡*], Yiyu Wang[†,§*], Zhan Li[¶], Ning Xu[‡] and Yuanqing Cheng[†,§]

[†]School of Integrated Circuit Science and Engineering, Beihang University, Beijing, China 100191
[‡]School of Computer and Artificial Intelligence, Wuhan University of Technology, Wuhan, Hubei 430070
[§]Shenzhen Institute of Beihang University, Shenzhen, China 518000
[¶]School of Mathematical Science, Beihang University, Beijing, China 100191
{zexiao852@gmail.com, wyy18735134197@buaa.edu.cn, yuanqing@buaa.edu.cn}

Abstract—With the slowdown of technology node scaling, 2.5D Chiplet technology has emerged as a promising approach to sustain Moore's Law. When routing clock signal on the passive interposer layer, transmission line effect must be considered, as signal rise/fall delay becomes comparable to propagation delay, significantly impacting signal integrity and system performance. Additionally, since active devices cannot be placed on the passive interposer, buffers can only be inserted in the chiplets mounted on the interposer. This paper investigates the problem of clock network synthesis for a 2.5D Chiplet with a passive interposer. Firstly, we propose to use a transmission line model to evaluate inter-chip clock skew and delay accurately. Then, we propose a clock network synthesis method considering transmission line effect and impedance matching. Finally, we propose a buffer insertion method with minimum clock wire detouring on passive interposer in order to optimize the inter-chiplet clock network. Experimental results show that our fast transmission line model can calculate clock skew more accurately compared to conventional Elmore model and only results in 2.4% error compared to HSPICE simulations. The clock network syntheses on several 2.5D Chiplet benchmarks show that our proposed algorithm is effective to construct zero-skew clock network for a 2.5D Chiplet in terms of clock skew and buffer area.

Index Terms—Transmission line model, clock network synthesis, passive interposer, buffer insertion, clock skew, 2.5D Chiplets

I. INTRODUCTION

In the post-Moore era, technology node scaling is approaching the physical limit in terms of integration density and economic returns, making traditional system-on-chip (SoC) design methodologies [1] unsustainable. To address this challenge, the semiconductor industry has been exploring new integration technologies, where 2.5D Chiplet emerges as a promising solution. 2.5D Chiplet integrates multiple small chips (referred to as "chiplets") on an interposer, forming a tightly integrated system that enhances performance and energy efficiency while mitigating the skyrocketing fabrication costs of large-scale chips, but it also introduces unique system-level design challenges.

*These authors contributed equally to this work. This work is supported in part by the Natural Science Foundation of China under Grant No. 92373205 and Shenzhen Science and Technology Program under Grant No. SGDX20230116093303006 and KJZD20231023100201003.

In the 2.5D Chiplet design [2], the clock transition time is comparable to interconnect transmission delay, leading to underdamped signal propagation where transmission line effect cannot be ignored. In previous research such as [3] [4] [5] [6] many interconnect optimizations or clock tree synthesis methods for multi-chip modules (MCMs) were proposed that considered transmission line modeling [7], however, most of these studies focused on multi-chip module technology, where different modules are loosely coupled and may not require a high quality global clock network. As for a 2.5D Chiplet, it require a high speed clock network to synchronize many chiplets. Recently, Murali *et al.* proposed clock synthesis method for the 2.5D Chiplet with EDA vendor tools [8], which also considered transmission line effect, but the buffer insertion strategy was not clearly discussed in the paper.

This paper presents a clock synthesis technique for a 2.5D Chiplet that accounts for transmission line effect and signal reflection, alongside a novel buffer insertion strategy for a passive interposer. Our main contributions can be summarized as follows.

- We propose a transmission line model to effectively calculate inter-chiplet clock delay induced by distributed interconnect parameters and develop an impedance matching technique during clock tree synthesis to avoid signal reflections at branching points, thereby optimizing the quality and stability of clock signal transmission.
- We propose a clock tree buffer insertion method for the passive interposer. By inserting buffers on nearby chiplets closest to the optimal buffer insertion positions, we can effectively optimize clock slew rate and clock skew with minimum wire detouring.
- Compared with HSPICE simulation, our transmission line model has delay error less than 2.4%, which is more accurate than Elmore delay model. Buffer insertion reduces skew by 65.7% and delay by 25.1%, with minimal increases in wire length and power compared to relevant clock synthesis method.

The structure of this paper is arranged as follows: Section

II introduces 2.5D Chiplet integration technology and the research motivation. Section III presents the formulation of the problem. Section IV details the technical workflow of the proposed clock network synthesis method. Section V shows the experimental results and comparisons with related work. Finally, Section VI concludes this paper.

II. PRELIMINARIES

A. 2.5D Chiplet integration

As shown in Fig. 1, 2.5D chiplet integration [9] involves mounting multiple chiplets onto a substrate and connecting different chiplets to C4 bumps through through-silicon vias (TSV) [10] in the interposer, and inter-chiplet interconnects are implemented in the redistribution layer (RDL). This integration approach represents a trade-off between traditional 2D chip design and more advanced 3D integration technology. Compared to traditional multi-chip modules (MCMs), 2.5D Chiplet integration offers greater design flexibility, better performance scalability, and higher production efficiency.

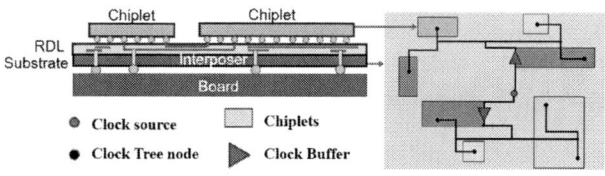

Fig. 1: An illustration of a 2.5D Chiplet and the clock network between chiplets.

B. A motivating example

To evaluate Elmore model accuracy for 2.5D Chiplet clock design and illustrate the necessity of considering transmission line effect, we present a synthesized example inspired by [11]. As shown in Fig. 2, each chiplet which has its local clock network on chip, and each local clock network is connected on a passive interposer to form the complete clock network.

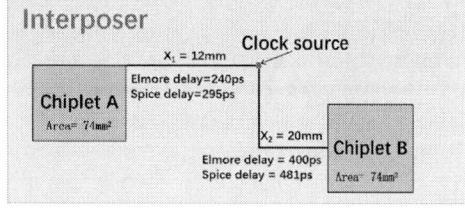

Fig. 2: An illustrative example for delay estimation of the inter-chiplet clock signal.

Chiplet A is $12mm$ from the clock driver, and chiplet B is $20mm$ away, and interposer interconnect parameters are: unit resistance $R = 0.1\Omega/\mu m$, unit capacitance $C = 0.2\text{fF}/\mu m$, and unit inductance $L = 5\text{fH}/\mu m$ [12]. Clock signal delays for both chiplets are obtained using the Elmore delay model [13] and HSPICE simulation.

TABLE I: Comparison of delay results between the Elmore delay model and HSPICE simulation.

Chiplet	Elmore Delay(ps)	HSPICE Delay (ps)	Delay Difference
Chiplet A	240	295	45 ps
Chiplet B	400	481	81 ps

Table I compares delay results of Elmore model and HSPICE simulation, revealing a 19.5% error of Elmore model in interposer delay estimation. This highlights significant inaccuracies in the Elmore model, necessitating the consideration of transmission line effect for delay estimation to construct high quality inter-chiplet clock network.

III. PROBLEM FORMULATION

Consider a 2.5D Chiplet system with multiple chiplets $C = \{c_1, \ldots, c_k\}$ integrated on a $W \times H$ passive interposer, where inter-chiplet clock interconnects route through the RDL layer on the interposer. Given interconnect parameters (unit resistance α, unit capacitance β, unit inductance γ) and n clock sinks with locations $S(x_i, y_i), i = 1 \ldots n$. We need to construct an inter-chiplet buffered clock network with buffer locations constrained by chiplet locations on the passive interposer, i.e., buffers cannot reside directly on the interposer but must be placed within some chiplets. Moreover, the clock network should meet the given clock skew and capacitive load constraints, which can be defined as follows:

$$\begin{cases} t_{\text{skew}} = \max\{t_i - t_j\} \\ (x_{B_m}, y_{B_m}) \in C \\ C_{\text{load}}(x_i, y_j) \leq C_{\text{max}} \quad \forall(x_i, y_j) \in S \end{cases} \quad (1)$$

where t_i and t_j are the clock delays from the source to sinks i and j, (x_{B_m}, y_{B_m}) represents the buffer insertion position. When the capacitive load $C_{\text{load}}(x_i, y_j)$ between nodes i and j exceeds the predefined threshold C_{max}, a buffer is inserted to improve clock skew rate. The objective is to minimize clock skew and wirelength of the buffered inter-chiplet clock network.

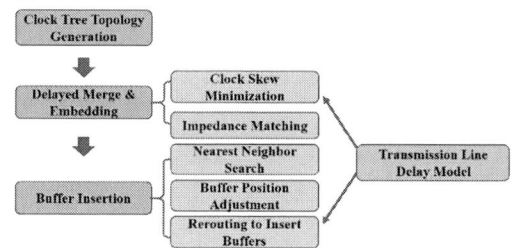

Fig. 3: The workflow of 2.5D Chiplet clock network synthesis.

IV. OUR PROPOSED METHOD FOR 2.5D CHIPLET CLOCK TREE SYNTHESIS

As illustrated in Fig. 3, our 2.5D interconnect clock network design can be divided into two phases. First, a bottom-up merging approach builds the clock network topology on the passive interposer. Using a transmission line model, we merge clock tree segments—optimizing delays/skews of subtrees with an admittance approximation method to obtain subtree merging points. Then, clock tree detailed routing and buffer insertion are performed based on chiplet positions. Some clock wire detouring may be introduced during buffer insertion to locate buffers into optimal nearby chiplets.

A. Clock tree construction

Considering transmission line effects, we adopt the second-order RCL transfer function approximation from [4] to model clock interconnect delays. The transfer function below approximates the response of any clock tree leaf node relative to its parent node:

$$H(s) = \frac{V_{leaf}(s)}{V_{parent}(s)} = \frac{1}{1 + b_1 s + b_2 s^2 + b_3 s^3 + \ldots + b_n s^n} \quad (2)$$

The transfer function $H(s) = \frac{V_{leaf}(s)}{V_{parent}(s)}$ models clock signal propagation on the transmission line, with its series expansion $\sum_{n=1}^{\infty} b_n s^n$ capturing $b_1 s$ for RC delay, $b_2 s^2$ for inductance and distributed impedance, and $b_n s^n$ ($n \geq 3$) for signal attenuation, reflection, and resonance; we adopt the transmission line model from [5] to estimate inter-chip delays in the passive interposer, thereby improving clock delay accuracy.

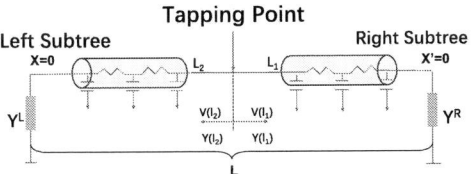

Fig. 4: Zero-skew merging based on the transmission line model.

Fig. 4 shows a clock internal node merging example based the transmission line model (shunt conductance assumed zero). Considering a length-l transmission line terminated by admittance $Y(0)$ at $x = 0$. We derive the near-end ($x = l$) admittance and transfer function $\frac{V_{x=l}(s)}{V_{x=0}(s)}$ to obtain the first-order admittance approximation $Y_1(0)$. The zero-skew clock network merging process in the figure uses (2) to approximate far-end admittance via first-moment matching.

$$Y_1(x) = Y_1(0) + \beta x \quad (3)$$

The value of the transfer coefficient b_1 in (2) can be determined by the following equation:

$$b_1(x) = b_1(0) + \frac{\alpha \beta x^2}{2} + \alpha Y_1(0) x \quad (4)$$

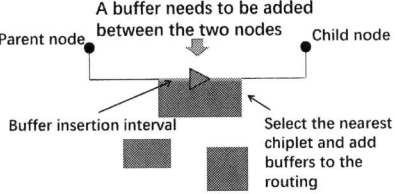

Fig. 5: An illustration of determining the buffer insertion interval.

We calculate the admittances Y^L and Y^R of the left and right subtrees. The ratio r of the wire lengths between the left and right subtrees is determined such that the first moment $H(s)$ is matched.

$$r = \frac{\frac{\alpha \beta}{2} l^2 + \alpha l Y_1^L(0) + (b_1^R(0) - b_1^L(0))}{\alpha \beta l^2 + \alpha l (Y_1^L(0) + Y_1^R(0))} \quad (5)$$

where b_1^L and b_1^R are the transfer function coefficients for the left and right subtrees, respectively. After determining the merging points with the bottom-up order, we perform top-down hierarchical clock embedding to generate the final clock tree structure based on DME algorithm [14].

B. Buffer insertion

Our buffer insertion strategy is as follows: Buffers are inserted when the capacitive load of subtree exceeds a specified threshold as mentioned in Section III. Due to the constraints of the passive interposer (no active components), we have to place buffers in nearby chiplets. To minimize delay and power consumption, as illustrated in Fig. 5, when a buffer needs to be inserted between a segment of interconnects, we adopt a nearest neighbor strategy based on Manhattan distance to select the chiplet closest to the interconnect segment as a buffer insertion candidate. At this time, through wire detouring, the interconnect between clock nodes is made to partially overlap with the edge of the candidate chiplet, and this overlapping region serves as the buffer insertion interval. After determining the buffer insertion interval, the specific position of the buffer within the insertion interval is determined through clock skew and admittance balancing between interconnect segments, which is detailed in the next paragraph.

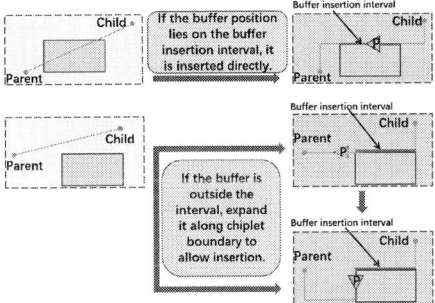

Fig. 6: Buffer insertion and wire detour scenarios: (a) Clock segment does not overlap with all chiplets. (b) Clock segment overlaps with a chiplet.

After determining the candidate chiplet, we utilize the obstacle avoidance method from [15] to align the interconnect with the chiplet edge through wire detouring. At this stage, we need to determine the specific insertion position of the buffer within this candidate interval. Using (6), we select the optimal buffer insertion point (denoted as p) where the terminal admittances are matched. Based on the transmission line model, we derive the insertion point formula:

$$r' = \frac{Y_R}{Y_L + Y_R} \quad (6)$$

where Y_L/Y_R denotes left/right admittances and r' the wire length ratio, to determine the buffer location. As shown in Fig. 6, if the calculated insertion point p falls within the insertion interval the buffer is inserted directly as shown in Fig. 6; otherwise, the insertion interval is expanded along the chiplet boundary via wire detouring until p lies on the expanded buffer insertion interval as shown in Fig.6. With this buffer insertion strategy, we can optimize buffer placement

979-8-3315-9813-6/25 $31.00 © 2025 IEEE

TABLE II: Comparisons between the clock tree constructed using the transmission line (TL) model and the clock tree constructed using the Elmore delay model.

Metrics	Delay (ps)		Skew (ps)		WL (mm)		Power (mW)		Buffers	
	Elmore	TL	Elmore	TL	Elmore	TL	Elmore	TL	Elmore	TL
CASE1	177.1	129.3	45.2	5.9	54.5	61.4	4.38	4.99	5	5
CASE2	364.7	284.0	55.4	17.5	85.9	92.8	5.67	6.12	12	9
CASE3	469.8	363.7	62.1	22.6	142.0	153.4	6.85	7.35	16	11
CASE4	761.2	528.1	54.3	29.5	181.7	212.1	8.33	8.72	23	17
Average Reduction (%)	25.6		66.2		-11.3		-8.5		20.6	

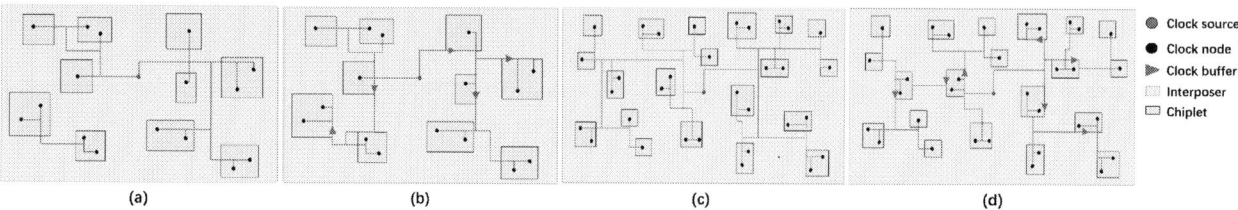

Fig. 7: CASE1/CASE2 clock topologies before/after buffer insertion: (a) before buffer insertion in CASE1; (b) after buffer insertion in CASE1; (c) before buffer insertion in CASE2; (d) after buffer insertion in CASE2.

with minimum extra wirelength, while meeting clock skew and slew rate constraints.

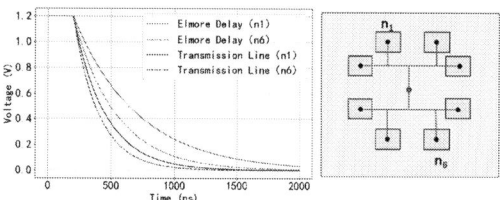

Fig. 8: Clock waveforms of some nodes in clock trees based on transmission line or Elmore model in CASE0.

V. EXPERIMENTAL RESULTS

A. Experiment Setup

In our experiments, the clock interconnect has unit resistance $R = 0.1\Omega/\mu m$, unit capacitance $C = 0.2\text{fF}/\mu m$, and unit inductance $L = 5\text{fH}/\mu m$ [12]. We chose CLKBUF1 from FreePDK45 library with $5.4fF$ input capacitance and $46.5ps$ intrinsic delay, and manually generated four 10 mm × 10 mm 2.5D Chiplet designs with randomly distributed clock sinks across chiplets. Table III lists the benchmarks used in our experiments. HSPICE [12] was used for clock skew and delay validations.

TABLE III: Benchmarks used in our experiments

Name	CASE0	CASE1	CASE2	CASE3	CASE4
sink_number	8	16	32	64	128
chiplets	8	10	22	45	92

TABLE IV: Comparisons before and after buffer insertion.

CASES	Before buffer insertion		Buffers	After buffer insertion	
	Skew (ps)	WL (mm)		Skew (ps)	WL (mm)
CASE1	12.7	59.3	5	5.9	61.4
CASE2	24.2	84.2	7	17.5	92.8
CASE3	38.0	128.0	11	22.6	153.4
CASE4	45.7	163.7	16	29.5	212.1

B. Comparisons with clock tree constructed by Elmore delay model

This section compares transmission line and Elmore-based clock trees. Waveform comparisons of CASE0 clock nodes show the former reduces delay significantly, validating its accuracy in delay estimation.

For test cases in Table III, we built clock trees using transmission line and Elmore models. Table II shows the former reduces delay by 25.1% and skew by 65.7%. Additionally, compared to Elmore delay based clock network, the transmission line based clock network consumes 9% more power due to 8.5% extra wirelength. Although the buffer count can be reduced, the power savings by clock buffers cannot compensate the interconnect power due to extra wirelength. As the scale of chiplet design increases, our method can show benefits not only on clock delay and skew reductions but also on overall power consumption.

C. Comparisons before and after buffer insertion

We validated our buffer insertion method via on four test cases. Table IV shows clock skew can be reduced by 38.7% on average, while wirelength increases by 16.5% after buffer insertion due to wire detouring to chiplets. Fig.7 illustrates clock topologies before and after buffer insertion of CASE1 and CASE2. It shows that buffer insertion improves clock skew effectively, but wire detouring on passive interposer must be considered.

VI. CONCLUSION

We propose a method for constructing an inter-chiplet clock tree on a passive interposer based on a transmission line model. This method uses a second-order approximation model suitable for more accurate delay estimation to recursively build the clock tree, while employing a rerouting approach to reasonably insert buffers for the passive interposer based 2.5D Chiplet. Our method reduces clock skew by 65.7% over Elmore delay based method, with minimal wirelength and power overheads.

979-8-3315-9813-6/25 $31.00 © 2025 IEEE

REFERENCES

[1] Fan Sang, Jaehyuk Lee, Xiaokuan Zhang, and Taesoo Kim. PORTAL: Fast and Secure Device Access with ARM CCA for Modern ARM Mobile System-on-Chips (SoCs). In *Proceedings of the IEEE Symposium on Security and Privacy (S&P)*, 2025.

[2] Shixin Chen, Hengyuan Zhang, Zichao Ling, Jianwang Zhai, and Bei Yu. The Survey of 2.5D Integrated Architecture: An EDA perspective. In *ASPDAC '25: Proceedings of the 30th Asia and South Pacific Design Automation Conference*, pages 285–293, 2025.

[3] J. S. H. Wang and W. W.-M. Dai. Optimal Design of Self Damped Transmission Lines for Multi-chip Modules. In *Proc. ICCAD*, pages 594–598, 1994.

[4] M. Sriram and S. M. Kang. *Physical Design of Multi-Chip Modules*. Kluwer Academic Publishers, Boston, MA, 1994.

[5] D. Lehther and S. S. Sapatnekar. Clock Tree Synthesis for Multi-Chip Modules. In *Proc. ICCAD*, pages 50–53, 1996.

[6] P Gerlach, C Linder, and K.-H Becks. Multi Chip Modules technologies. *Nuclear Instruments and Methods in Physics Research Section A: Accelerators, Spectrometers, Detectors and Associated Equipment*, 473(1–2):102–106, 2001.

[7] Mouataz Billah Mesmouli, Abdelouaheb Ardjouni, Ioan-Lucian Popa, Hicham Saber, Faten H Damag, Yasir A Madani, and Taher S Hassan. Qualitative properties of nonlinear neutral transmission line models and their applications. *Axioms*, 14(4):269, 2025.

[8] Gauthaman Murali, Heechun Park, Eric Qin, Hakki Mert Torun, Majid Ahadi Dolatsara, Madhavan Swaminathan, Tushar Krishna, and Sung Kyu Lim. Clock delivery network design and analysis for interposer-based 2.5-D heterogeneous systems. *IEEE Transactions on Very Large Scale Integration (VLSI) Systems*, 29(4):605–616, 2021.

[9] Jinwoo Kim et al. Architecture, Chip, and Package Codesign Flow for Interposer-Based 2.5-D Chiplet Integration Enabling Heterogeneous IP Reuse. *IEEE Transactions on Very Large Scale Integration (VLSI) Systems*, 28(11):2424–2437, 2020.

[10] Sooyong Choi, Sooman Lim, Muhamad Mukhzani Muhamad Hanifah, Paolo Matteini, Wan Yusmawati Wan Yusoff, and Byungil Hwang. An introductory overview of various typical lead-free solders for tsv technology. *Inorganics*, 13(3):86, 2025.

[11] Samuel Naffziger, Kevin Lepak, Milam Paraschou, and Mahesh Subramony. 2.2 AMD chiplet architecture for high-performance server and desktop products. In *2020 IEEE International Solid-State Circuits Conference-(ISSCC)*, pages 44–45. IEEE, 2020.

[12] Samuel H Russ. *Signal Integrity*. Springer, 2022.

[13] A.I. Abou-Seido, B. Nowak, and C. Chu. Fitted Elmore delay: a simple and accurate interconnect delay model. *IEEE Transactions on Very Large Scale Integration (VLSI) Systems*, 12(7):691–696, 2004.

[14] Ting-Hai Chao, Yu-Chin Hsu, Jan-Ming Ho, and A.B. Kahng. Zero skew clock routing with minimum wirelength. *IEEE Trans. Circuits Syst. II, Analog Digit. Signal Process.*, 39(11):799–814, November 1992.

[15] Z. Qian, C. Weng, and R. Suaya. Generalized Impedance Boundary Condition for Conductor Modeling in Surface Integral Equation. *IEEE Transactions on Microwave Theory and Techniques*, 55(11):2354–2364, 2007.

Fault Modeling and Testing of Spin-Orbit Torque-Based Multipillar Memory Cell

Arshid Nisar
Univ. Grenoble Alpes, CNRS, CEA,
Grenoble INP, SPINTEC, France
arshid-nisar.laway@cea.fr

Lorena ANGHEL
Univ. Grenoble Alpes, CNRS, CEA,
Grenoble INP, SPINTEC, France
lorena.anghel@phelma.grenoble-inp.fr

Gregory DI PENDINA
Univ. Grenoble Alpes, CNRS, CEA,
Grenoble INP, SPINTEC, France
gregory.dipendina@cea.fr

Abstract—**Multipillar spintronic memory has emerged as a promising candidate for next-generation magnetoresistive random access memory (MRAM), offering advantages such as low power consumption, high integration density, and suitability for in-memory computing and neuromorphic applications. However, the intricate fabrication of multipillar structures increases susceptibility to manufacturing defects, necessitating robust and efficient testing methodologies. In this work, we analyze a single-cell spin-orbit torque (SOT)-based multipillar MRAM device under various resistive bridges, open defects, and transistor stuck-on/stuck-open faults. Corresponding fault models such as stuck-at faults (SAFs), transition faults (TFs), coupling faults (CFs), and read disturbance faults (RDFs) are proposed. A dedicated test algorithm is developed to detect these faults using a multilevel sensing scheme. It offers full coverage of all SAFs, TFs, CFs, and RDFs, and enables concurrent testing of multiple MTJs to significantly reduce test time. The proposed approach provides a robust design for testing framework for emerging multilevel SOT-MRAM architectures.**

Index Terms—**Defects, Fault modeling, Magnetic tunnel junction, Memory testing, Multipillar-MRAM, Spintronics**

I. INTRODUCTION

Magnetic random-access memory (MRAM) has emerged as a promising candidate among next-generation memory technologies due to its non-volatility, high endurance, low power consumption, fast operation, and CMOS compatibility. In particular, spin-transfer torque (STT) and spin-orbit torque (SOT) MRAM have gained significant attention. However, STT-MRAM suffers from limited switching speed and high energy consumption due to a large incubation delay, while SOT-MRAM faces challenges in integration density due to the need for an additional access transistor. To address these challenges, a multi-pillar SOT device was reported, leveraging the advantages of unidirectional SOT and STT writing mechanisms to achieve low-energy operation [1], [2]. This device has demonstrated significant potential for high-density memory and in-memory computing applications [3], [4]. However, the complex and still-maturing manufacturing processes make these technologies susceptible to various defects, impacting their quality and reliability. Therefore, precise fault models that characterize the faulty behavior of these defects, along with effective test methods for their detection, are essential for ensuring robust and reliable MRAM technology.

Wu et al. proposed a device-aware defect and fault modeling framework, analyzing resistive and pinhole defects [5], and introduced a systematic fault analysis for intermediate-state defects using silicon data, non-linear approach, and Verilog-A models [6], [7]. Taghipour et al. developed a low-cost design-for-testability (DFT) scheme to enhance the detection of hard-to-detect faults by introducing a voltage mismatch in the sense amplifier [8]. Nair et al. conducted detailed defect injections based on magnetic device and layout characteristics to derive unique fault models and efficient test algorithms [9], [10]. Chintaluri et al. presented a comprehensive analysis of fault models, considering parametric variations, intra-cell and inter-cell defects, and data pattern dependencies [11]. Yoon et al. studied resistive and capacitive defects, proposing a novel MBIST architecture for testing read, write, and retention characteristics under variations and magnetic coupling [12]. Yuan et al. introduced and characterized back-hopping defects, providing corresponding fault models and test solutions [13]. Previous studies have primarily focused on single-bit STT-MRAMs. Given the potential of multi-pillar SOT-MRAM for high-density memory and advanced applications like in-memory and neuromorphic computing, there is a critical need for comprehensive testing methodologies, which remain largely unexplored in this domain.

In this work, a spin-orbit torque-based multi-pillar memory device is analyzed for various electrical defects, including resistive bridges between nodes, open defects, and transistor stuck-on and stuck-open faults. Corresponding fault models, such as stuck-at faults, transition faults, coupling faults, and read disturb faults are proposed. Additionally, a test algorithm is developed to detect these faults efficiently. A multilevel sensing scheme is utilized for simultaneous testing of multiple memory elements within the multi-pillar device, significantly reducing overall testing time.

II. MULTIPILLAR SOT-MRAM AND ITS OPERATION

The multi-pillar SOT-MRAM consists of four magnetic tunnel junctions (MTJs), with their free layers placed on SOT channel layer as shown in Fig.1. Each MTJ is composed of a free layer and a fixed layer, separated by a thin MgO tunnel barrier. The magnetization of the free layer can be altered, while the pinned layer remains fixed. The resistance of the device depends on the relative orientation of the magnetization between the free and pinned layers: a parallel (P) orientation results in a low resistance state ("0"), while an anti-parallel (AP) orientation leads to a high resistance state ("1").

979-8-3315-9813-6/25 $31.00 © 2025 IEEE

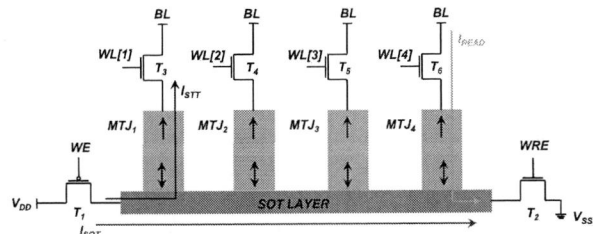

Fig. 1. Structure of the multi-pillar SOT-MRAM device.

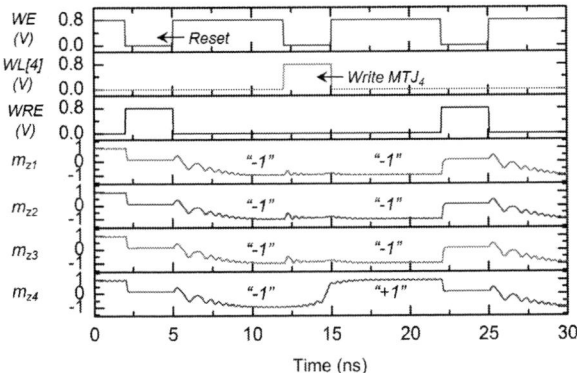

Fig. 2. Transient behavior showing reset and write operations (write "1" to MTJ_4). Here, $m_{zi} = +1$ (-1) represents the P(AP) magnetization orientation of ith MTJ in the multipillar device

The write operation is performed in two steps: First, all bits are switched to the '0' or parallel (P) state in a single write operation by passing current through the SOT layer via the SOT mechanism. This is achieved by activating transistors T_1 and T_2 while deactivating all word line (WL) transistors. To write '1' or the anti-parallel (AP) state for a specific bit, the corresponding WL and T_1 transistors are activated, allowing current to flow from the free layer to the pinned layer, switching the device via the STT mechanism. Fig. 2 illustrates the transient behavior during the reset operation, where all MTJs are switched to the "0" state using the SOT current. Subsequently, a "1" is written to MTJ_4 through the STT mechanism. The read operation can be performed by passing a read current through the intended MTJ by activating the corresponding $WL[n]$ and transistor T_2, followed by sensing the output using a sense amplifier [14]. The sense amplifier compares the cell current or voltage with a predefined reference to determine the stored "0" or "1" state.

III. DEFECTS AND PROPOSED FAULT MODELS

In this section, various defects, corresponding fault models, and proposed sensitization methodology are discussed, showing the impact of these defects on the functionality of the multi-pillar SOT-MRAM device. The fault behavior of the multipillar device is analyzed by injecting three types of defects: *1) Bridge defects*, which represent unintended connections between two nodes; *2) Open defects*, which signify interruptions in a connection; and *3) Transistor stuck-on and stuck-open defects*, where a transistor remains permanently in

the ON or OFF state, regardless of the applied gate voltage. These three types of defects have been specifically discussed here as they are the most prevalent issues arising from fabrication complexities in modern memory technologies. Fig. 3 shows possible defects and defect locations in multipillar SOT-MRAM. R_B, R_O, T_{ON}, and T_{OFF} denote possible bridge, open, transistor stuck-on, and transistor stuck-off defects, respectively.

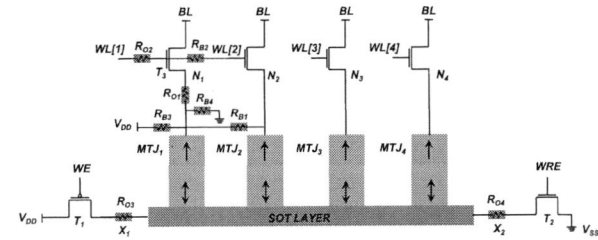

Fig. 3. Multipillar SOT-MRAM with representative bridge and open defects modeled as resistances R_B and R_O, respectively (only shown for MTJ_1 and MTJ_2).

A. Proposed Methodology

In this work, we only focus on specific target states corresponding to four MTJs in the multipillar: "0000, 0001, 0011, 0111, and 1111", corresponding to "P.P.P.P, P.P.P.AP, P.P.AP.AP, P.AP.AP.AP, and AP.AP.AP.AP". These states provide sufficient coverage for evaluating various defects while ensuring a reliable read operation with adequate sensing margin. The read operation is performed by selecting all four MTJs in the multi-pillar device simultaneously, rather than reading them individually, enabling the retrieval of all bits in a single read operation. This is achieved by activating transistor T_2 and all WL transistors, allowing the read current to flow through the multiple MTJs in the multi-pillar structure. A multilevel voltage sense amplifier with four reference levels as shown in Fig. 4 is used to distinguish the states "0000, 0001, 0011, 0111, and 1111". In this way, all the five different states can be detected using only one read operation. To assess read reliability, 100 Monte Carlo simulations are conducted for each state, considering 3% variation in free layer thickness, oxide thickness, MTJ diameter, and tunnel magneto-resistance (TMR). Fig. 5 shows statistical distribution of read voltages corresponding to "0000, 0001, 0011, 0111, and 1111". The results show clearly separated voltage levels with no overlap, ensuring distinct state identification. The sense margin can be further improved by using a pre-sense amplifier [12], enabling reliable read operation. The simulation framework is implemented using 28nm CMOS technology for the access transistors and the peripheral sense amplifier design and a modified SOT-assisted STT-MTJ model, as reported in [15], [16]. The simulation parameters used for the multi-pillar device are summarized in Table I.

B. Fault models

In context to R_B, R_O, T_{ON}, and T_{OFF} the primary fault models considered in this work include *1) Stuck-at Fault*

979-8-3315-9813-6/25 $31.00 © 2025 IEEE

(SAF): This can lead to faulty outputs such as SAF0000, SAF0001, SAF0011, SAF0111, or SAF1111, depending on the location of the fault. Such faults are typically observed in the presence of both open and bridge defects. *2) Transition Fault (TF)*: A transition fault occurs when one or more MTJs

TABLE I
PARAMETER VALUES FOR MULTIPILLAR SOT-MTJ DEVICE [15], [17]

Parameter	Value
Free layer thickness (t_f)	1.1 nm
Saturation magnetization (M_s)	$8\times10^5 A/m^2$
Anisotropy Field (H_K)	$8\times10^4 A/m^2$
Tunnel magneto resistance (TMR)	200%
MTJ diameter	50 nm
Length, Width, and Thickness of SOT layer	280nm, 60nm, and 3nm
Resistivity of SOT layer	200 ($\mu\Omega$.cm)

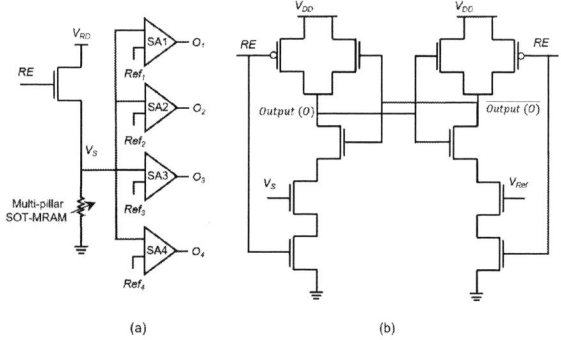

Fig. 4. (a) Design of multilevel sense amplifier. It compares its input against 4 references at once to detect all the states in single read operation. (b) Schematic of sense amplifier circuit.

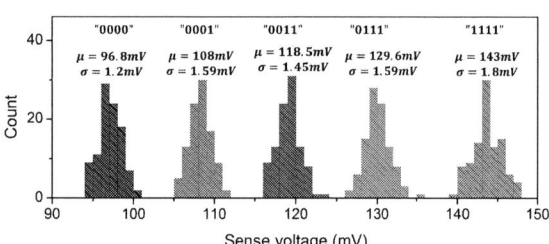

Fig. 5. Statistical distribution of the read voltage (V_s) across the multipillar SOT-MRAM device for the states "0000", "0001", "0011", "0111", and "1111".

in the multi-pillar cell fail to switch to the intended state, either from parallel (P) to anti-parallel (AP) or vice versa, during a write operation. *3) Coupling Fault (CF)*: In a coupling fault, a write operation intended for a specific MTJ affects the state of neighboring MTJs within the multi-pillar device. In this study, coupling faults are considered only for transitions from '0' to '1' (P to AP), as the reverse transition (1 to 0) resets all MTJs to '0' simultaneously. *4) Read Disturb Fault (RDF)*: A read disturb fault occurs when the read operation unintentionally disturbs the stored data in memory cell. The various open and bridge defects and their corresponding fault manifestations are discussed as follows:

a) Faults due to open defects

This includes defects at locations N_1 to N_4 (R_{O1} for MTJ_1), $WL[1]$ to $WL[4]$ (R_{O2} for MTJ_1), X_1 (R_{O3}), and X_2 (R_{O4}). The parameters in parentheses denote the injected defects modeled as resistances, as illustrated in Fig. 3. Table II summarizes the potential open defects, their respective resistance values (defect sizes), and the corresponding fault models induced in the multipillar SOT-MRAM device. The defect sizes listed correspond to the worst-case scenarios, wherein at least one memory state exhibits faulty behavior due to the defect. To illustrate this further, let us consider an open fault occurring at any node N_n (represented by R_{O1} in the case of MTJ_1). An open fault at N_n results in a TF during the $0 \rightarrow 1$ transition, as the required STT switching current cannot flow through the affected device. Additionally, the open fault increases the overall resistance of the cell, thereby raising the readout voltage for each state. This behavior manifests as SAFs 0001, 0011, 0111, and 1111 for the intended states 0000, 0001, 0011, and 0111, respectively. The state 1111 does not exhibit this fault due to all MTJs already being in the AP state. Similar behavior is observed corresponding to the open defect in WL. TFs can be sensitized by executing the full sequence: $w0000, r0000, w0001, r0001, w0011, r0011, w0111, r0111,$ and $w1111, r1111$ while SAFs can be detected by performing $w0000, r0000, w0001, r0001, w0011, r0011,$ and $w0111, r0111$. Here, w and r denote write and read operations of a particular state, respectively. Fig. 6 illustrates the occurrence of both SAF and TF when a resistance of $3k\Omega$ (R_{O4}) is introduced at node X_2.

b) Faults due to bridge defects

This includes defects between any pair of N nodes (R_{B1} between MTJ_1 and MTJ_2), pair of WL lines (R_{B2} between $WL[1]$ and $WL[2]$), node N and V_{DD} (R_{B3} between N_1 and V_{DD}), and node N and V_{SS} (R_{B4} between N_1 and V_{SS}). Table I summarizes the potential bridge defects, their respective defect sizes, and the corresponding fault models induced in the multipillar SOT-MRAM device. Consider the case of a bridge defect between any pair of nodes, such as $N-N$ or $WL-WL$. This can lead to TF during $0 \rightarrow 1$ switching. This occurs when two nodes are unintentionally shorted, causing the STT write current to split between multiple MTJs. As a result, the intended MTJ does not receive sufficient current to complete the switching operation. If the write current significantly exceeds the critical switching threshold, it may unintentionally switch the bridged MTJs through the STT mechanism and the preceding MTJs via the SOT mechanism, resulting in a CF. To sensitize TFs, the complete write-read sequence $w0000, r0000 \rightarrow w1111, r1111$ should be executed while CFs are detected by performing $w0000, r0000, w0001, r0001, w0011, r0011,$ and $w0111, r0111$. Furthermore, if T_1 is stuck on, unintended SOT current flows through the channel during read operation, elevating the voltage and causing an incorrect interpretation of the state, typically detected as SAF1111. This also results in resetting all MTJ_s to "0000" during read operation resulting in RDF. This fault can be identified by performing a write

TABLE II
OPEN AND BRIDGE DEFECTS, THEIR DEFECT SIZES, AND CORRESPONDING FAULT MODELS IN MULTIBIT SOT-MRAM

Open Defect Location	Defect Size	Fault Models	Bridge Defect Location	Defect Size	Fault Models
N_n (R_{O1})	$R_{O1} \geq 1.6k\Omega$	SAF, TF	T_1 ON	-	SAF, RDF
	$1k\Omega \leq R_{O1} < 1.6k\Omega$	TF	T_2 ON	-	TF
X_2 (R_{O4})	$1k\Omega \leq R_{O4} < 4k\Omega$	SAF	N-N (R_{B1})	$R_{B1} \leq 3k\Omega$	TF, CF
	$R_{O4} \geq 4k\Omega$	TF, SAF	WL-WL (R_{B2})	$R_{B2} \leq 300k\Omega$	TF, CF
X_1 (R_{O3})	$R_{O3} \geq 2k\Omega$	TF	N-V_{DD} (R_{B3})	$R_{B3} \leq 10k\Omega$	TF, SAF
WL_n (R_{O2})	$R_{O2} \geq 9M\Omega$	SAF, TF		$10k\Omega < R_{B3} \leq 50k\Omega$	SAF
	$400k\Omega \leq R_{O2} < 9M\Omega$	TF	N-V_{SS} (R_{B4})	$R_{B4} \leq 10k\Omega$	SAF

operation that sets at least one MTJ to the "1" state, followed by two consecutive read operations. The second read helps determine whether an unintended reset has occurred due to the stuck-on fault, as all $MTJs$ would be erroneously reset to "0" during the read phase. A stuck-on T_2 allows STT current to bypass the MTJ and flow directly through the channel, preventing a proper $0 \rightarrow 1$ transition, resulting in a TF.

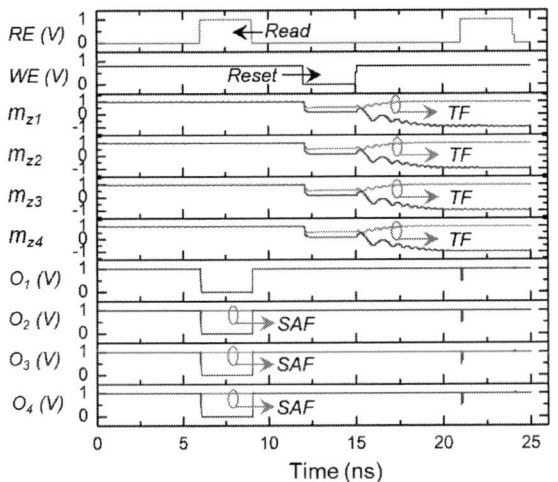

Fig. 6. Transient response showing the occurrence of SAF and TF (highlighted in magenta) when a 4kΩ resistance (R_{O4}) is introduced at node X_2. The ideal behavior is shown in blue for comparison.

This fault can be sensitized by executing $w0000, r0000, w0001, r0001$, where the failure to observe the expected state change confirms the issue.

IV. TEST ALGORITHM

A test algorithm is proposed to detect the discussed faults in a single-cell multi-pillar SOT-MRAM, as follows:

$$(w0000, r0000, w0001, r0001, w0011, r0011,$$
$$w0111, r0111, w1111, r1111, w0000, r0000)$$

SAF0000 is detected by r_2, r_3, r_4, and r_5, SAF0001 is detected by r_1, r_3, r_4, r_5, r_6, and r_7, SAF0011 is detected by r_1, r_2, r_4, r_5, r_6, and r_7, SAF0111 is detected by r_1, r_2, r_3, r_5, r_6, and r_7, and SAF1111 is detected by r_1, r_2, r_3, r_4, r_7. TFs for "0000", "0001", "0011", "0111" are detected by r_1, r_2, r_3, r_4, r_5, r_6, and r_7. CFs are detected by r_2, r_3, r_4, r_5, and r_6. RDF is detected at step r_6 by performing two successive read operations. Here, r_n denotes the nth read

operation in the proposed test algorithm. For example, r_1 corresponds to $r0000$, r_2 corresponds to $r0001$, and so on. The fault dictionary of the algorithm is presented in Table III. Each fault type is mapped against the set of read operations r_1-r_7, with '1' indicating that the fault is detectable during the corresponding read. Here, CF $i \leftrightarrow j$ represents CF between MTJ_i and MTJ_j in the multipillar device. From Table III, it is evident that the proposed test algorithm achieves complete coverage of all SAFs, TFs, CFs, and RDFs in the single-cell multi-pillar SOT-MRAM, ensuring reliable fault detection across the device.

TABLE III
FAULT DICTIONARY OF THE PROPOSED TEST ALGORITHM

Fault Type	$r1$	$r2$	$r3$	$r4$	$r5$	$r6$	$r7$
SAF0000	0	1	1	1	1	1	0
SAF0001	1	0	1	1	1	1	1
SAF0011	1	1	0	1	1	1	1
SAF0111	1	1	1	0	1	1	1
SAF1111	1	1	1	1	0	0	1
TF0000	1	0	0	0	0	0	1
TF0001	0	1	0	0	0	0	0
TF0011	0	0	1	0	0	0	0
TF0111	0	0	0	1	0	0	0
TF1111	0	0	0	0	1	1	0
CF $1 \leftrightarrow 2$	0	0	0	1	1	1	0
CF $1 \leftrightarrow 3$	0	0	1	0	1	1	0
CF $1 \leftrightarrow 4$	0	1	0	0	1	1	0
CF $2 \leftrightarrow 3$	0	0	1	1	0	0	0
CF $2 \leftrightarrow 4$	0	1	0	1	0	0	0
CF $3 \leftrightarrow 4$	0	1	1	0	0	0	0
RDF	0	0	0	0	0	1	0

V. CONCLUSION

This paper presented comprehensive fault modeling for a SOT-based multi-pillar memory cell, emphasizing bridge, open, and transistor-level defects. Corresponding fault models including stuck-at, transition, coupling faults, and read disturb faults are analyzed in detail. To ensure effective detection, a test algorithm is proposed, leveraging a multilevel sensing scheme for efficient and simultaneous readout of multiple bits. Simulation results demonstrate full coverage of SAFs, TFs, CFs, and RDFs, confirming the viability of the proposed approach for reliable defect detection in high-density SOT-MRAM architectures. This work addresses the critical reliability challenges of multilevel spintronic based memories which are now increasingly adopted in neuromorphic and in-memory computing and provides a necessary design for testing framework for their fault analysis.

979-8-3315-9813-6/25 $31.00 © 2025 IEEE

REFERENCES

[1] Z. Wang, L. Zhang, M. Wang, Z. Wang, D. Zhu, Y. Zhang, and W. Zhao, "High-density nand-like spin transfer torque memory with spin orbit torque erase operation," *IEEE Electron Device Letters*, vol. 39, no. 3, pp. 343–346, 2018.

[2] Y. Wu, K. Garello, W. Kim, M. Gupta, M. Perumkunnil, V. Kateel, S. Couet, R. Carpenter, S. Rao, S. Van Beek, K. Vudya Sethu, F. Yasin, D. Crotti, and G. Kar, "Voltage-gate-assisted spin-orbit-torque magnetic random-access memory for high-density and low-power embedded applications," *Phys. Rev. Appl.*, vol. 15, p. 064015, Jun 2021. [Online]. Available: https://link.aps.org/doi/10.1103/PhysRevApplied.15.064015

[3] J. Doevenspeck, K. Garello, S. Rao, F. Yasin, S. Couet, G. Jayakumar, A. Mallik, S. Cosemans, P. Debacker, D. Verkest, R. Lauwereins, W. Dehaene, and G. Kar, "Multi-pillar sot-mram for accurate analog in-memory dnn inference," in *2021 Symposium on VLSI Technology*, 2021, pp. 1–2.

[4] Y. Zhao, J. Yang, B. Li, X. Cheng, X. Ye, X. Wang, X. Jia, Z. Wang, Y. Zhang, and W. Zhao, "Nand-spin-based processing-in-mram architecture for convolutional neural network acceleration," *Science China Information Sciences*, vol. 66, no. 4, p. 142401, 2023.

[5] L. Wu, S. Rao, M. Taouil, G. C. Medeiros, M. Fieback, E. J. Marinissen, G. S. Kar, and S. Hamdioui, "Defect and fault modeling framework for stt-mram testing," *IEEE Transactions on Emerging Topics in Computing*, vol. 9, no. 2, pp. 707–723, 2019.

[6] L. Wu, S. Rao, M. Taouil, E. J. Marinissen, G. S. Kar, and S. Hamdioui, "Characterization, modeling, and test of intermediate state defects in stt-mrams," *IEEE Transactions on Computers*, vol. 71, no. 9, pp. 2219–2233, 2021.

[7] L. Wu, M. Taouil, S. Rao, E. J. Marinissen, and S. Hamdioui, "Electrical modeling of stt-mram defects," in *2018 IEEE International Test Conference (ITC)*. IEEE, 2018, pp. 1–10.

[8] S. Taghipour, M. Kamal, R. N. Asli, A. Afzali-Kusha, and M. Pedram, "Lchc-dft: A low-cost high-coverage design-for-testability technique to detect hard-to-detect faults in stt-mrams in the presence of process variations," *IEEE Transactions on Device and Materials Reliability*, vol. 22, no. 4, pp. 477–487, 2022.

[9] S. M. Nair, R. Bishnoi, M. B. Tahoori, G. Tshagharyan, H. Grigoryan, G. Harutyunyan, and Y. Zorian, "Defect injection, fault modeling and test algorithm generation methodology for stt-mram," in *2018 IEEE International Test Conference (ITC)*. IEEE, 2018, pp. 1–10.

[10] S. M. Nair, R. Bishnoi, M. B. Tahoori, H. Grigoryan, and G. Tshagharyan, "Variation-aware fault modeling and test generation for stt-mram," in *2019 IEEE 25th International Symposium on On-Line Testing and Robust System Design (IOLTS)*. IEEE, 2019, pp. 80–83.

[11] A. Chintaluri, H. Naeimi, S. Natarajan, and A. Raychowdhury, "Analysis of defects and variations in embedded spin transfer torque (stt) mram arrays," *IEEE Journal on Emerging and Selected Topics in Circuits and Systems*, vol. 6, no. 3, pp. 319–329, 2016.

[12] I. Yoon, A. Chintaluri, and A. Raychowdhury, "Emacs: Efficient mbist architecture for test and characterization of stt-mram arrays," in *2016 IEEE International Test Conference (ITC)*. IEEE, 2016, pp. 1–10.

[13] S. Yuan, M. Taouil, M. Fieback, H. Xun, E. J. Marinissen, G. S. Kar, S. Rao, S. Couet, and S. Hamdioui, "Device-aware test for back-hopping defects in stt-mrams," in *2023 Design, Automation & Test in Europe Conference & Exhibition (DATE)*. IEEE, 2023, pp. 1–6.

[14] A. Nisar, S. Dhull, S. Mittal, and B. K. Kaushik, "Sot and stt-based 4-bit mram cell for high-density memory applications," *IEEE Transactions on Electron Devices*, vol. 68, no. 9, pp. 4384–4390, 2021.

[15] Z. Wang, W. Zhao, E. Deng, J.-O. Klein, and C. Chappert, "Perpendicular-anisotropy magnetic tunnel junction switched by spin-hall-assisted spin-transfer torque," *Journal of Physics D: Applied Physics*, vol. 48, no. 6, p. 065001, 2015.

[16] K. Jabeur, G. Di Pendina, G. Prenat, L. D. Buda-Prejbeanu, and B. Dieny, "Compact modeling of a magnetic tunnel junction based on spin orbit torque," *IEEE Transactions on Magnetics*, vol. 50, no. 7, pp. 1–8, 2014.

[17] S. S. Parkin, C. Kaiser, A. Panchula, P. M. Rice, B. Hughes, M. Samant, and S.-H. Yang, "Giant tunnelling magnetoresistance at room temperature with mgo (100) tunnel barriers," *Nature materials*, vol. 3, no. 12, pp. 862–867, 2004.

Exploring Enhancements to 1T1C FeMFET Bitcells with a Versatile DTCO Methodology

Rosario Pronsato[*1], Antoine Cauquil[*1], Pascal Vivet[†], Jean Coignus[‡], Damien Deleruyelle[*], Cédric Marchand[*], Lioua Labrak[*], Ian O'Connor[*]

[*]Ecole Centrale de Lyon, INSA Lyon, CNRS, Universite Claude Bernard Lyon 1, CPE Lyon, INL,
UMR5270, 69130 Ecully, France [†]CEA-LIST, Grenoble, France
[‡]CEA-LETI, Univ. Grenoble Alpes, Grenoble, France, [1]Equal Contribution

Abstract—Non-volatile in-memory computing (iMC) has emerged as an energy-efficient paradigm well suited to AI workloads. Its implementation using 1T1C FeMFETs (Ferroelectric Memory Field Effect Transistors), a best-in-class emerging non-volatile memory technology that integrates BEOL ferroelectric devices with FEOL transistors, is of particular interest. This interest stems from their potential to enable large-scale multiply-accumulate (MAC) operations in both digital and analog domains. However, realizing tangible performance benefits requires comprehensive cross-layer exploration of both design and technology parameters, extending up to accelerator level. In this work, we propose a bitcell-level multi-objective optimization methodology to identify and extract optimal sizing solutions that provide tractable trade-offs between key performance indicators (KPI). We further demonstrate how this approach facilitates cross-stack exploration of accelerator architectures. Results are presented as Pareto fronts spanning 2-4 KPIs: a 2-KPI problem illustrates the methodology, while a 4-KPI problem represents a realistic design scenario. Comparison is made between 130nm and 28nm technologies demonstrating a decrease in the average of write energy and area up to 24X and 30X respectively.

Index Terms—DTCO, Emerging Technology, Ferroelectric, FeMFET, Bitcell, Optimization, in-Memory-Computing

I. INTRODUCTION

The rapid adoption of AI-driven applications, particularly large language models (LLMs) and edge AI, demands unprecedented computational power to handle increasingly complex models. However, physical miniaturization limits and the Von Neumann bottleneck create critical hardware challenges. Shrinking transistor sizes faces fundamental physical barriers, limiting traditional hardware scaling for both datacenter LLMs and resource-constrained edge devices. The latter further struggle with thermal and power budgets, making conventional architectures inefficient for deploying billion-parameter models. For edge AI, this bottleneck exacerbates latency and power inefficiencies, hindering real-time applications such as autonomous robotics or on-device generative AI. To address these limitations, designers must explore innovative computing paradigms and leverage emerging technologies – such as ferroelectric materials – to fundamentally rearchitect AI hardware. Trends [1], [2] show a shift to memory-centric specialized processing units, leveraging performance with data reuse and avoiding energy consuming fetch instructions [3]. Also known as in-Memory Computing (iMC), this paradigm uses memory arrays as processing units to limit data movement and perform multiple operations in parallel.

The development of AI accelerators spans multiple abstraction levels – from material properties to full-system architecture – creating a complex design space where parameter interactions and their system-wide performance impacts become difficult to anticipate. This multidimensional optimization challenge obscures critical tradeoffs between competing design objectives across the hardware stack.

Thus, it is necessary to develop a methodology that utilizes design-technology co-optimization (DTCO), to understand the impact of low-level parameters on high-level performance, and to explore the impact of high-level specifications on low-level requirements. By analyzing performance across the design stack, it is possible to evaluate the final system, exploring more effectively different architectures, interconnections, mapping of a workload, or technologies.

While [4], [5] demonstrate automated memory-circuit optimization, their tools are specific to one technology or target isolated metrics. Other studies related to iMC present a system-level methodology to leverage the mapping efficiency [6] or the customizability of the architecture for better exploration [7], but it abstracts away lower-level effects, resulting in lower accuracy when predicting actual circuit performance.

In this work, we bring the following contributions:

- a simulator-agnostic optimization framework that integrates heuristic algorithms with SPICE-like simulations, ensuring broad compatibility across different circuit design environments
- a case study of a 1T1C FeMFET bitcell with performance, power and area considerations at circuit level

This paper is structured as follows. In Section II, we introduce the basics of the FeMFET technology. In Section III we discuss how we use heuristic algorithms to perform design-technology co-optimization of the bitcell. In Section IV, different multi-objective optimizations are performed and the results are presented, and we conclude in Section V by discussing this approach and its benefits.

II. FEMFET BITCELL

Ferroelectric materials present compelling advantages for iMC by introducing non-volatility to conventional CMOS technologies through leveraging a hysteretic polarization-field response. This study focuses on ferroelectric capacitors (Fe-CAPs) integrated in the back-end-of-line (BEOL), which retain polarization after voltage removal. By directly connecting the FeCAP to a MOSFET gate in a 1T1C FeMFET configuration as seen in Figure 1, the remanent polarization modulates the channel surface potential, translating the material's hysteretic response into a threshold voltage shift.

We provide exploration of FeMFET devices based on STMicroelectronics proprietary 130nm [8] and GlobalFoundries

979-8-3315-9813-6/25 $31.00 © 2025 IEEE

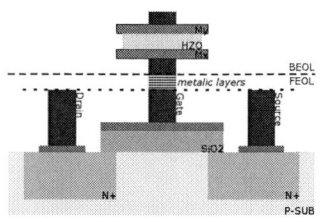

(a) FeMFET cross section

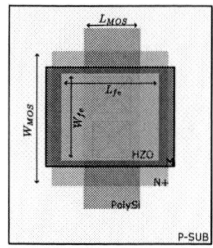

(b) FeMFET layout

Fig. 1: N-type FeMFET technology representation

28nm [9] PDK alongside with BEOL CEA-LETI process [10]. As simulating FeCAP is proven to be challenging regarding convergence, we fitted LETI's FeCAP measurement on a Preisach-based Verilog-A compact model [11]. This approach does not take into account the transient response of the ferroelectric layer, but this is valid as long as the voltage used to program the FeMFET polarizes the FeCAP substantially above its coercive field, since the latency of the operation can then be approximated by the transient response of the capacitive load [12].

The bitcell studied in this work consists of a FeMFET. Many bitcells can be organized within an array to act as a complete memory [13]. Sharing the same drain and source by column; and gate by row, it is possible to control them individually. To write data, a programming voltage ($\pm V_W$) is applied on the row while the column signals are grounded to avoid drain-source conduction. For the read operation, a read voltage ($V_{R_{ds}}$) is applied to a row while a small differential voltage is applied to the target column. With this principle, it is possible to perform vector-matrix multiplication [14], [15] in one cycle between an input vector and the content of the array. The results are sensed as current per column.

Instead of focusing on the optimization of a full array that faces many challenges, not to mention disturbs on the half selected bitcell when programming [13] or the use of high voltages that reduces the endurance of the devices [16], we propose to focus on a single bitcell offering a certain level of abstraction through the extraction of key performance indicators (KPIs).

TABLE I: Key Performance Indicator descriptions

KPI	Description of extraction
Write latency L_W [sec]	Time between 10% and 90% response of the voltage across the FeCAP to an input step.
Write energy E_W [J]	Energy is the integral of the power (P_W) of current and voltage gate during the write response time.
Current ratio I_{LVT}/I_{HVT}	Ratio of the drain source current when the FeMFET is programmed with a positive (LVT) and a negative (HVT) writing pulse.
Memory window MW [V]	Difference of threshold voltages when the FeMFET is programmed in LVT and HVT states.
Threshold voltage V_T [V]	Voltage to be applied on the gate of the FeMFET to measure a drain-source current of 100nA under 100mV voltage difference between drain-source.
Area [m^2]	Surface occupied by the stack of the FeCAP and the MOSFET.

Table I provides a summary of the KPIs associated with the memory characteristics of interest and how to measure them. Furthermore, Figure 2 illustrates the complete set of testbench configurations employed for the experimental extraction of these KPIs, illustrating the methodology used in our evaluation process.

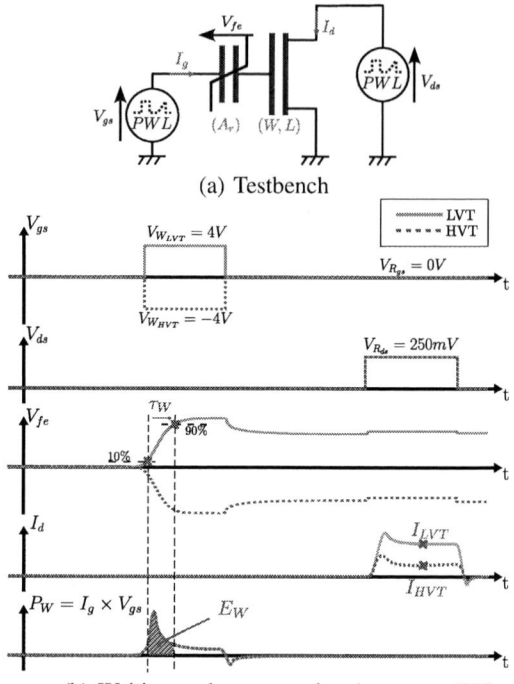

(a) Testbench

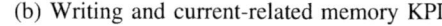

(b) Writing and current-related memory KPI

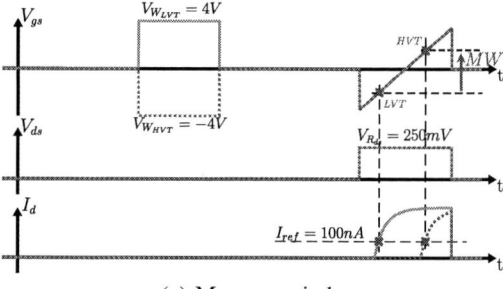

(c) Memory window

Fig. 2: KPI extraction protocols

III. DTCO PLATFORM

To perform design technology co-optimization of the bitcell we used evolutionary optimization algorithms. This kind of algorithms excel at navigating high-dimensional design spaces with competing objectives (e.g., power, performance, area). Unlike gradient-based methods that require differentiable models, EAs can optimize black-box circuit simulators and empirical silicon data. In principle, these methods are a stochastic search procedure with a population-based approach. This methodology copies the biological mechanism of evolution, i.e. it gathers its best population members, then mutates and recombines them from generation to generation to find an optimal design or until a certain criterion is met within specific constraints [17].

The exploration of the FeMFET bitcell is formulated as a multi-objective optimization problem, due to the need to study different KPIs that must be either minimized or maximized. It consists of three parts: the definition of the variables composing the design space, the objective functions, and the constraints. For our application, the design variables are transistor *length* (L), *width* (W), and the *area ratio* (A_r) of the FeCAP over the MOSFET gate area.

979-8-3315-9813-6/25 $31.00 © 2025 IEEE

TABLE II: Design variables for optimization

	Unit	Boundaries					
		G01		G02		GF	
		Min	Max	Min	Max	Min	Max
Geometric Parameters	Length	m	130n	3μ	500n	3μ	30n	3μ
	Width	m	150n	3μ	800n	3μ	80n	3μ
	Area ratio	–	0.01	1	0.01	1	0.01	1

The heuristic search starts with the initialization of a population. That is, proposing a first group of devices with different values of L, W and A_r. Individuals are diverse, i.e. meaningfully different from each other to ensure a wide range of potential solutions and avoid premature convergence. The proposed candidates are evaluated on the basis of their performance metrics. We have chosen KPIs relevant for this device, as previously presented in Table I.

This evaluation step leads to the selection of candidates for further genetic operations, a process to cross them and introduce random variations to enhance diversity. An iteration of the evolutionary algorithm is called an *epoch*, or a generation. This iterative process continues until a solution meets the desired criteria or after a chosen number of generations.

The constraints and boundaries of this multi-objective optimization problem are mainly related to physical limitations. The technology used has minimum sizes, as shown in Table II. In addition, we put limitations on the FeCAP area. Since ferroelectric grains in $Hf_{0.5}Zr_{0.5}O_2$ (HZO) thin films [18] are mainly of 10-15nm diameter [19], a minimal area of $8fm^2$ represents a realistic constraint for exploration. That is, $W \cdot L \cdot A_r > 8fm^2$ for the optimization algorithm.

We propose a multi-objective optimization platform that uses real-time simulation results to evaluate the device candidates. In the next section, we detail the algorithm used, the tool that connects the simulator to the evolutionary algorithm, and the calibration of the platform.

IV. SIMULATION RESULTS AND DISCUSSIONS

The proposed optimization was implemented using the PyMOO Python library [20], which offers single- and multi-objective optimization algorithms. We simulate with Spectre® – a Spice-like simulator. In order to obtain a unified exploration platform, we use Python-Skill Bridge [21] to interface PyMOO with the command interface window of the simulator.

The Non-dominated Sorting Genetic Algorithm II (NSGA-II) [22] was selected for this multi-objective optimization problem due to its well-established efficiency in handling Pareto-optimal solutions. It's computational efficiency and scalability makes it suitable for problems with complex non-linear tradeoffs, as demonstrated in prior literature [23]–[25]. A key advantage of its implementation in modern frameworks like PyMOO is the flexibility to tailor parameters. Population size, number of offsprings, and variation operators can be chosen to fit the specific computational budget and convergence characteristics of a given problem.

Three different explorations were carried out, two with two objectives and one with four. The first optimization aims to maximize the current ratio I_{LVT}/I_{HVT} while minimizing the writing energy consumption. Then, we perform an exploration that seeks to maximize the memory window (MW) also by minimizing write enrgy (E_W). We use these two scenarios to calibrate the algorithm, tune the genetic operators, and

choose the algorithm parameters. The Pareto-optimal solutions were generated using 100 generations, a population size of 50, and 20 offspring per generation. Afterwards, we performed a 4-objective exploration maximizing I_{LVT}/I_{HVT} and minimizing the area footprint, the write latency, and the read latency. We used and compared three technologies: two regular transistors with different gate oxide thickness from the HCMOS9A PDK [8], i.e GO1 for the thinnest and GO2 for the thicker one; and the super low voltage threshold transistor from the 28SLP PDK [9], referred as to GF in the following.

A. Calibration

Figure 3 shows the results of the first exploration carried out maximizing I_{LVT}/I_{HVT} while minimizing E_W. Figure 3a shows the set of best solutions found by the algorithm, while 3b shows the Pareto front obtained that corresponds to these devices. At the top-left of the Pareto front we observe an infeasible region, while any solution below the curve is in the non-optimal region. The optimization direction is pointed with an arrow.

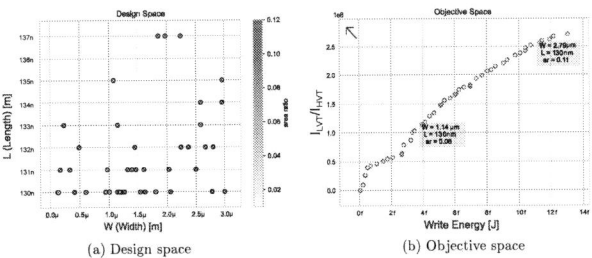

(a) Design space   (b) Objective space

Fig. 3: 2-objectives optimization between I_{LVT}/I_{HVT} and E_w varying L, W, and A_r.

This exploration corresponds to the GO1 transistor. As expected, the solutions go toward minimal length with variations in width. The tradeoff involves maximizing the current ratio by using wider transistors, which increases the gate capacitance and, therefore, the energy required to charge it.

Figure 4 shows a similar 2-KPI optimization to the previous exploration. However, the current ratio is replaced by the memory window. Here, the design space shows a trend for minimal length and width devices. This is due to the short-channel phenomenon that reduces the threshold voltage of the MOSFET [26]. Two mechanisms can explain this: (1) charge sharing between the drain/source and depletion region, and (2) drain-induced barrier lowering when a drain voltage is applied while measuring the threshold voltage.

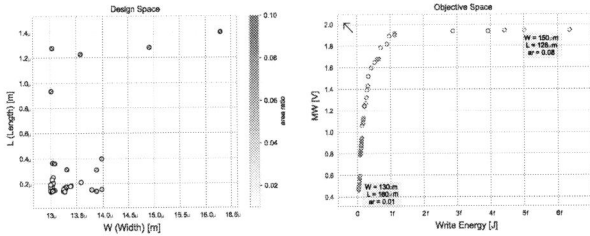

Fig. 4: 2-objectives optimization between MW and E_W varying L, W, and A_r.

This example also highlights the importance of setting boundaries in the study as it can bias the output result. For instance, the voltage ramp used in this example goes from -1V to 1V, thus limiting the MW to 2V.

B. Optimization example

A 4-KPI optimization is performed by maximizing I_{LVT}/I_{HVT}, and minimizing the E_W and L_W, as well as the area of the device. We also conducted the tests between technologies with GO1, GO2 and GF.

In figure 5, the design space shows different design trends. First in both GO1 and GO2 we can see that most solutions are devices with minimal channel lengths; however, as the maximum width evaluated in the simulation is approached, the trend shifts toward longer channels with reduced FeCAP dimensions. The 28nm technology goes against this distribution with more candidates at smaller areas.

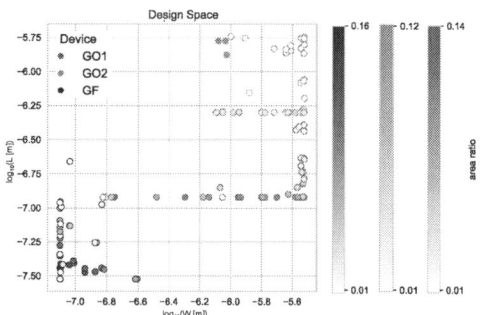

Fig. 5: Design Space of the 4-objectives optimization

We propose a *pairplot* representation of the objective space in Figure 6. It displays the pairwise tradeoffs between the objective functions. The diagonal plots highlight the distribution of the KPI as referred to the X-axis. This set of plots reveals that the 28nm technology solutions have the smallest areas and, consequently, the smallest write energy. They represent a decrease of 30.2X and 24X respectively, with respect to the other technologies. These solutions also tend to have the largest current ratio out of the three, 27.5X compared to GO1 on average. Off diagonal plots show clear groups between each technology, emphasizing the improvement provided between each technology shift.

While the exploration is thorough, the simulation also achieves fast convergence and extraction of results. The simulation times are given in Table III - these are results from sequential runs executed on 16 threads of a 2.4 GHz Xeon Silver 4210r server.

TABLE III: Simulation and optimization time

Simulation	Execution Time
I_{LVT}/I_{HVT} extraction	2.03 - 2.85 seconds
E_W and L_W extraction	2.16 - 2.58 seconds

Optimization program	Number of simulations	
2-KPI	4060	153 minutes
4-KPI	4060	164 minutes

Each dot in the pairplot of Figure 6 has an associated solution in the design space (Figure 5). This information gives the designer a wide range of options for a given application. However, we want to propagate this information to higher levels to ease cross-layer exploration of accelerator architectures, i.e. to pass on low-level performance choices to higher

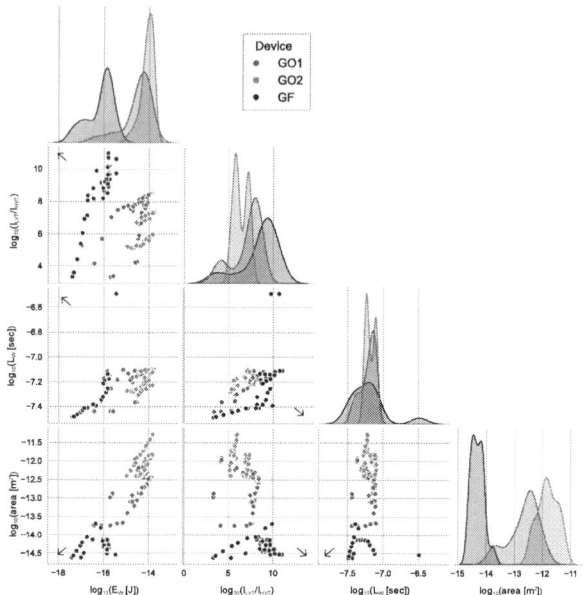

Fig. 6: Scatterplot matrix (pairplot) of 4-KPI optimization

architectural levels. To implement this, we must define the mechanism for upward data propagation, balancing generality, accuracy, and scalability. A structured data-base in the form of a look-up table is an interesting approach. In this way, we can store the result of the co-optimization process across design points, and use a method of interpolation, regression or ML-based models to generalize between sampled points.

V. CONCLUSION AND FUTURE WORK

This work demonstrates the interest of a co-optimization methodology applied to a FeMFET bitcell. It focuses on the multi-objective optimization of memory-related KPIs with respect to width, length and area ratio variations. The calibration of two different memory characteristics resulted in expected tradeoffs. Compared to the first set, an additional experiment on a less tangible optimization problem with 4 KPIs, resulted in an output that offers tradeoffs on the sizing of the bitcell. The use of a stochastic multi-objective optimization provided an efficient mean to explore a complex design space, enabling the efficient discovery of high-performing design solutions within a feasible computational timeframe.

Both case studies highlighted the importance of the optimization problem definition as it sets limits on some parameters. While this study focuses on FeCAPs with areas greater than $8fm^2$ to align with fabrication constraints, the proposed methodology is inherently scalable. This scalability provides a pathway to explore co-optimization between material properties (e.g., film thickness) and circuit-level metrics (e.g., write energy, latency).

ACKNOWLEDGMENT

We would like to express our sincere gratitude to CEA-LETI for providing real case values of ferroelectric capacitor. This work was partially funded by Agence Nationale de la Recherche under grant no. 22-PEEL-0013 (France 2030 - PEPR Electronique - CHOOSE) and by the European Commission under grant no. 101135656 (HORIZON-CL4-2023-DIGITAL-EMERGING-01-CNECT Ferro4EdgeAI).

REFERENCES

[1] B. Murmann, "Mixed-signal computing for deep neural network inference," *IEEE Transactions on Very Large Scale Integration (VLSI) Systems*, vol. 29, no. 1, pp. 3–13, 2021.

[2] W. Wan, R. Kubendran, C. Schaefer, S. B. Eryilmaz, W. Zhang, D. Wu, S. Deiss, P. Raina, H. Qian, B. Gao, S. Joshi, H. Wu, H.-S. P. Wong, and G. Cauwenberghs, "A compute-in-memory chip based on resistive random-access memory," *Nature*, vol. 608, no. 7923, p. 504–512, Aug. 2022. [Online]. Available: http://dx.doi.org/10.1038/s41586-022-04992-8

[3] A. Pedram, S. Richardson, M. Horowitz, S. Galal, and S. Kvatinsky, "Dark memory and accelerator-rich system optimization in the dark silicon era," *IEEE Design Test*, vol. 34, no. 2, pp. 39–50, 2017.

[4] F. García-Redondo, L. Verschueren, S. Rao, P. Pandey, D. Abdi, P. Weckx, S. Couet, M. García-Bardon, and G. Hellings, "A novel dtco-driven 1t1r bitcell for sub-10ns stt-mram llc macros at n12 node," in *2024 IEEE European Solid-State Electronics Research Conference (ESSERC)*, 2024, pp. 665–668.

[5] K. Touloupas, N. Chouridis, and P. P. Sotiriadis, "Local bayesian optimization for analog circuit sizing," in *2021 58th ACM/IEEE Design Automation Conference (DAC)*, 2021, pp. 1237–1242.

[6] K. Mishty and M. Sadi, "System and design technology co-optimization of sot-mram for high-performance ai accelerator memory system," *IEEE Transactions on Computer-Aided Design of Integrated Circuits and Systems*, vol. 43, no. 4, pp. 1065–1078, 2024.

[7] Y. Cai, Y. Gao, Z. Wang, L. Bao, L. Liang, Q. Zheng, C. Wang, and R. Huang, "Device-architecture co-optimization for rram-based in-memory computing," in *2023 IEEE 15th International Conference on ASIC (ASICON)*, 2023, pp. 1–4.

[8] STMicroelectronics, "HCMOS9A process design kit for the 130nm high speed CMOS technology node," Proprietary, accessed under NDA, 2025, provided by STMicroelectronics through Europractice.

[9] GlobalFoundries, "28SLP process design kit for the 28nm HKMG technology node," Proprietary, accessed under NDA, 2025, provided by GlobalFoundries through Europractice.

[10] R. Alcala, M. Materano, P. D. Lomenzo, L. Grenouillet, T. Francois, J. Coignus, N. Vaxelaire, C. Carabasse, S. Chevalliez, F. Andrieu, T. Mikolajick, and U. Schroeder, "Beol integrated ferroelectric hfo-based capacitors for feram: Extrapolation of reliability performance to use conditions," *IEEE Journal of the Electron Devices Society*, vol. 10, pp. 907–912, 2022.

[11] B. Jiang, Zurcher, Jones, Gillespie, and Lee, "Computationally efficient ferroelectric capacitor model for circuit simulation," in *1997 Symposium on VLSI Technology*, 1997, pp. 141–142.

[12] N. Gong, X. Sun, H. Jiang, K. S. Chang-Liao, Q. Xia, and T. P. Ma, "Nucleation limited switching (nls) model for hfo2-based metal-ferroelectric-metal (mfm) capacitors: Switching kinetics and retention characteristics," *Applied Physics Letters*, vol. 112, no. 26, p. 262903, 06 2018. [Online]. Available: https://doi.org/10.1063/1.5010207

[13] Z. Jiang, Z. Zhao, S. Deng, Y. Xiao, Y. Xu, H. Mulaosmanovic, S. Duenkel, S. Beyer, S. Meninger, M. Mohamed, R. Joshi, X. Gong, S. Kurinec, V. Narayanan, and K. Ni, "On the feasibility of 1t ferroelectric fet memory array," *IEEE Transactions on Electron Devices*, vol. 69, no. 12, pp. 6722–6730, 2022.

[14] S. De, F. Müller, N. Laleni, M. Lederer, Y. Raffel, S. Mojumder, A. Vardar, S. Abdulazhanov, T. Ali, S. Dünkel, S. Beyer, K. Seidel, and T. Kämpfe, "Demonstration of multiply-accumulate operation with 28 nm fefet crossbar array," *IEEE Electron Device Letters*, vol. 43, no. 12, pp. 2081–2084, 2022.

[15] T. Soliman, R. Olivo, T. Kirchner, C. D. l. Parra, M. Lederer, T. Kämpfe, A. Guntoro, and N. Wehn, "Efficient fefet crossbar accelerator for binary neural networks," in *2020 IEEE 31st International Conference on Application-specific Systems, Architectures and Processors (ASAP)*, 2020, pp. 109–112.

[16] A. Wang, R. Chen, Y. Yun, J. Xu, and J. Zhang, "Review of ferroelectric materials and devices toward ultralow voltage operation," *Advanced Functional Materials*, vol. 35, no. 7, p. 2412332, 2025. [Online]. Available: https://advanced.onlinelibrary.wiley.com/doi/abs/10.1002/adfm.202412332

[17] T. Bartz-Beielstein, J. Branke, J. Mehnen, and O. Mersmann, "Evolutionary algorithms," *WIREs Data Mining and Knowledge Discovery*, vol. 4, no. 3, p. 178–195, Apr. 2014. [Online]. Available: http://dx.doi.org/10.1002/widm.1124

[18] S. J. Kim, J. Mohan, S. Summerfelt, and J. Kim, "Ferroelectric hf0.5zr0.5o2 thin films: A review of recent advances," *JOM*, vol. 71, 09 2018.

[19] M. H. Park, Y. H. Lee, H. J. Kim, T. Schenk, W. Lee, K. D. Kim, F. P. G. Fengler, T. Mikolajick, U. Schroeder, and C. S. Hwang, "Surface and grain boundary energy as the key enabler of ferroelectricity in nanoscale hafnia-zirconia: a comparison of model and experiment," *Nanoscale*, vol. 9, pp. 9973–9986, 2017. [Online]. Available: http://dx.doi.org/10.1039/C7NR02121F

[20] J. Blank and K. Deb, "pymoo: Multi-objective optimization in python," *IEEE Access*, vol. 8, pp. 89 497–89 509, 2020.

[21] "Python-skill bridge," https://github.com/unihd-cag/skillbridge, 2023, accessed: 2025-03-26.

[22] K. Deb, A. Pratap, S. Agarwal, and T. Meyarivan, "A fast and elitist multiobjective genetic algorithm: Nsga-ii," *IEEE Transactions on Evolutionary Computation*, vol. 6, no. 2, p. 182–197, Apr. 2002. [Online]. Available: http://dx.doi.org/10.1109/4235.996017

[23] D. Malyna, J. Duarte, M. Hendrix, and F. van Horck, "Multi-objective optimization of power converters using genetic algorithms," in *International Symposium on Power Electronics, Electrical Drives, Automation and Motion, 2006. SPEEDAM 2006.*, 2006, pp. 713–717.

[24] I. Canturk and N. Kahraman, "Comparative analog circuit design automation based on multi-objective evolutionary algorithms: An application on cmos opamp," in *2015 38th International Conference on Telecommunications and Signal Processing (TSP)*. IEEE, Jul. 2015, p. 1–4. [Online]. Available: http://dx.doi.org/10.1109/TSP.2015.7296478

[25] C. Tang, X. Chen, Y. Luo, and Y. Zeng, "Automatic optimal design method for circuit sizing based on cnn surrogate model assisted differential evolution algorithm," *IEEE Access*, vol. 12, p. 136238–136247, 2024. [Online]. Available: http://dx.doi.org/10.1109/ACCESS.2024.3462952

[26] B. Razavi, *Design of Analog CMOS Integrated Circuits (17.2)*, 1st ed. USA: McGraw-Hill, Inc., 2000.

Non-Volatile Ferroelectric-AND (FeAND) Memory Cell Design

Basile Darne, Miqueas Filsinger, Alberto Bosio, Damien Deleruyelle, Ian O'Connor, Bertrand Vilquin,
Cédric Marchand

Centrale Lyon, INSA Lyon, CNRS,
Université Claude Bernard Lyon 1, CPE Lyon,
INL, UMR5270, 69130 Ecully, France
basile.darne@ec-lyon.fr

Abstract—Ferroelectric memory devices have emerged as a promising class of non-volatile memory technologies, offering a unique combination of high-speed operation, low power consumption, and good endurance compared to conventional flash memory. These devices leverage the bistable polarization states of ferroelectric materials to store data, enabling non-volatile retention while maintaining fast read/write capabilities. The discovery of hafnium-based ferroelectric materials that are fully CMOS compatible and exhibit robust ferroelectricity at nanoscale dimensions has further enhanced their integration and scalability potential. For IoT devices, which require non-volatile state retention under constrained power budgets and frequent interruptions, we propose a novel FeAND memory cell designed to serve as a non-volatile backup for volatile memory. Unlike conventional ferroelectric memories that rely on current sensing, our design directly outputs a voltage signal, eliminating the need for sensing circuits. The cell exhibits a logical AND-like behavior, enabled by an innovative read scheme based on a CMOS inverter. The cell can function as both a non-volatile memory element and a logic gate where one input is permanently stored as a polarization state. This dual functionality enables novel Computing-in-Memory architectures by embedding logic operations directly within the memory array. We validate our design using Cadence Spectre simulations with the GlobalFoundries 28SLP technology.

I. INTRODUCTION

In the context of the Internet of Things (IoT), the power budget of computing units is constrained and a constant power supply is not guaranteed. Power failures can erase volatile memory, causing reliability and performance issues. A common solution is to back up the computational state onto non-volatile memory (NVM). The concept of non-volatile computing [1] is thus a promising approach to preserve and restore system state. This approach relies on embedded memories, integrated on the same silicon as the logic circuit, enhancing data transfer between compute units and memory [2].

Conventional embedded non-volatile memory (eNVM) technology is based on embedded Flash with floating gate transistors. It has improved microcontroller performance in power, density, and cost [3], [4], but faces challenges scaling below the 28 nm nodes [5]. As a result, emerging non-volatile memory technologies have been proposed to replace embedded Flash [6]. Among these, ferroelectric materials are promising for energy efficiency and latency [3], [5], [7]. Ferroelectric

memories show energy/bit values between 1 fJ and 1 pJ, and latency around 10 ns, comparable to embedded SRAM or DRAM. They also offer a 10-year retention time and endurance above 10^{10} cycles [5].

Initially, perovskite materials like $Pb(Zr, Ti)O_3$ (PZT) were considered in the late 1980s [8], offering low programming energy (1 pJ/bit) and high endurance ($> 10^{14}$). However, their lack of CMOS compatibility [9] limited their practical use beyond the 130 nm. The discovery of ferroelectric properties in the orthorhombic HFO_2 [10] in 2011 renewed the interest in ferroelectrics [11], [12] given hafnium-based materials are CMOS compatible and show good endurance, latency, and retention [13].

These materials can thus be integrated into electronic devices to form new types of memory. Several integration methods have been demonstrated, each with its own trade-offs.

The simplest form employs ferroelectric capacitors (FeCaps) in a metal-ferroelectric-metal (MFM) structure. FeCaps can be integrated at the back-end-of-line on a transistor's drain, forming ferroelectric-RAM (FeRAM), similar to DRAM cells. These devices have shown strong performance : >10^{15} endurance cycles, 15 ns latency, and data retention of 10 years at 85°C [14]. FeRAMs are advantageous for their performance and decoupled scaling of transistors and FeCaps. However, they suffer from a destructive readout operation requiring a write-back, which consumes additional endurance cycles. Still, their high endurance is sufficient to tolerate this behavior for most eNVM applications.

Another type of ferroelectric device consists in integrating a ferroelectric layer directly on the gate of a transistor, forming a ferroelectric field effect transistor (FeFET) with a metal-ferroelectric-semiconductor (MFS) structure on the gate. The integration of the ferroelectric material is performed in the front-end-of-line of the process. This device allows a nondestructive read operation, tackling the issue of FeRAMs, and such integration also opens up new applications such as non-volatile logic operators. However, they also come with a reduced endurance ($\sim 10^5$) compared to FeRAM cells [15], [16]. This is due to the higher voltages that must be applied in order to write FeFETs compared to FeRAMs, which lead to interface degradation and charge injection [15].

A hybrid approach is the ferroelectric-metal-FET (FeM-

FET) with a metal-ferroelectric-metal-insulator-semiconductor structure [5], [6], which combines the ease of integration of back-end-of-line FeRAM with the behavior of FeFETs. These devices consist in a FeCap (MFM structure) connected to the gate of a transistor (metal-insulator-semiconductor structure). They solve the main integration drawback of FeFETs which is their poor scalability, because the FeMFET approach allows to size independently the FeCap and the transistor, which is crucial for the behavior optimization of such devices as described in section III-B. This also allows to potentially add metal connections to other devices between the FeCap and the gate of the transistor, opening up new possibilities of circuit design.

Ferroelectric-based devices revolve around a common operating mechanism [17], [18]. Information is written as a ferroelectric polarization state by applying a voltage, which generates an electric field within the ferroelectric material. The unique property of ferroelectric materials is based on the hysteretic Polarization-Electric field curve (see Figure 1 a)). These materials can therefore retain a positive or a negative polarization state depending on the applied electric field (or voltage) history. This allows to alter the charge distribution inside the channel of the transistor, enabling either to accumulate electrons (and reduce the threshold voltage of the device) or to accumulate positive charges (which will increase the threshold voltage of the device). Depending on the polarization state, the device can exhibit low threshold voltage state (low-V_T or LVT) or a high threshold voltage state (high-V_T or HVT) as shown in Figure 1 (b), that can be current-sensed to distinguish the two memory states.

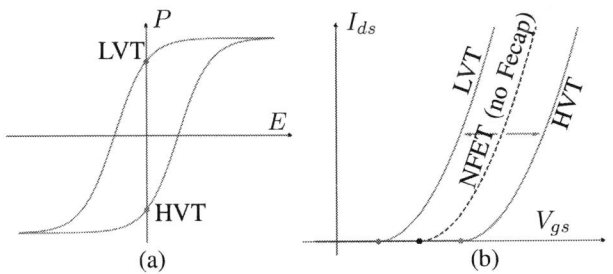

Figure 1: N-FeFET behavior (a) Hysteresis loop (b) Threshold voltage shift

This work focuses on the FeMFET approach and proposes a proof of concept of a new ferroelectric-based memory cell, where the read operation is performed through voltage sensing. Within the scope of this work, we will put the emphasis on the theoretical ideal behavior of such a cell, and we will provide simulation results that matches this behavior. The purpose is to validate the viability of the circuit, and device optimization and performance assessment remain to be performed. The targeted theoretical behavior of the circuit will be described in section II. Next, section III presents the steps that we chose to follow in order to match the ideal behavior. Finally, Cadence Spectre simulation results are detailed in section IV.

II. PROPOSED CIRCUIT

Whereas most of the work conducted around ferroelectric memory devices in the literature uses current sensing for readout operation [15], we propose a new voltage-based readout architecture called the Ferroelectric-AND (FeAND) as shown in Figure 2.

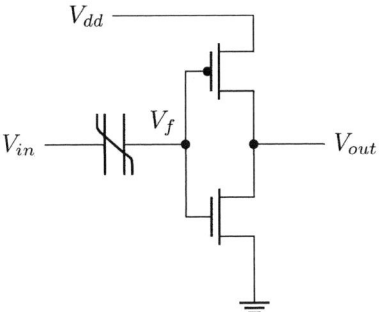

Figure 2: Proposed FeAND memory cell

The circuit is inspired by the FeMFET approach, where a FeCap is connected to the gate of a regular transistor. The novelty of this circuit lies in connecting the FeCap not to a single transistor but to a CMOS inverter (one NMOS, one PMOS). The input V_{in} writes the data, while V_{dd} triggers readout. The output V_{out} reflects the stored state: high for logic '1', low for logic '0'. The FeCap serves as a non-volatile storage element and the inverter is used as a circuit that converts the stored information into a voltage, either logic '1' or logic '0'.

This circuit supports two different operation modes which are summarized in table I. Simultaneous read and write is undefined (when both V_{dd} is high and a writing voltage is applied at V_{in}).

Table I: Operation control of the memory cell

V_{in}	V_{dd}	Operation
$V_{in,p} > 0$	'0'	Write '1'
$V_{in,m} < 0$	'0'	Write '0'
'0'	'1'	Read
$V_{in,p}$ or $V_{in,m}$	'1'	Undefined

The memory state is stored as a polarization state in the FeCap, which is written by applying a writing voltage at V_{in}. By convention, a positive polarization corresponds to a memory state $mem = 1$ while a negative polarization corresponds to $mem = 0$. The polarization determines the voltage on the floating node V_f, which can reach 2 different levels defined in Figure 3: V_{fp} if $mem = 1$ and V_{fm} if $mem = 0$.

The information stored as V_{fp} or V_{fm} is read using the inverter, which must be designed in such a way that the lowest value drives the output to logic '1' and the highest value of V_f sets the output to '0'. This idea is represented on Figure 3.

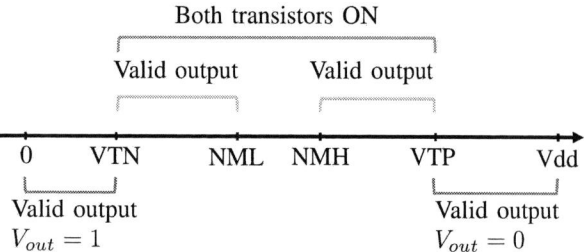

Legend:
V_{fp}: V_f when $mem = 1$ and $V_{dd} = 1$
V_{fm}: V_f when $mem = 0$ and $V_{dd} = 1$
$VT_N = V_{th,n}$: NMOS threshold
$VT_P = V_{dd} + V_{th,p}$: PMOS threshold

Figure 3: Device behavior as a function of V_f

The challenging aspect of this circuit's design is that we want to design the inverter in such a way that V_{fm} and V_{fp} are outside of the inverter's linear region, so as to avoid the configuration when both transistors are ON. This is why we want to make sure V_{fm} and V_{fp} are outside of the regions between VT_N and the low noise margin (NML), and between the high noise margin (NMH) and VT_P. Even though in these regions the output is a valid logical value, both transistors are ON. This increases the power consumption and is not desirable. Therefore, we want $V_{fm} < VT_N$ and $V_{fm} < VT_P$ (NMOS OFF and PMOS ON, $V_{out} = 1$) and $V_{fp} > VT_N$ and $V_{fp} > VT_P$ (NMOS ON and PMOS OFF, $V_{out} = 0$), or vice versa. This ensures that one of the two transistors of the inverter will always be blocked when data is read, thus limiting power consumption.

Thus, we will aim at respecting the following design conditions:

$$\begin{cases} V_{fm} < VT_N \text{ and } V_{fm} < VT_P \\ V_{fp} > VT_N \text{ and } V_{fp} > VT_P \end{cases}$$

or

$$\begin{cases} V_{fp} < VT_N \text{ and } V_{fp} < VT_P \\ V_{fm} > VT_N \text{ and } V_{fm} > VT_P \end{cases}$$

These conditions ensure a robust and low-power readout of the stored bit, avoiding the inverter's linear region and targeting Always ON and Always OFF transistors.

III. DESIGN METHODOLOGY

The purpose of this work is to propose a simulation that validates the concept of the device described previously. This is why we are not yet assessing nor optimizing the performance of proposed device. Two separate sets of parameters will allow to tune the behavior of the device. The writing pulse applied at V_{in} defines the polarization states and the levels V_{fp} and V_{fm}. The sizes of the components affect V_{fm} and V_{fp}, but also the values of VT_N and VT_P. For this work,

the transistor technology is fixed and we will focus on thin oxide regular threshold transistors from the Global Foundries 28 nm technology. In this work a Presiach-based Verilog-A compact model was used for the FeCap. The thickness of the ferroelectric layer was set to 10 nm and the coercive field to 1.2 MV/cm resulting in a coercive voltage of 1.2 V [19].

A. Writing pulse V_{in}

In order to ensure that the FeCap is correctly programmed we set a long writing time $T_{write} = 3\mu s$, and we apply high voltages (in absolute value) at V_{in} in order to maximize the polarization difference between the two memory states. We will set $V_{in,m} = -5V$ the value of the negative voltage applied in order to write a negative polarization state in the FeCap, and $V_{in,p} = +4V$ to obtain a positive polarization state. These values will remain fixed in order to simplify the exploration, but further optimization could allow to improve latency and energy consumption.

B. Component sizing

The sizes of the NMOS and PMOS affect the values VT_N and VT_P which are crucial for the behavior of the correct behavior of the circuit as highlighted in section II.

The areas of the FeCap, the NMOS and the PMOS also affect their respective capacitances and the capacitive coupling of the circuit. This determines the way voltages V_{in} and V_{dd} affect the value of V_f.

The FeCap capacitance can be expressed as $C_{fecap} = \frac{A_{fe}}{T_{fe}}\epsilon_0\epsilon_{fe}$, with A_{fe} is the FeCap area, T_{fe} the thickness of the ferroelectric layer and ϵ_{fe} the dielectric permittivity of the ferroelectric material. Similarly, we have the gate oxide capacitance of the NMOS $C_N = \frac{A_P}{T_{ox,N}}\epsilon_0\epsilon_{ox}$ and of the PMOS $C_P = \frac{A_P}{T_{ox,P}}\epsilon_0\epsilon_{ox}$, with A_N the area of the NMOS and A_P the area of the PMOS.

The design space thus consists in the following parameters: the width W_N and length L_N of the NMOS, the width W_P and length L_P of the PMOS, and a_{ratio} the area ratio between the FeCap and the CMOS inverter such as $A_{fecap} = a_{ratio} \cdot (A_N + A_P) = a_{ratio} \cdot (W_N L_N + W_P L_P)$.

The exploration methodology we propose consists of two steps. First, we conduct the optimization of the CMOS inverter alone, without the FeCap, this first step allows to set the values of VT_N and VT_P. Once the inverter is adjusted, we conduct a second exploration to size the FeCap, which allows to set the values of V_{fp} and V_{fm}.

IV. RESULTS

The memory cell is tested with the following operating scheme:

1) Write '1' to the memory, no read, read, no read
2) Write '0' to the memory, no read, read, no read

In order to perform the test, we must apply V_{in} and V_{dd} pulses as described in table II. Figure 4 illustrates the V_{in} and V_{dd} signals used for the simulation. We then extract the simulated V_f (floating node voltage) and V_{out} signals.

979-8-3315-9813-6/25 $31.00 © 2025 IEEE

Table II: Test signals applied on the device

	Write '1'	No read	Read	No read
V_{in}	$V_{in,p}$	0	0	0
V_{dd}	0	0	1	0
	Write '0'	No read	Read	No read
V_{in}	$V_{in,m}$	0	0	0
V_{dd}	0	0	1	0

The parameters of the applied pulses are presented in table III, and table IV summarizes the design parameters used to simulate the circuit.

Table III: Definition of pulses applied on V_{in} and on V_{dd}

$V_{in,p}$	+4V
$V_{in,m}$	−5V
T_{write}	$3\mu s$
V_{dd}	1V
T_{read}	$6\mu s$

Table IV: Design parameters

L_n	30nm	W_n	80nm
L_p	75nm	W_p	200nm
a_{ratio}	0.5		

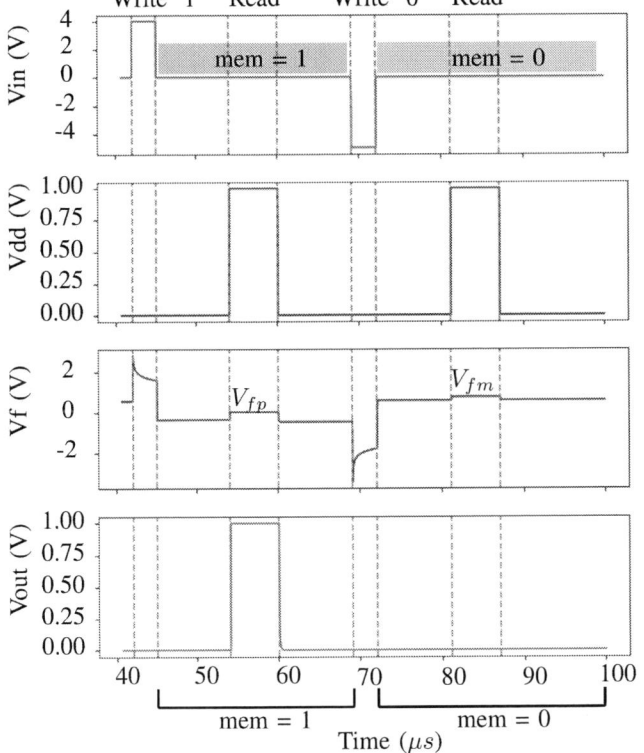

Figure 4: FeAND simulation results

Simulation results presented in figure 4 show that when $mem = 1$ and $V_{dd} = 1$, we get $V_{out} = 1$. In a similar way, when $mem = 0$ and $V_{dd} = 1$ we obtain $V_{out} = 0$. This device thus displays the behavior of an AND logic gate having mem and V_{dd} as inputs, as illustrated in table V.

Table V: Logical behavior of the FeAND memory cell

mem	V_{dd}	V_{out}
0	0	0
0	1	0
1	0	0
1	1	1

Regarding the floating node voltage, V_f, several observations can be made:

- Depending on the value of mem, V_f is either high or low as expected. The components have been sized in such a way that V_{fp} is low enough to generate a '1' at the output and V_{fm} is high enough to generate a '0'. We have $V_{fp} = 0.022V$, $VT_N = 0.31V$, $VT_P = 0.64V$ and $V_{fm} = 0.76V$.
- For a given mem value, the value of V_f reaches a different value during a read pulse, compared to the stable value when V_{in} and V_{dd} are '0'. This slight increase of V_f during a reading pulse is is due to the capacitive coupling between the FeCap and the PMOS and NMOS transistors, but it is small enough and doesn't disturb the stored polarization state.
- During the writing operations, the pulses applied at V_{in} also cause V_f to reach a high absolute value (around 3V) which is an issue for the gate dielectric integrity of the transistors. Tackling this challenge requires to reduce the values of the programming voltages. [20], [21].

The proposed operating scheme is very promising for low-power applications because the only energy consumption contributions come from applying the writing V_{in} and reading V_{dd} pulses, no energy is consumed when the transistors switch (because when they switch, V_{dd} is kept at '0'). Evaluation and reduction of energy consumption is an optimization challenge that remains to be conducted.

V. CONCLUSION AND FUTURE WORKS

We have introduced a functional memory cell in which the data is written by applying a positive or negative voltage, and the data is read as a voltage output. This innovative readout mechanism allows to avoid the use of sense amplifiers that convert the output current into a voltage making it easier to integrate the proposed FeAND memory into mixed volatile and non-volatile memories. The proposed cell presents several advantages, among which the voltage memory readout, the possibility to scale down the size of transistors without impacting the ferroelectric behavior of the FeCap, and the low-energy operating scheme (no switch energy consumption, only write and read energy contributions).

However, several challenges remain to be solved. First, extensive design space exploration remains to be performed in order to identify the design parameters that optimize latency and energy consumption. Second, in order to fulfill the conditions on the maximum voltage sustained by the gate of the transistors, the value of V_f must be reduced during writing operations. Finally, a variability analysis must be conducted in order to assess the robustness of the proposed circuit.

REFERENCES

[1] Y. Liu, Z. Li, H. Li, Y. Wang, X. Li, K. Ma, S. Li, M.-F. Chang, S. John, Y. Xie, J. Shu, and H. Yang, "Ambient energy harvesting nonvolatile processors: from circuit to system," in *Proceedings of the 52nd Annual Design Automation Conference*, ser. DAC '15. New York, NY, USA: Association for Computing Machinery, Jun. 2015, pp. 1–6. [Online]. Available: https://doi.org/10.1145/2744769.2747910

[2] E. Marinissen, B. Prince, D. Keltel-Schulz, and Y. Zorian, "Challenges in embedded memory design and test," in *Design, Automation and Test in Europe*, Mar. 2005, pp. 722–727 Vol. 2, iSSN: 1558-1101. [Online]. Available: https://ieeexplore.ieee.org/abstract/document/1395663

[3] R. Strenz, "Review and Outlook on Embedded NVM Technologies – From Evolution to Revolution," in *2020 IEEE International Memory Workshop (IMW)*, May 2020, pp. 1–4, iSSN: 2573-7503. [Online]. Available: https://ieeexplore.ieee.org/abstract/document/9108121

[4] ——, "Embedded Flash technologies and their applications: Status & outlook," in *2011 International Electron Devices Meeting*, Dec. 2011, pp. 9.4.1–9.4.4, iSSN: 2156-017X. [Online]. Available: https://ieeexplore.ieee.org/abstract/document/6131521

[5] A. I. Khan, A. Keshavarzi, and S. Datta, "The future of ferroelectric field-effect transistor technology," *Nature Electronics*, vol. 3, no. 10, pp. 588–597, Oct. 2020, publisher: Nature Publishing Group. [Online]. Available: https://www.nature.com/articles/s41928-020-00492-7

[6] A. Keshavarzi, K. Ni, W. Van Den Hoek, S. Datta, and A. Raychowdhury, "FerroElectronics for Edge Intelligence," *IEEE Micro*, vol. 40, no. 6, pp. 33–48, Nov. 2020. [Online]. Available: https://ieeexplore.ieee.org/document/9207822/

[7] K. Ni, S. Dutta, and S. Datta, "Ferroelectrics: From Memory to Computing," in *2020 25th Asia and South Pacific Design Automation Conference (ASP-DAC)*, Jan. 2020, pp. 401–406, iSSN: 2153-697X. [Online]. Available: https://ieeexplore.ieee.org/abstract/document/9045150

[8] J. F. Scott and C. A. Paz de Araujo, "Ferroelectric Memories," *Science*, vol. 246, no. 4936, pp. 1400–1405, Dec. 1989, publisher: American Association for the Advancement of Science. [Online]. Available: https://www.science.org/doi/abs/10.1126/science.246.4936.1400

[9] J. Müller, P. Polakowski, S. Mueller, and T. Mikolajick, "Ferroelectric Hafnium Oxide Based Materials and Devices: Assessment of Current Status and Future Prospects," *ECS Journal of Solid State Science and Technology*, vol. 4, no. 5, p. N30, Feb. 2015, publisher: IOP Publishing. [Online]. Available: https://iopscience.iop.org/article/10.1149/2.0081505jss/meta

[10] T. S. Böscke, J. Müller, D. Bräuhaus, U. Schröder, and U. Böttger, "Ferroelectricity in hafnium oxide thin films," *Applied Physics Letters*, vol. 99, no. 10, p. 102903, Sep. 2011. [Online]. Available: https://doi.org/10.1063/1.3634052

[11] T. Mikolajick, U. Schroeder, and S. Slesazeck, "The Past, the Present, and the Future of Ferroelectric Memories," *IEEE Transactions on Electron Devices*, vol. 67, no. 4, pp. 1434–1443, Apr. 2020. [Online]. Available: https://ieeexplore.ieee.org/abstract/document/9032362

[12] X. Xu, Z. Luo, H. Sun, Y. Xu, L. Gao, and Z. Yu, "A review of hafnium-based ferroelectrics for advanced computing," *Solid-State Electronics*, vol. 225, p. 109053, Apr. 2025. [Online]. Available: https://www.sciencedirect.com/science/article/pii/S0038110124002028

[13] S. Zhou, L. You, A. Chaturvedi, S. A. Morris, J. S. Herrin, N. Zhang, A. Abdelsamie, Y. Hu, J. Chen, Y. Zhou, S. Dong, and J. Wang, "Anomalous polarization switching and permanent retention in a ferroelectric ionic conductor," *Materials Horizons*, vol. 7, no. 1, pp. 263–274, 2020, publisher: Royal Society of Chemistry. [Online]. Available: https://pubs.rsc.org/en/content/articlelanding/2020/mh/c9mh01215j

[14] J. Okuno, T. Kunihiro, K. Konishi, H. Maemura, Y. Shuto, F. Sugaya, M. Materano, T. Ali, M. Lederer, K. Kuehnel, K. Seidel, U. Schroeder, T. Mikolajick, M. Tsukamoto, and T. Umebayashi, "High-Endurance and Low-Voltage operation of 1T1C FeRAM Arrays for Nonvolatile Memory Application," in *2021 IEEE International Memory Workshop (IMW)*, May 2021, pp. 1–3, iSSN: 2573-7503. [Online]. Available: https://ieeexplore.ieee.org/abstract/document/9439595

[15] S. Dunkel, M. Trentzsch, R. Richter, P. Moll, C. Fuchs, O. Gehring, M. Majer, S. Wittek, B. Muller, T. Melde, H. Mulaosmanovic, S. Slesazeck, S. Muller, J. Ocker, M. Noack, D.-A. Lohr, P. Polakowski, J. Muller, T. Mikolajick, J. Hontschel, B. Rice, J. Pellerin, and S. Beyer, "A FeFET based super-low-power ultra-fast embedded NVM technology for 22nm FDSOI and beyond," in *2017 IEEE International Electron Devices Meeting (IEDM)*. San Francisco, CA, USA: IEEE, Dec. 2017, pp. 19.7.1–19.7.4.

[16] H. Mulaosmanovic, E. T. Breyer, S. Dünkel, S. Beyer, T. Mikolajick, and S. Slesazeck, "Ferroelectric field-effect transistors based on HfO2: a review," *Nanotechnology*, vol. 32, no. 50, p. 502002, Sep. 2021, publisher: IOP Publishing. [Online]. Available: https://dx.doi.org/10.1088/1361-6528/ac189f

[17] J. Ajayan, P. Mohankumar, D. Nirmal, L. M. I. L. Joseph, S. Bhattacharya, S. Sreejith, S. Kollem, S. Rebelli, S. Tayal, and B. Mounika, "Ferroelectric Field Effect Transistors (FeFETs): Advancements, challenges and exciting prospects for next generation Non-Volatile Memory (NVM) applications," *Materials Today Communications*, vol. 35, p. 105591, Jun. 2023. [Online]. Available: https://www.sciencedirect.com/science/article/pii/S2352492823002817

[18] S. L. Miller and P. J. McWhorter, "Physics of the ferroelectric nonvolatile memory field effect transistor," *Journal of Applied Physics*, vol. 72, no. 12, pp. 5999–6010, Dec. 1992. [Online]. Available: https://doi.org/10.1063/1.351910

[19] Bo Jiang, Zurcher, Jones, Gillespie, and Lee, "Computationally Efficient Ferroelectric Capacitor Model For Circuit Simulation," in *Symposium on VLSI Technology*. Kyoto, Japan: IEEE, 1997, pp. 141–142.

[20] S. Lancaster, S. Slesazeck, and T. Mikolajick, "On the thickness scaling of ferroelectric hafnia," *IEEE Transactions on Materials for Electron Devices*, pp. 1–13, 2024.

[21] D. I. Han, H. Choi, D. H. Lee, S. H. Kim, J. Lee, I. Jeon, C. H. Jung, H. Lim, and M. H. Park, "Strategies for Reducing Operating Voltage of Ferroelectric Hafnia by Decreasing Coercive Field and Film Thickness," *Advanced Physics Research*, p. 2400194, Mar. 2025.

On the Possibility of Relying Solely on FeMFET Variability for PUF Implementations

Miqueas Filsinger, Antoine Cauquil, Damien Deleruyelle, David Navarro, Ian O'Connor, Cédric Marchand
Univ Lyon, Ecole Centrale de Lyon, CNRS,
INSA Lyon, Université Claude Bernard Lyon 1,
CPE Lyon,
INL, UMR5270, 69130 Ecully, France
miqueas.filsinger@ec-lyon.fr

Abstract—The promising features introduced by the integration of ferroelectric devices into conventional integrated circuit fabrication processes have spurred extensive research into device reliability, non-volatile memory circuits and system-level applicability. Their low-power operation makes them particularly suitable for Internet of Things applications, and their intrinsic memory properties position them as strong candidates for non-volatile memory technologies and In-memory Computing. For such data-intensive applications, the need to ensure secure data storage, processing and transmission has motivated the adaptation of classic hardware security strategies to this emerging in-memory computing paradigm, demonstrating high effectiveness with ferroelectric designs. However, a variability analysis from a design perspective remains unexplored towards either implementing security primitives based on identity, e.g. Physical Unclonable Functions, or based on stochasticity, e.g. True Random Number Generators.

In this work, we focus on the variations expected in a commercial 28 nm process and its compatibility with memory cell design, in view of the implementation of a Physical Unclonable Function in a ferroelectric memory array. In particular we show that while obtaining a sufficient output variability, which can be used for fingerprinting the device, a fair current ratio is maintained, allowing memory array implementations.

I. INTRODUCTION

In recent years, ferroelectric technology based on hafnium oxide has gained considerable attention as a possible solution to address many applications involving the use of memories in general [1]. The low energy operation and the continuous scaling down of device area of Ferroelectric-Metal Field Effect Transistors (FeMFETs) [2], [3], make them attractive for the electronics design environment [4]. These devices are used in various domains, ranging from edge computing and Internet of Things (IoT) to neural networks, where their non-volatile properties are particularly beneficial. As a result, its imminent application makes the implementation of security primitives crucial to ensure secure communication, storage, and operation of ferroelectric memory-based systems. Interestingly, the same ferroelectric memory arrays designed for data storage can be dual-operated as Physical Unclonable Functions (PUFs), reusing existing hardware for security purposes.

A large body of work exists in recent literature regarding the implementation of PUFs as an effective security solution in ferroelectric technology [5], [6], [7]. The idea behind these devices is to leverage random variations at the fabrication

stage, in order to give every instance (chip) a unique identity. Each device includes the same security primitive, which is later tested by the design house in order to store a record of every corresponding input-output, called the challenge-response pair (CRP) library. At the time of use, the authenticator will interrogate the device in order to compare its CRPs with the ones stored in the cloud [8]. If they match, then access is granted to the user.

A valid implementation requires proof that the design satisfies certain properties [9], [10]:

- Physical Uniqueness: for the same challenge applied to different instances, different responses are desired. Also, different challenges must produce different responses.
- Reliability: for any experimental condition in a given range of operation, challenge-response pairs must remain stable.
- Unclonability: even if an attacker has physical access to the PUF, he must not be able to reproduce an exact copy of it.
- Unpredictability: attackers should not be able to predict, under any circumstance, the response to any challenge.

Such characteristics imply an abstract relationship with the entropy source — that is, the source of random variations. This disconnection between output metrics and design requirements makes achieving a successful PUF a delicate task throughout the entire design process. For instance, while a cell can be engineered to enhance variability, numerous other factors may still influence the cell's stochastic variation, potentially introducing bias that only becomes evident after fabrication.

At the same time, designers must be given a certain degree of flexibility to ensure that a fair trade-off between performance, reliability, and stochasticity can always be achieved.

In this work, we tackle the entropy source: how random variations in FeMFETs can enable the design of PUFs using a 28 nm HKMG commercial design kit. We aim to investigate how much variability can be effectively harnessed from the technology.

II. BACKGROUND

A. FeMFET basics

The FeMFET consists of a conventional Field-Effect Transistor (FET) whose gate is connected to a back-end-of-line

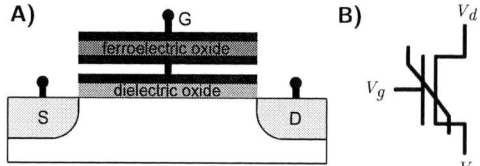

Figure 1: A) Cross-section view of a FemFET. B) schematic symbol.

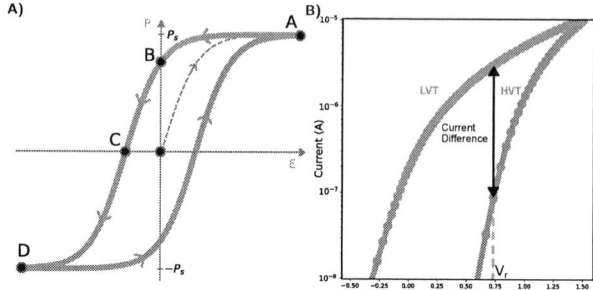

Figure 2: A) Typical hysteretic Polarization-Electric Field curve of a ferroelectric capacitor. B) Characteristic response of the FeMFET, drain-source current vs. gate voltage. For both, the red curve corresponds to the positive polarization state (i.e. LVT state), and the blue path to the negative one (i.e. HVT state). An example of state differentiation is given.

(BEOL) Ferroelectric Capacitor (FeCap) as shown in Figure 1A. The ferroelectric material can be characterized by two specific properties [11] which combine to lend a non-volatile behavior to the transistor:

- The material exhibits spontaneous polarization, meaning it possesses a natural electric dipole moment even in the absence of an applied electric field.
- The mean dipole moment can be oriented in one of two directions.

To illustrate this, the typical polarization curve is shown in Figure 2A. When a ferroelectric material is initially unpolarized and a positive electric field is applied, the polarization increases as electric dipoles align with the field direction. This continues until the material reaches saturation polarization (point A), beyond which further increases in the field no longer result in significant changes in polarization. When the external electric field is removed, the polarization does not return to zero. Instead, the material retains a non-zero polarization known as the remanent polarization (point B), indicating the spontaneous alignment of dipoles even in the absence of an external field. To bring the net polarization back to zero, a reverse electric field must be applied. The magnitude of this reverse field at which the net polarization cancels is called the coercive field (point C). If the reverse field continues to increase, the polarization switches direction, eventually reaching the negative saturation polarization (point D). Upon cycling the field again in the positive direction, the polarization follows a different trajectory than during the initial sweep. This hysteresis behavior illustrates the bistable nature of ferroelectric materials, a key characteristic exploited in FeCaps and non-volatile memory devices.

It is clear then that this behavior can be exploited to encode digital identity, a digital '0' or '1' in a unique transistor.

To perform a write operation in the conventional way, both drain and source are grounded, and a positive voltage V_p is applied to the gate [12], [13]. This causes charge to flow through the ferroelectric layer inducing a positive polarization state. Conversely, if a negative V_p is applied, then a reversed polarization state is induced. These two different states produce a shift in the transistor's response, virtually changing the threshold voltage to High Threshold Voltage (HVT) and Low Threshold Voltage (LVT).

If a read operation is performed, the current I_{ds} is measured in order to distinguish the LVT state from the HVT state. A constant reading voltage V_d is applied to the drain, the source

remains grounded ($V_s = 0V$), and a voltage V_r is applied to the gate. On Figure 2B, we represent the current differentiation between the two states with this gate voltage variation.

B. Modeling

In the Preisach model, originally developed to describe magnetic hysteresis phenomena, but also applicable to ferroelectric materials, the overall polarization is represented as the sum of the polarities of individual "hysterons", which are formal abstractions of ferroelectric domains. Each hysteron exhibits the same hysteretic step response but differs in the width and center of its hysteresis window, as described by a distribution of their energies. In practice, when simulating a large number of hysterons, the total polarization can be adequately approximated through a hyperbolic function [14] as:

$$P(t) = 2\alpha_i P_s \tanh\left(w(v(t) \pm V_c)\right) + \beta_i ,$$

where P_s corresponds to the saturation polarization, V_c to the coercive voltage, and $w = \frac{t}{2V_c} \ln \frac{Ps+Pr}{Ps-Pr}$, with t the thickness of the ferroelectric layer. At the same time, α_i and β_i are scale and offset constants, needed to correctly describe the trajectories in non-saturated loops [15].

Although this approximation is valid, in scenarios where only a few grains are present in the ferroelectric layer, such as in the FeCap sizes presented in this work, the dominant switching behavior is rather abrupt and stochastic [16]. This is in line with Nucleation Limited Switching theory [17], where the polarization reversal (switching) is no longer deterministic, but described by a Poisson process. Thus, the probability of having a switching event is dependent on the pulse programming time and amplitude. This behavior can be leveraged for implementing Random Number Generators, where a dynamic noise source can be used to encrypt important information. In this work, however, we focus on the potential of relying on the fabrication variability of these devices on a stable operation regime, in order to obtain a static noise source.

III. METHODOLOGY AND RESULTS

A commercial 28 nm design kit was selected to enable a realistic evaluation of ferroelectric integration in advanced CMOS nodes. It is mandatory to ensure that a sufficient proportion of the input voltage is applied across the ferroelectric capacitor for proper inter-layer programming conditions, while also ensuring that the voltage applied to the transistor gate is sufficiently small to respect gate oxide integrity. At the same time, the ferroelectric capacitor must be large enough to allow a gate current to polarize the FET. For that, a sufficiently low capacitance ratio $CR = \frac{C_{fe}}{C_{mos}}$ was selected by tuning the area ratio A_r, i.e. the ratio between the area of the FeCap and the area of the transistor [18]. It was observed that a satisfactory behavior was achieved using thin oxide transistors and a $CR \approx 5\%$. This strategy is effective because, even though the capacitance ratio does not fully capture the real dynamic interaction between the two elements, it provides a useful approximation that holds for any transistor of the same oxide thickness.

For the design of each FeMFET, ferroelectric capacitors were chosen to be as small as possible. To the best of our knowledge, the smallest diameters used in fabrication range between 60 nm $\leq \varnothing \leq 100$nm [19]. For the transistor sizing, a design exploration over 10^3 points was performed in order to capture the best I_{HVT}/I_{LVT} relation, as shown in Figure 3.

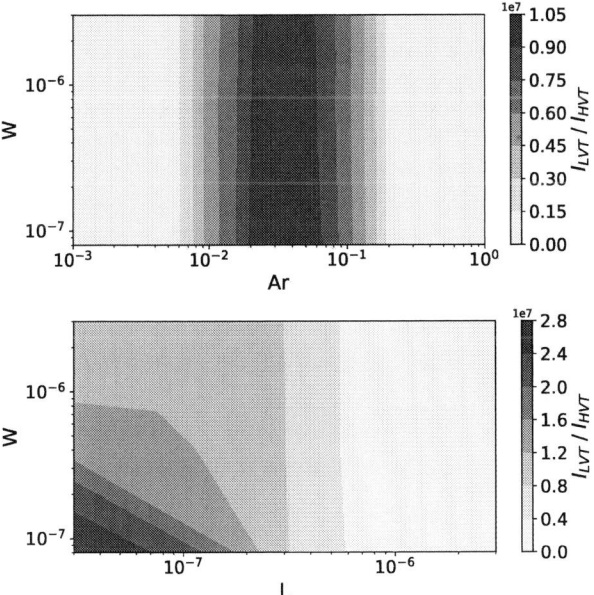

Figure 3: Design space exploration of FeMFET, optimizing I_{HVT}/I_{LVT} ratio in terms of transistor width, length, and area ratio $A_r = A_{\text{FeCap}}/A_{\text{transistor}}$

For the following simulations, in each design point a pre-polarization operation is carried out prior to a sequence of two program-read pulses, each lasting $t_P = t_R = 50\ \mu$s, longer than required to stabilize the FeMFET response, as shown in Figure 4.

A design case is stated for both ferroelectric sizes as shown in Table I.

Table I: Design parameters for both FeCap sizes.

	$\varnothing = 60$nm	$\varnothing = 100$ nm
W	750 nm	
L	250 nm	
V_p	3V	2.4V
V_r	150mV	120mV

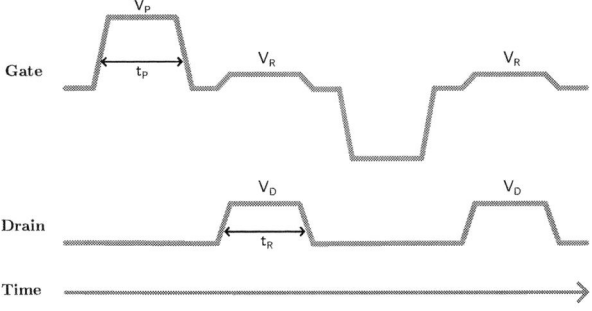

Figure 4: Simulated voltage waveforms at the gate and drain of the FeMFET under design

To explore variability impact, Monte-Carlo simulations were conducted, taking into account transistor and FeCap area variability. Different intra-chip scenarios were studied, highlighting the local variations of 500 FeMFETs bitcells on 50 randomly sampled global variation points. The Monte-Carlo analysis results are shown in Figure 5. From an intra-chip perspective (i.e. local variation), dispersion of the LVT state current demonstrates advantageous circuit fingerprinting behavior. As the LVT current of each bitcell ranges around the same local mean, differential circuits with high enough discrimination capabilities can be used to produce challenges by comparing the content of two equally programmed bitcells. However, the global mean variations are likely to strongly bias the response of the aforementioned challenge in the case of extra-chip comparison. This principle can be seen as two non-overlapping bell curves on Figure 5.

Also, during different simulations it was noted that a bias in the LVT state read current naturally induces a similar bias in the HVT state read current, meaning that both states are correlated upon variations. This implies that cell-to-cell differences can be characterized in only one write-read cycle, i.e. either LVT or HVT state, instead of comparing both. Therefore, the primary output metric for each evaluation was the current value for an LVT state programmed FeMFET, measured at the midpoint of the read pulse, i.e. $I_d(t = t_0 + \frac{t_R}{2})$.

On the other hand, FeCap variations do not introduce significant variations in the output but induce a linear relation $\sigma/\mu(I_{LVT}) = 1.78\ \sigma_{\text{Area}}$. This, and the known stability of the

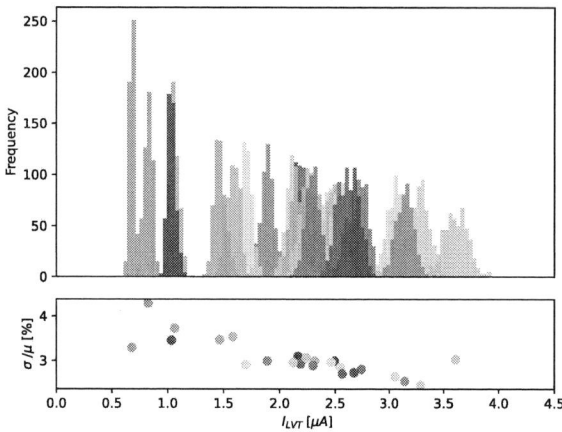

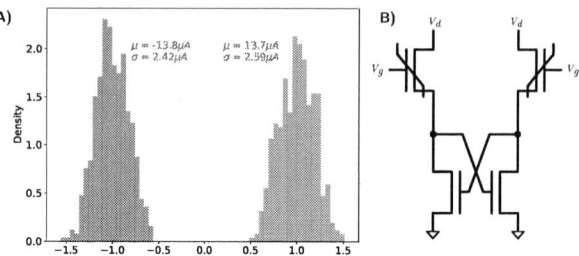

Figure 5: Histogram of LVT current for 500 FeMFETs for various Monte Carlo runs. Below, I_{LVT} relative deviation.

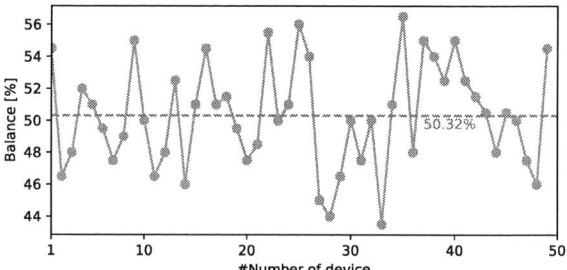

Figure 6: A) Relative current difference between left and right FemFET B) Latch circuit for identifying variability in the reading stage

fabrication process, makes us think FeCap variability can be negligible compared with the transistor variability.

Observing the global variations, a significant I_{LVT} dispersion becomes evident. This can be interpreted as a potential problem for memory array applications, where fixed current ratios are expected. Simulations indicate a worst-case current ratio of $I_{LVT}/I_{HVT} \approx 100$, and a minimum $I_{LVT} \approx 50$ nA, suggesting some room for state differentiation with this transistor size, but leaving out multi-level memory applications.

To assess the feasibility of cell-to-cell detection, a simple latch circuit was implemented [5] to amplify the mismatch between the LVT state current of each cell, (Figure 6B). When both FeMFETs are being programmed, the cross coupled nodes of the circuits are brought to ground, then both devices are read, latching the circuit on one wing. Noise aside, it saturates in one direction due to current mismatch that comes from process variations of the memory cells. From this given methodology we conducted 1000 local variation runs in Figure 6A. Where the blue bell distribution represents the iterations where the circuit latched on the left side and in red the ones that latches to the right wing. Another experiment highlights the balance of the setup in Figure 7, this time the experiment is repeated 50 times with global variations. In the end, we measured a distribution close to 50% for both sides (here, 50, 32%). This balance suggests appealing characteristics for PUF applications as after process, the fingerprint of the circuit will not be biased to a given wing of the comparator circuit.

IV. CONCLUSION AND FUTURE WORK

By using a Preisach model fitted with experimentally obtained parameters we observed the behavior of a single cell, and the impact of FeCap and transistor variability in the read-write scheme. Through a proof of concept, we observed that the positive feedback of latch circuits would be sufficient for

Figure 7: Percentage of latch circuits that converge to the same side in 200 latch circuits for 50 Monte-Carlo Simulations.

identifying FeMFET mismatch. Via the emulation of intra-chip Monte Carlo simulations, current relative variations were observed, varying in the range of 2%-4%, which sets a reference for the design of any variability detection scheme. It would be interesting in the future to explore the impact of layout parasitics, noise and memory perturbations in the dual operation of memory arrays as PUFs.

ACKNOWLEDGMENT

This work has received the financial support of the French Agence Nationale de la Recherche (ANR), through project ANR-23-CE24-0015-01 (ECHOES) and by the European Commission under grant no. 101135656 (HORIZON-CL4-2023-DIGITAL-EMERGING-01-CNECT Ferro4EdgeAI).

REFERENCES

[1] *Embedded Memories for Nano-Scale VLSIs*. Springer US, 2009, pp. 279–328. [Online]. Available: http://dx.doi.org/10.1007/978-0-387-88497-4

[2] J. F. Scott, *Ferroelectric Memories*. Springer Berlin Heidelberg, 2000. [Online]. Available: http://dx.doi.org/10.1007/978-3-662-04307-3

[3] H. Dahlberg, O. Kaatranen *et al.*, "Memory state dynamics in beol fefets: Impact of area ratio on analog write mechanisms and charging," *IEEE Access*, vol. 13, pp. 9923–9930, 2025.

[4] A. Bosio, M. Cantan *et al.*, "Emerging technologies: Challenges and opportunities for logic synthesis," in *2021 24th International Symposium on Design and Diagnostics of Electronic Circuits & Systems (DDECS)*, 2021, pp. 93–98.

[5] S. Lim, J. Hwang *et al.*, "Design of physically unclonable function using ferroelectric fet with auto write-back technique for resource-limited iot security," *IEEE Internet of Things Journal*, vol. 11, no. 16, p. 27676–27686, Aug. 2024. [Online]. Available: http://dx.doi.org/10.1109/JIOT.2024.3399482

[6] H. Shao, Y. Zhou *et al.*, "A novel fefet array-based puf: Co-optimization of entropy source and crp generation for enhanced robustness in iot security," in *2023 International Electron Devices Meeting (IEDM)*. IEEE, Dec. 2023. [Online]. Available: http://dx.doi.org/10.1109/IEDM45741.2023.10413787

[7] T. Li, X. Guo *et al.*, "Demonstration of high-reconfigurability and low-power strong physical unclonable function empowered by fefet cycle-to-cycle variation and charge-domain computing," *Nature Communications*, vol. 16, no. 1, Jan. 2025. [Online]. Available: http://dx.doi.org/10.1038/s41467-024-55380-x

[8] R. Pappu, B. Recht *et al.*, "Physical one-way functions," *Science*, vol. 297, no. 5589, p. 2026–2030, Sep. 2002. [Online]. Available: http://dx.doi.org/10.1126/science.1074376

[9] P. Jimenez, R. Cardoso *et al.*, "Complexity assessment of analog and digital security primitives signals using the disentropy of autocorrelation," 2024. [Online]. Available: https://arxiv.org/abs/2402.17488

[10] S. Joshi, S. P. Mohanty, and E. Kougianos, "Everything you wanted to know about pufs," *IEEE Potentials*, vol. 36, no. 6, p. 38–46, Nov. 2017. [Online]. Available: http://dx.doi.org/10.1109/mpot.2015.2490261

[11] R. Waser, *Nanoelectronics and information technology: advanced electronic materials and novel devices*. John Wiley & Sons, 2012.

[12] J. Merkel, "Fefet process integration and characterization," *Journal of the Microelectronic Engineering Conference*, vol. 25, no. 1, 2019. [Online]. Available: https://repository.rit.edu/ritamec/vol25/iss1/20

[13] J. F. McGlone, "Ferroelectric hfo$_2$ thin films for fefet memory devices," *Journal of the Microelectronic Engineering Conference*, vol. 22, no. 1, 2016. [Online]. Available: https://repository.rit.edu/ritamec/vol22/iss1/24

[14] A. D. Gaidhane, R. Dangi *et al.*, "A computationally efficient compact model for ferroelectric switching with asymmetric nonperiodic input signals," *IEEE Transactions on Computer-Aided Design of Integrated Circuits and Systems*, vol. 42, no. 5, p. 1634–1642, May 2023. [Online]. Available: http://dx.doi.org/10.1109/TCAD.2022.3203956

[15] B. J. et al, "Computationally Efficient Ferroelectric Capacitor Model For Circuit Simulation," in *Symposium on VLSI Technology*. Kyoto, Japan: IEEE, 1997, pp. 141–142.

[16] H. Mulaosmanovic, J. Ocker *et al.*, "Switching kinetics in nanoscale hafnium oxide based ferroelectric field-effect transistors," *ACS Applied Materials & Interfaces*, vol. 9, no. 4, p. 3792–3798, Jan. 2017. [Online]. Available: http://dx.doi.org/10.1021/acsami.6b13866

[17] S. Zhukov, Y. A. Genenko *et al.*, "Dynamics of polarization reversal in virgin and fatigued ferroelectric ceramics by inhomogeneous field mechanism," *Phys. Rev. B*, vol. 82, p. 014109, Jul 2010. [Online]. Available: https://link.aps.org/doi/10.1103/PhysRevB.82.014109

[18] Z. Zheng, L. Jiao *et al.*, "Beol-compatible mfmis ferroelectric/anti-ferroelectric fets—part i: Experimental results with boosted memory window," *IEEE Transactions on Electron Devices*, vol. 71, no. 3, pp. 1827–1833, 2024.

[19] S. Martin, C. Jahan *et al.*, "Hf$_{0.5}$Zr$_{0.5}$O$_2$ feram scalability demonstration at 22nm fdsoi node for embedded applications," in *2024 IEEE International Electron Devices Meeting (IEDM)*. IEEE, Dec. 2024, p. 1–4. [Online]. Available: http://dx.doi.org/10.1109/IEDM50854.2024.10873378

TSPC-Based Low-Power High-Resolution CMOS Phase Frequency Detector

Dhandeep Challagundla[1], Venkata Krishna Vamsi Sundarapu[1], Ignatius Bezzam[2], Riadul Islam[1]

[1]*University of Maryland, Baltimore County, [2]Rezonent Inc.*

vd58139@umbc.edu, vsundar1@umbc.edu, i@rezonent.us, riaduli@umbc.edu

Abstract—**Phase Frequency Detectors (PFDs) are essential components in Phase-Locked Loop (PLL) and Delay-Locked Loop (DLL) systems, responsible for comparing phase and frequency differences and generating up/down signals to regulate charge pumps and/or, consequently, Voltage-Controlled Oscillators (VCOs). Conventional PFD designs often suffer from significant dead zones and blind zones, which degrade phase detection accuracy and increase jitter in high-speed applications. This paper addresses PFD design challenges and presents a novel low-power True Single-Phase Clock (TSPC)-based PFD. The proposed design eliminates the blind zone entirely while achieving a minimal dead zone of 40 ps. The proposed PFD, implemented using TSMC 28 nm technology, demonstrates a low-power consumption of $4.41\mu W$ at 3 GHz input frequency with a layout area of $10.42\mu m^2$.**

Index Terms—**Phase frequency detectors, delay-locked loops, phase-locked loops, power consumption, dead zone.**

I. INTRODUCTION

Phase-Locked Loops (PLLs) and Delay-Locked Loops (DLLs) are fundamental clock synchronization circuits widely used in high-speed digital and communication systems. As shown in Figure 1, a DLL (Figure 1(a)) employs a voltage-controlled delay line to align the output phase with the reference clock, ensuring precise delay matching without frequency synthesis. In contrast, a PLL (Figure 1(b)) utilizes a Voltage-Controlled Oscillator (VCO) to generate a frequency-locked output that tracks the reference clock, enabling frequency multiplication and jitter minimization [1], [2].

A crucial component in both PLL and DLL architectures is the Phase Frequency Detector (PFD). The PFD compares the phase and frequency difference between the reference and feedback signals, generating "Up" and "Down" control signals that regulate the charge pump and loop filter, ultimately adjusting the delay line in DLLs or the VCO in PLLs. The accuracy and speed of the PFD directly impact the loop's lock time, jitter performance, and overall system stability, making its design optimization critical.

Different types of PFDs are designed to meet specific performance requirements. Latch-based PFDs [3] are preferred for low-power applications but suffer from dead zone limitations. In contrast, tri-state PFDs help mitigate dead zone effects and have a linear phase-detecting range [4], [5].

This research was funded in part by National Science Foundation (NSF) award number: 2138253, Rezonent Inc. award number: CORP0061, and UMBC Startup Fund.

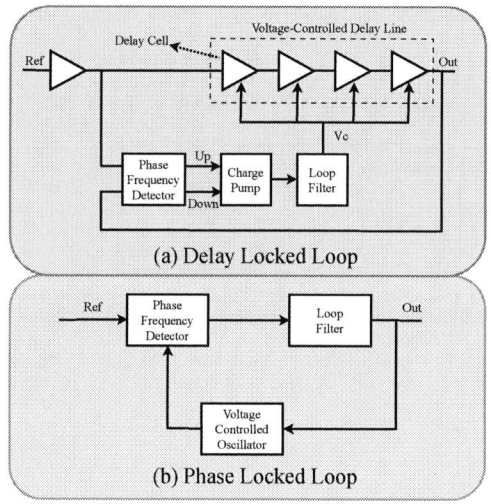

Fig. 1. Generalized block diagram of (a) a delay-locked loop and (b) a phase-locked loop [6].

PFD design improvements primarily focus on reducing the dead zone and minimizing phase noise while balancing power consumption and operating frequency. Various techniques, such as reset-path removal and the use of True Single-Phase Clock (TSPC) [6]–[9] logic, offer promising solutions, but each comes with trade-offs. The selection of an appropriate PFD architecture depends on the specific application requirements, particularly in high-performance PLLs where jitter and phase accuracy are critical considerations.

This work proposes a novel TSPC-based PFD circuit designed to enhance power efficiency and reduce dead-zone effects, thereby improving the locking characteristics of both PLL and DLL architectures in advanced low-power applications. The key contributions of this research are:

- A novel low-power, high-resolution PFD circuit based on TSPC logic.
- The dead zone of the proposed PFD is primarily determined by the setup time of the TSPC latch, exhibiting a minimal dead zone of 40 ps.
- The proposed design achieves 2.5× lower power consumption compared to [10], and a 72.5% reduction in power relative to [11].

979-8-3315-9813-6/25 $31.00 © 2025 IEEE

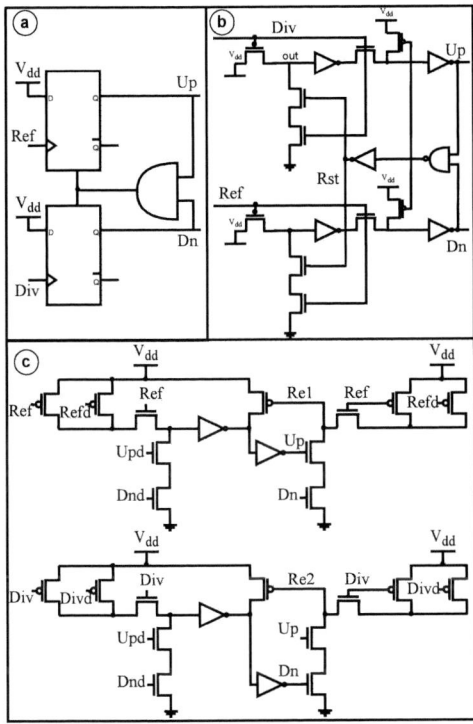

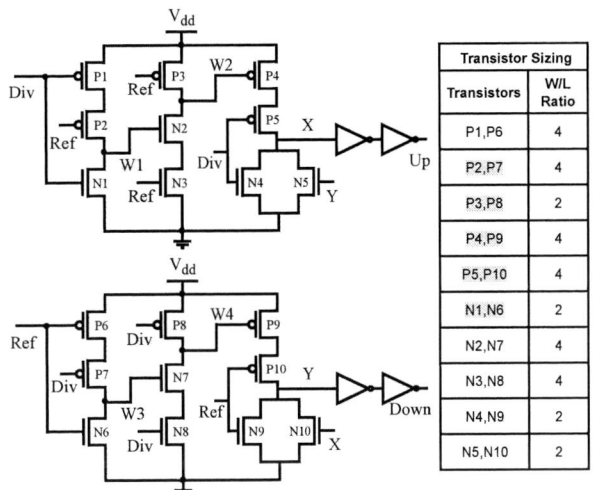

Transistor Sizing	
Transistors	W/L Ratio
P1,P6	4
P2,P7	4
P3,P8	2
P4,P9	4
P5,P10	4
N1,N6	2
N2,N7	4
N3,N8	4
N4,N9	2
N5,N10	2

Fig. 3. The proposed TSPC-based phase frequency detector circuit is designed using 20 transistors with symmetric design to generate Up and $Down$ signals.

Fig. 2. (a) Traditional PFD using D latch design, (b)a pass transistor based fast PFD [13] (c) existing dead zone free PFD [10] using 22 transistors.

II. BACKGROUND

Traditional PFDs suffer from a large dead zone in their phase detection characteristics at a steady state, leading to increased jitter in the locked state [12]. This dead zone occurs when the phase difference between the input signals is too small for the PFD to detect, causing erroneous phase detection and potentially locking the PLL to an incorrect phase. Additionally, conventional tristate charge pumps used in PLLs introduce issues such as charge sharing, charge injection, and clock feedthrough when the Up and $Down$ signals switch states, further degrading the performance by introducing phase noise.

Another limitation of conventional PFD designs is the blind zone, which occurs when the input signals are close to $\pm 360°$. This can lead to instability in phase detection, further complicating the lock-in process of the PLL.

Various design modifications have been explored to minimize the dead zone [10], [11], [14]–[20] while maintaining power efficiency [21], [22] and high-speed operation [23]–[25]. One common approach involves using D flip-flops and an "AND" gate to generate the Up and $Down$ signals [6], [26]. While this method is simple, it suffers from high power consumption at increased operating frequencies and requires a large transistor count, leading to higher area utilization.

Figure 2 illustrates various PFD architectures. Figure 2(a) shows the conventional PFD design employing D–latches

and a reset logic gate, which suffers from a considerable dead zone due to latch timing limitations. Figure 2(b) [13] presents a PFD optimized for low signal-to-noise ratio (SNR) environments, offering improved sensitivity but with limited dead zone mitigation. Figure 2(c) depicts an advanced dead zone-free PFD architecture, as proposed in [10], utilizing 22 transistors and incorporating inverters for faster reset paths to minimize the dead zone.

An alternative design aims to eliminate the dead zone entirely [15] by removing the reset path in the PFD. This results in a more linear phase detection characteristic, enabling precise phase error correction. Such an approach has demonstrated a maximum operating frequency of 4 GHz with high power consumption.

Another notable design utilizes a D flip-flop combined with gate diffusion input (GDI) logic technique to optimize power efficiency [11]. This PFD offers significantly lower power consumption while achieving operating frequencies of up to 3.33 GHz. However, it suffers from a high blind zone of $2\pi/10$ around $\pm 360°$, which impacts performance under certain operating conditions. Moreover, previous works have demonstrated their designs at higher technology nodes. In contrast, this work presents a novel TSPC-based PFD circuit implemented in 28nm technology, achieving zero blind zone, minimal dead zone, and low power consumption. The TSPC latch is advantageous as it operates with a single clock phase, eliminating the need for complementary clocks. Its dynamic structure enables faster switching and a lower transistor count, making it well-suited for high-frequency, low-power PFDs.

III. PROPOSED PHASE FREQUENCY DETECTOR CIRCUIT

The PFD is a crucial component in PLLs and DLLs and is responsible for detecting both phase and frequency differences between the reference clock ("Ref") and the feedback clock ("Div"). The proposed PFD, shown in Figure 3, is

979-8-3315-9813-6/25 $31.00 © 2025 IEEE

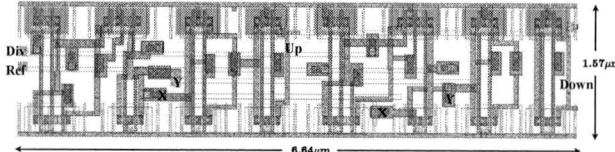

Fig. 4. The layout of the proposed TSPC-based phase frequency detector circuit occupies $10.42\,\mu m^2$ and uses metal 1 to metal 3 layers.

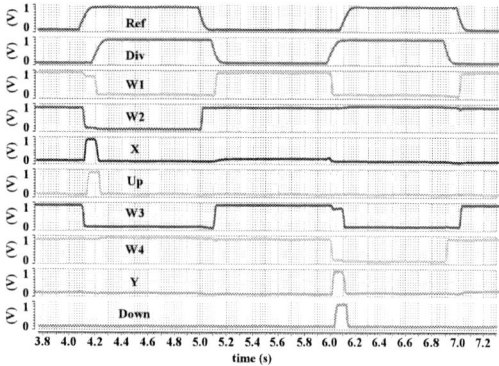

Fig. 5. Simulation results of the proposed PFD circuit producing appropriate up and down signals according to the phase difference of "Ref" and "Div" signals.

implemented using a CMOS-based dynamic logic design to achieve high-speed operation with minimal power consumption. The circuit consists of two symmetrical dynamic TSPC flip-flop [27]-like structures, each dedicated to generating Up and $Down$ signals based on the relative timing of "Ref" and "Div." These signals serve as inputs to a charge pump, ultimately controlling the loop filter and frequency synthesis mechanism.

The PFD operates by comparing the phase and frequency of "Ref" and "Div" and producing an appropriate control signal. When the reference clock leads the divided clock, the Up signal is asserted, prompting the charge pump to increase the control voltage of the VCO or delay line. Conversely, when the feedback clock leads the reference clock, the $Down$ signal is activated, instructing the system to slow down. The circuit ensures that Up and $Down$ signals are never active simultaneously, preventing ambiguous states and reducing phase jitter. Additionally, an internal reset mechanism is implemented to suppress excessive pulses, thereby improving the overall locking stability of the loop.

The circuit schematic of the proposed PFD is shown in Figure 3, and the corresponding simulation waveform is described in Figure 5. When both "Div" and "Ref" signals are "0," transistors P1 and P2 are "ON," which charges the wire W1 to V_{dd} and, in turn, activates N2. As "Ref" transitions to "1" while "Div" remains "0," node X is driven to "1." Since N2 remains "ON" and N3 also turns "ON" ("Ref" = "1"), wire W2, the input to P4 discharges. Simultaneously, as "Div" remains "0," P5 stays "ON," allowing X to charge and maintain its state at "1." With X = "1," transistor N10 becomes "ON," discharging node Y and setting Y to "0." This is shown in the first cycle of the simulation waveform, Figure 5.

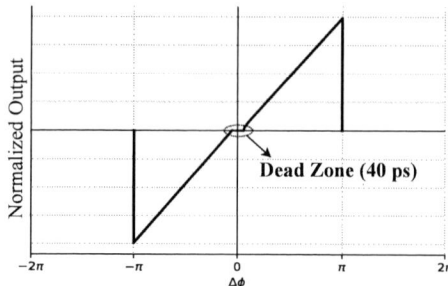

Fig. 6. The transfer characteristics of the proposed TSPC-based phase frequency detector circuit demonstrate a low dead zone of 40 ps.

In the second cycle, as shown in Figure 5, when "Div" transitions to "1" while "Ref" remains "0," transistors N8 and N7 turn "ON," discharging the gate input of P9. This causes Y to be set to "1," which subsequently turns N5 "ON," driving X to "0." This operational sequence ensures that the circuit accurately detects the phase difference between the "Ref" and "Div" signals, enabling reliable phase frequency detection essential for high-speed clock synchronization applications.

Figure 4 shows the layout of the proposed PFD circuit. It has a standard cell height of $1.57\mu m$ and an area of $10.42\mu m^2$ and is constructed using Metal 1 to Metal 3 layers.

IV. RESULTS AND DISCUSSION

A. Experimental Setup

The proposed TSPC-based PFD circuit was designed using TSMC 28nm CMOS technology. The schematic and layout designs were implemented in Cadence Virtuoso, and transient simulations were conducted using the Cadence Spectre simulator. Simulations were performed with a 1 GHz to 3 GHz clock frequency, and power consumption values were extracted directly from the Spectre tool. Since 28 nm implementations of conventional PFDs were not readily available, we used Dennard's power scaling law [28] to compare the reference circuits. Additionally, transfer characteristic simulations were performed to determine the dead zone of the proposed circuit. To assess design robustness, process, voltage, and temperature (PVT) variation analyses were also conducted.

B. Transfer Characteristics

The transfer characteristics plot shown in Figure 6 is obtained by simulating the Ref and Div signals with a phase difference ranging from $-\pi$ to π. The plot illustrates a normalized output response, demonstrating the expected phase detection behavior. Notably, the frequency divider operation observes a dead zone of 40 ps. This dead zone is relatively low due to the dependence of the PFD on the TSPC latch, where the setup time of the flip-flop introduces this phase-detection limitation. The minimized dead zone ensures improved phase tracking accuracy and enhances the overall performance of the PLL system.

TABLE I
COMPARISON OF VARIOUS PFD CIRCUITS

	This Work	[15]	[10]	[11]	[13]	[29]	[30]
Process Technology (nm)	28	180	28	28	28	28	65
Dead Zone (ps)	40	0	0	0	70	45	–
Power Consumption (uW)	4.41	201.6*	11.04**	15.44**	2.92**	21.9**	422.4*
Transistor Count	20	8	22	16	54	22	14
Max Frequency (GHz)	3	4	2.9	3.3	0.8	1	2.2

* Scaled using Dennard's scaling law [28] to match the 28nm technology.
** Simulated circuits in-house using 28nm technology.

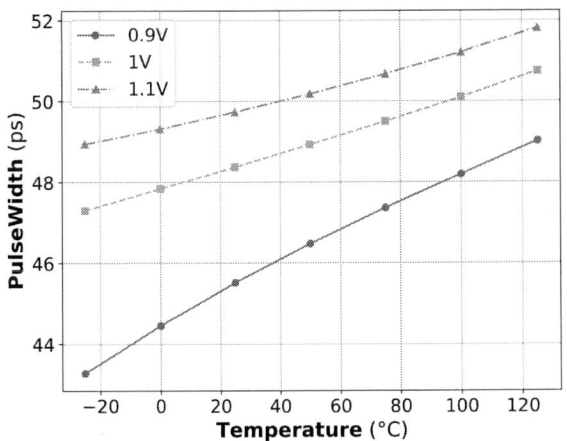

Fig. 7. The temperature and voltage variations show minimal changes across temperatures ranging from $-25°C$ to $125°C$ and $\pm10\%$ change in supply voltage.

C. Comparison with Previous Works

Table I presents a comparative analysis of various PFD designs in terms of process technology, dead zone, power consumption, transistor count, and maximum operating frequency. The proposed TSPC-based PFD, implemented in TSMC 28 nm technology, demonstrates a low dead zone of 40 ps while consuming only $4.41\mu W$ of power at 3 GHz. We performed simulations of comparative circuits under identical conditions as those used for the proposed circuit to measure their power consumption. To ensure a fair comparison for remaining circuits, we applied Dennard's power scaling law [28] to scale down the power consumption results from larger technology nodes to the 28 nm technology node.

Compared to [15], the proposed design achieves 99.3% reduction in power consumption. Additionally, the proposed PFD design consumes $2.5\times$ lower power than [10] and $3.5\times$ lower than [11]. While these designs have zero dead zones, they have high power consumption. The proposed TSPC-based PFD design exhibits a higher operating frequency and a lower dead zone of 40 ps compared to [13] and [29], which have values of 70 ps and 45 ps, respectively.

D. PVT Variation Analysis

Figure 7 shows the temperature-voltage variation analysis for a nominal phase difference of $\pm0.1\pi$ at 1 GHz, where the ideal pulse width is 50 ps. We considered seven different tem-

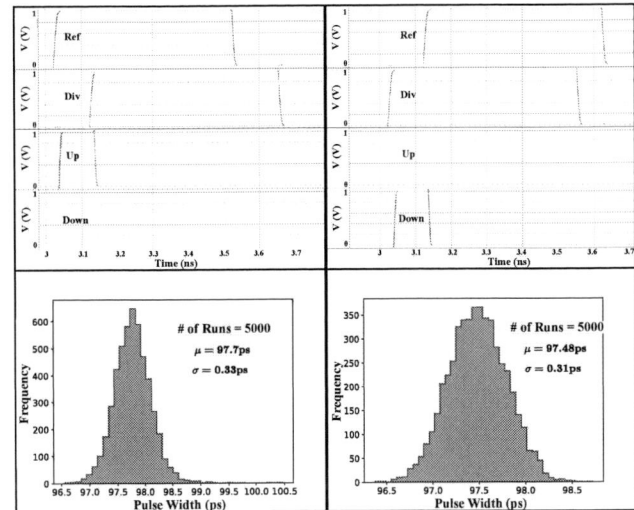

Fig. 8. The process variation analysis shows a very similar mean pulse width value when considering 5000 samples of 3σ deviations in $\pm10\%$ change in length of all transistors of the proposed PFD circuit.

peratures ranging from $-25°C$ to $125°C$ and $\pm10\%$ changes in supply voltages. The average pulse width of "Up" and "Down" is measured at each combination of temperature and voltage considering a nominal phase difference of 0.1π. It is observed that the proposed PFD circuit generates $1.1\times$ higher pulse width on average with a higher supply voltage of 1.1 V than 0.9 V. Moreover, the proposed circuit generates only 5.5% higher average pulse width at $100°C$ when compared to $25°C$. This analysis showcases a maximum change of 44 ps to 52 ps average pulse width when considering extreme temperatures of $-25°C$ at 0.9 V and $125°C$ at 1.1 V, respectively.

Figure 8 presents the process variation analysis of the proposed PFD circuit, based on 5000 samples of the "Up" and "Down" signal pulse widths, considering $\pm10\%$ length variation of all transistors under 3σ deviations. This analysis is performed considering a phase difference of $\pm0.2\pi$ at 1 GHz. The analysis yielded a mean pulse width of 97.7 ps with a standard deviation of 0.33 ps for the "Up" signal. Meanwhile, the "Down" pulse width exhibits a very similar mean pulse width of 97.48 ps and a standard deviation of only 0.31 ps.

V. CONCLUSION

In this work, we proposed a novel low-power TSPC-based PFD optimized for high-speed and low-power operation. The proposed design was implemented in TSMC 28 nm CMOS technology and achieved a dead zone of 40 ps with zero blind zones, demonstrating improved phase detection accuracy. Our design achieved a low power consumption of $4.41\mu W$ at 3 GHz frequency, with a layout area of $10.42\mu m^2$. Comparison with prior works shows high power savings ranging from 61% when compared with [10] to 80% when compared with [29]. Additionally, PVT variation analysis conducted using 5000 samples of $\pm10\%$ length variation with 3σ deviations demonstrated the robustness of the proposed PFD circuit with a mean average pulse width of 97 ps at a nominal phase difference of $\pm0.2\pi$ around $0°$.

REFERENCES

[1] B. Razavi, "A Simple Precharged CMOS Phase Frequency Detector," 2003.

[2] R. Islam and M. R. Guthaus, "HCDN: Hybrid-mode clock distribution networks," *IEEE Transactions on Circuits and Systems I: Regular Papers*, vol. 66, no. 1, pp. 251–262, 2018.

[3] A. Koithyar and T. Ramesh, "A faster phase frequency detector using transmission gate–based latch for the reduced response time of the PLL," *International Journal of Circuit Theory and Applications*, vol. 46, no. 4, pp. 842–854, 2018.

[4] C.-W. Hsu, K. Tripurari, S.-A. Yu, and P. R. Kinget, "A sub-sampling-assisted phase-frequency detector for low-noise PLLs with robust operation under supply interference," *IEEE Transactions on Circuits and Systems I: Regular Papers*, vol. 62, no. 1, pp. 90–99, 2014.

[5] J. A. Crawford, "Frequency synthesizer design handbook," 1994.

[6] J. M. Rabaey, A. Chandrakasan, and B. Nikolic, *Digital integrated circuits*. Prentice hall Englewood Cliffs, 2002, vol. 2.

[7] R. Islam, "High-speed energy-efficient soft error tolerant flip-flops," Ph.D. dissertation, Concordia University, 2011.

[8] D. Challagundla, I. Bezzam, and R. Islam, "Design automation of series resonance clocking in 14-nm FinFETs," *Circuits, Systems, and Signal Processing*, vol. 42, no. 12, pp. 7549–7579, 2023.

[9] R. Islam, D. Challagundla, and I. Bezzam, "System and methods of reducing wideband series resonant clock skew," Oct. 10 2024, US Patent App. 18/627,479.

[10] M. Divya and K. Sundaram, "A novel replica technique based dead zone free phase frequency detector and a self-cascode current-splitting charge pump for a low-spur low power phase-locked loop architecture," *AEU-International Journal of Electronics and Communications*, vol. 176, p. 155098, 2024.

[11] A. Rezaeian, G. Ardeshir, and M. Gholami, "A low-power and high-frequency phase frequency detector for a 3.33-GHz delay locked loop," *Circuits, Systems, and Signal Processing*, vol. 39, pp. 1735–1750, 2020.

[12] J. Sharma, T. Varma, and D. Boolchandani, "A brief review of the various phase-frequency detector architectures," in *2021 IEEE International Symposium on Smart Electronic Systems (iSES)*. IEEE, 2021, pp. 74–78.

[13] M. Mansuri, D. Liu, and C.-K. Yang, "Fast frequency acquisition phase-frequency detectors for GSa/s phase-locked loops," in *Proceedings of the 27th European Solid-State Circuits Conference*. IEEE, 2001, pp. 333–336.

[14] T. Azadmousavi, M. Azadbakht, E. Najafi Aghdam, and J. Frounchi, "A novel zero dead zone PFD and efficient CP for PLL applications," *Analog Integrated Circuits and Signal Processing*, vol. 95, pp. 83–91, 2018.

[15] J. Sharma, R. Ahmad, A. Yadav, T. Varma, and D. Boolchandani, "Design and optimization of phase frequency detector through Taguchi and ANOVA statistical techniques for fast settling low power frequency synthesizer," *Integration*, vol. 96, p. 102162, 2024.

[16] N. M. Ismail and M. Othman, "CMOS phase frequency detector for high speed applications," in *2009 International Conference on Microelectronics-ICM*. IEEE, 2009, pp. 201–204.

[17] Y. Liang and C. C. Boon, "A 40 GHz CMOS PLL With- 75-dBc Reference Spur and 121.9-fs rms Jitter Featuring a Quadrature Sampling Phase-Frequency Detector," *IEEE Transactions on Microwave Theory and Techniques*, vol. 70, no. 4, pp. 2299–2314, 2022.

[18] K. Park, M. Shim, H.-G. Ko, B. Nikolić, and D.-K. Jeong, "Design techniques for a 6.4–32-Gb/s 0.96-pJ/b continuous-rate CDR with stochastic frequency–phase detector," *IEEE Journal of Solid-State Circuits*, vol. 57, no. 2, pp. 573–585, 2021.

[19] M. Divya and K. Sundaram, "Dead zone-less low power phase frequency detector, independent of duty cycle variations for charge pump phase locked loop," *Analog Integrated Circuits and Signal Processing*, vol. 114, no. 1, pp. 13–30, 2023.

[20] H. Ju, K. Lee, W. Jung, and D.-K. Jeong, "A 48Gb/S 2.4 pJ/B PAM-4 Baud-Rate Digital CDR with Stochastic Phase Detection Technique in 40nm CMOS," in *2021 IEEE Asian Solid-State Circuits Conference (A-SSCC)*. IEEE, 2021, pp. 1–3.

[21] B. Jamadi, J. Lee, and J. S. Walling, "A low-power phase frequency detector using sram cells in 22nm fd-soi," in *2024 22nd IEEE Interregional NEWCAS Conference (NEWCAS)*, 2024, pp. 218–222.

[22] K. Abdul Majeed and B. J. Kailath, "Low power PLL with reduced reference spur realized with glitch-free linear PFD and current splitting CP," *Analog Integrated Circuits and Signal Processing*, vol. 93, pp. 29–39, 2017.

[23] N. A. Badiger and S. Iyer, "Design & implementation of high speed and low power pll using gpdk 45 nm technology," *Journal of The Institution of Engineers (India): Series B*, vol. 105, no. 2, pp. 239–249, 2024.

[24] W.-H. Lee, J.-D. Cho, and S.-D. Lee, "A high speed and low power phase-frequency detector and charge-pump," in *Proceedings of the ASP-DAC'99 Asia and South Pacific Design Automation Conference 1999 (Cat. No. 99EX198)*. IEEE, 1999, pp. 269–272.

[25] N. Pradhan and S. K. Jana, "Design of phase frequency detector with improved output characteristics operating in the range of 1.25 MHz–3.8 GHz," *Analog Integrated Circuits and Signal Processing*, vol. 107, no. 1, pp. 101–108, 2021.

[26] P. Nagarajan, N. A. Kumar, J. A. Dhanraj, T. S. Kumar *et al.*, "Delay Flip Flop based Phase Frequency Detector for Power Efficient Phase Locked Loop Architecture," in *2022 International Conference on Electronics and Renewable Systems (ICEARS)*. IEEE, 2022, pp. 410–414.

[27] R. Islam, "Low-power resonant clocking using soft error robust energy recovery flip-flops," *Journal of Electronic Testing*, vol. 34, no. 4, pp. 471–485, 2018.

[28] R. Dennard, F. Gaensslen, H.-N. Yu, V. Rideout, E. Bassous, and A. LeBlanc, "Design of ion-implanted MOSFET's with very small physical dimensions," *IEEE Journal of Solid-State Circuits*, vol. 9, no. 5, pp. 256–268, 1974.

[29] A. Fathi, M. Mousazadeh, and A. Khoei, "High-speed, low power, and dead zone improved phase frequency detector," *IET Circuits, Devices & Systems*, vol. 13, no. 7, pp. 1056–1062, 2019.

[30] C.-W. Hsu, K. Tripurari, S.-A. Yu, and P. R. Kinget, "A sub-sampling-assisted phase-frequency detector for low-noise PLLs with robust operation under supply interference," *IEEE Transactions on Circuits and Systems I: Regular Papers*, vol. 62, no. 1, pp. 90–99, 2014.

EmFIA: A Novel Emulation-based Fault Injection Vulnerability Assessment Framework at RTL Level

Tanvir Rahman, Shuvagata Saha, Sujan Kumar Saha, Farimah Farahmandi, Mark Tehranipoor

Department of Electrical and Computer Engineering
University of Florida, Gainesville, FL, USA
{tanvir.rahman, sh.saha, sujansaha}@ufl.edu, {farimah,tehranipoor}@ece.ufl.edu

Abstract—**Fault-injection attacks (FIA) intentionally disrupt circuit behavior allowing adversaries to bypass safety mechanisms, disrupt system functionality, or extract sensitive information, thereby posing severe risks to the security and reliability of modern System-on-Chips (SoCs). However, pre-silicon security assessments targeting FIA predominantly rely on gate-level simulation or late-stage layout analysis, which are slow, limited in coverage, and often fail to capture realistic operating conditions—leaving critical vulnerabilities undetected until post-silicon stages. To address these limitations, we propose EmFIA, an emulation-driven register-transfer level (RTL) fault injection assessment framework designed to analyze security-critical vulnerabilities against FIA. EmFIA systematically analyzes security-critical signals in a design by modeling the faults in hardware emulation platform, inserting SystemVerilog assertions, and monitoring security property violations. Demonstrated on a RISC-V SoC and standalone AES-128 and RSA-128 cores, EmFIA enables rapid exploration of fault scenarios, achieving speedups of several orders of magnitude compared to exhaustive gate-level simulation. EmFIA provides designers with fast, property-aware security insight early in the design cycle, significantly strengthening hardware resilience prior to fabrication.**

Index Terms—**Fault Injection Assessment, Emulation, Pre-silicon, Focused Ion Beam(FIB) Fault Injection, Register-Transfer Level**

I. INTRODUCTION

Fault injection attacks (FIAs) pose severe threats to modern System-on-Chip (SoC) designs by corrupting safety-critical logic, bypassing privilege controls, or extracting cryptographic secrets. These attacks can be induced via voltage glitches, laser pulses, electromagnetic interference, or focused ion beams (FIB), targeting various abstraction levels of the system [1]–[3]. As a countermeasure, designers rely on pre-silicon vulnerability assessment to detect exploitable design-level weaknesses early in the development cycle. However, existing assessment flows remain limited to slow gate-level simulation or late-stage layout analyses [4], which often operate on abstracted or incomplete models and fail to replicate real hardware behaviors accurately. Traditional fault injection assessment techniques [5]–[7] are constrained by slow execution speed and minimal fault type & space coverage, making them unsuitable for large-scale SoC verification.

Recent state-of-the-art work has introduced Gate-level simulation frameworks [8] to improve scalability, but these still fall short in terms of speed and observability. Gate-level fault simulation provides exhaustive controllability but at the cost

of impractical runtime, especially for large SoCs [9]. Post-silicon analyses, while more realistic, are too late in the design lifecycle and offer limited design fixability [10]. The lack of fast, property-aware RTL methodologies capable of capturing hardware-software interactions under realistic fault conditions remains a key gap in the hardware security verification ecosystem.

To address these limitations, we introduce EmFIA, an emulation-driven fault injection assessment framework that operates at the RTL level. EmFIA focuses on stuck-at fault modeling inspired by FIB-based permanent faults that can disable security-critical signals in digital hardware. Unlike transient fault models, the stuck-at abstraction aligns with the physical persistence and spatial locality of FIB-induced damage, allowing accurate emulation of permanent circuit modifications. EmFIA enables the designer to inject these faults dynamically during execution using RTL emulation platforms, enabling high-throughput, cycle-accurate observation of fault effects.

EmFIA targets pre-identified security-sensitive signals that influence the confidentiality, integrity, and availability (CIA) of the system. By embedding SystemVerilog Assertions (SVAs) linked to security properties, the framework acts as a property-checking oracle that detects assertion violations during fault emulation. To achieve realistic emulation speed and fidelity, EmFIA leverages a commercial hardware emulation platform [11] that supports signal forcing and SVA monitoring with RTL granularity, avoiding the need for lower-level layout or gate netlist representations.

We demonstrate EmFIA on a RISC-V based SoC platform [12] and standalone cryptographic accelerators (AES-128 and RSA-128). In these case studies, EmFIA is able to uncover assertion violations caused by strategically injected stuck-at faults. Our results show that EmFIA offers orders of magnitude faster fault space exploration while preserving observability and maintaining semantic alignment with high-level design intent.

This paper makes the following contributions:

- We present EmFIA, a pre-silicon, RTL-level fault injection framework based on hardware emulation for scalable security assessment.
- We define a stuck-at fault model tailored to emulate FIB-style attacks, focusing on permanent signal manipulation rather than transient glitches.

979-8-3315-9813-6/25 $31.00 © 2025 IEEE

- We integrate property-aware fault detection using SVAs mapped to CIA-classified signals and demonstrate their effectiveness under RTL emulation.
- We validate EmFIA on complex SoC and crypto IPs, showing significantly faster fault analysis than simulation with enhanced coverage of early-cycle vulnerabilities.

The paper is organized as follows: in Section II, we describe our proposed framework along with the threat model and the fault model. Section III illustrates the implementation of our framework in an emulation platform, and the assessment results are shown in Section IV. Finally, we conclude our paper in Section V.

II. PROPOSED FRAMEWORK

A. Threat Model

We consider a pre-silicon adversary model where the attacker's goal is to violate the system's security guarantees by permanently altering the behavior of a hardware design. The adversary may exploit vulnerabilities in the RTL implementation to gain unauthorized access, bypass security checks, or leak cryptographic information. Unlike post-silicon adversaries, this threat model assumes access to RTL design representations and the ability to simulate or emulate faults before fabrication.

We focus on attacks that compromise the Confidentiality, Integrity, or Availability (CIA) of the system through permanent signal-level modifications that emulate physical phenomena such as Focused Ion Beam (FIB)-induced faults. The attacker can target security-sensitive registers, control paths, or interrupt signals with a high degree of precision, exploiting design flaws before tape-out. The attacker is assumed to know the ISA-level behavior of the SoC and may also have knowledge of the expected control or dataflow under benign conditions. However, the adversary does not modify the testbench or design-for-test hardware and relies solely on fault effects observable through RTL signal propagation.

B. Fault Model

Our fault model focuses on stuck-at faults representing permanent, controllable disruptions to a single-bit signal at the RTL. These faults correspond to FIB-style physical attacks [2], effectively fixing the logic level to '1' (stuck-at-1) or '0' (stuck-at-0). This abstraction avoids the need to model transient timing or electrical effects while maintaining realistic coverage for permanent hardware modifications.

We assume the attacker can inject faults at any time step, but the fault, once activated, persists throughout the emulation cycle unless explicitly released. To maintain tractability and emulation efficiency, we constrain the number of fault sites per experiment to one for our initial analysis. Faults are injected only at identified security-sensitive signals, determined through static analysis and prior security property classification. These include privilege check signals, memory access control flags, interrupt lines, and cryptographic state indicators.

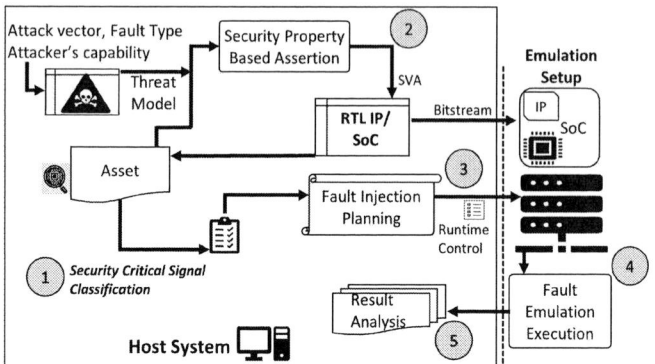

Fig. 1. EmFIA Framework

EmFIA uses direct cycle-accurate emulation with dynamic signal forcing to emulate stuck-at conditions without rerunning compilation, synthesis, or altering the hardware description.

C. Framework Description

The Emulation-based Fault Injection Assessment (EmFIA) framework is a five-stage pre-silicon assessment framework (Fig. 1) for emulation-based fault injection at the RTL level. It identifies security-sensitive components, instruments RTL with security assertions, schedules realistic fault scenarios, and observes fault-induced violations on a hardware emulation platform. Below, we outline each stage.

1) Security Signal Classification: EmFIA begins with manual or semi-automated classification of signals that uphold confidentiality, integrity, or availability (CIA). Targets include privilege bits, memory flags, cryptographic enables, resets, and FSM control paths. The designer may apply static analysis tools or architectural insights to flag signals at control-flow or data-path intersections.

2) Assertion Insertion: Next, SystemVerilog Assertions (SVAs) are added to encode expected security behavior (e.g., key immutability, valid privilege transitions). These enable real-time violation detection under faulted conditions. By inserting these assertions directly into the RTL, EmFIA enables on-the-fly detection of security violations triggered by fault effects.

3) Fault Injection Planning: For each protected signal, EmFIA schedules stuck-at fault (SA) campaigns using a FIB-inspired model (SA-0/SA-1). Each fault is mapped to a specific cycle and injected as a persistent logic-level override mimicking conductive material deposition on interconnects.

4) Fault Emulation Execution: EmFIA executes the fault campaigns using a cycle-accurate hardware emulation platform that supports dynamic signal forcing. During runtime, the emulator injects faults at pre-specified cycles and locations. Assertions are evaluated on-the-fly under realistic workloads or testbench. EmFIA supports high-throughput testing without recompilation.

5) Violation Logging and Analysis: Assertion violations, timing, and signal context are logged to trace security vulnerability. Designers can correlate fault sites with CIA violations and prioritize fixes or insert countermeasures.

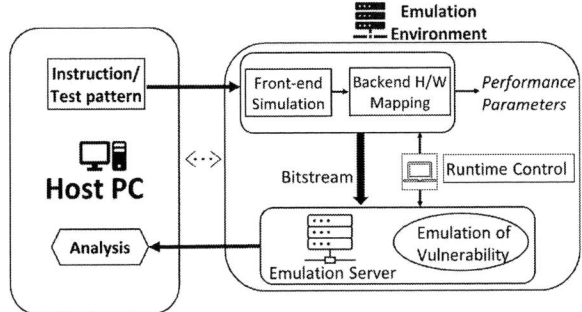

Fig. 2. Emulation environment setup.

This structured flow enables fast, property-aware RTL-level vulnerability assessment with broader coverage and speed than traditional simulation techniques.

III. IMPLEMENTATION

The EmFIA framework is implemented on the *Synopsys ZeBu Server* [11], a hardware-assisted emulation platform optimized for RTL design validation. This setup enables *cycle-accurate fault injection* while maintaining high throughput. The full emulation infrastructure consists of two main components: the *Host PC* and the *Emulation Environment*, illustrated in Fig. 2.

A. Host PC

The Host PC puts together the overall emulation process. It generates *SystemVerilog-based test patterns* and transmits them to the emulator. During post-emulation, the Host PC receives trace data and assertion logs, which are processed to reflect the timing, severity, and propagation effects of injected faults, supporting a structured assessment of design resilience under fault scenarios.

The Emulation Environment consists of the following stages:

1) Front-End Simulation and Hardware Mapping: RTL modules and testbenches are first compiled using *Synopsys VCS*, then mapped onto hardware using *Xilinx Vivado* to produce bitstreams. This ensures timing-accurate representation and resource-mapped behavior aligned with real hardware execution.

2) Bitstream Loading and Runtime Control: The generated bitstream is deployed on the emulator. A runtime controller manages the execution and applies *fault injection patterns* dynamically during emulation cycles using signal forcing.

3) Vulnerability Emulation: During runtime, *stuck-at faults* are applied to selected RTL signals. SystemVerilog assertions monitor the system for violations of security properties. The emulator logs any triggered assertions and abnormal behaviors under faulted conditions.

B. Data Flow and Analysis

Once emulation completes, *runtime data* is collected on the Host PC. Designers can examine which faults caused assertion violations, when they occurred, which security properties were affected and where in the design they originated. This enables direct observation of fault effects with accurate traceability and provides practical insight into the vulnerability of RTL components under *stuck-at fault* scenarios.

IV. RESULT ANALYSIS

We evaluated EmFIA on three benchmark designs: AES-128 encryption core [13], RSA-128 exponentiation core [14], and the NEORV32 RISC-V SoC [12]. For each design, we targeted RTL-level signals previously classified as security-critical [8] and inserted assertion-based security properties (Table I) to monitor confidentiality, integrity, and availability violations caused by emulated stuck-at faults.

A. Emulation Findings

As summarized in Table II, in AES-128, EmFIA detected violations of SP-1a (data integrity) and SP-2 (output availability) at cycle 56. These violations stemmed from injected faults in round control logic and key registers. In RSA-128, SP-3, SP-4, and SP-6 were violated due to improper state transitions during Montgomery exponentiation [15], revealing susceptibility in FSM transitions. Notably, SP-5 remained robust under all tested conditions, confirming the effectiveness of laddering-based security hardening. NEORV32 violations (SP-7, SP-8) reflected incorrect mode of operation and illegal PC values, representing potential privilege escalation paths.

B. Property-Driven Fault Validation: AES-128 Deeep Dive

To demonstrate EmFIA's capability beyond raw violation detection, we performed a two-stage *static & dynamic* investigation focused on SP-2 (cipher-text availability).

1) Static fan-in analysis: Using `Yosys Open SYnthesis Suite`, we traced the structural fan-in cone of `cipher_text` and `valid_out`. The cone comprises 41 nets across key schedule, round logic, and control FSM. Filtering for single-bit status/control lines produced 23 *local fault sites*. Nets were automatically ranked by topological level (depth): 11 belong to the final-round control path (e.g., `valid_shift2key`), 7 to intermediate round counters, and 5 to key schedule enables, providing guidance for dynamic fault injection ordering.

2) Single-site stuck-at campaign: Each of the 23 sites was forced to `stuck-at 0` and `stuck-at 1` over the full encryption cycles to emulate FIB based faults. 9 sites (all among the static ranks) *deterministically* violated SP-2 (Table III), indicating their critical role in handshake signaling. The remaining faults either masked internally or influenced other properties, revealing complex interdependencies between internal signals and security invariants.

3) Two-site interaction study: Although our threat model targets single-bit faults, EmFIA can flip several bits at once without re-synthesis. We examined the pair (`valid_sub2shift`, `valid_shift2key`). The combination *(0,1)* propagated to violate SP-2, whereas *(1,0)* did not. This asymmetry highlights nonlinear fault propagation effects, as the violation did not occur during single-site analysis, showcasing emulation's capability in the early detection of

TABLE I
LIST OF SECURITY PROPERTIES FOR VARIOUS IPS

SP	Design	Security Asset	Security Property Description	Violation Type
SP-1a	AES-128	Plaintext	The output ciphertext must be the valid encrypted form of the input	Data Integrity
SP-1b	RSA-128		Plaintext once the encryption process completes	
SP-2	AES-128	Ciphertext	Ciphertext should be available for use when the encryption completes	Availability
SP-3	RSA-128	Plaintext	Done signal should be asserted at the end of encryption	Availability, Integrity
SP-4	RSA-128	Operation Validity	Transition from FSM state "Start" to "State_1" must not be immediately followed by a transition from "State_1" to "Done" that bypasses an intermediate State "Check_condition"	Integrity
SP-5	RSA-128	Operation Validity	The FSM must not transition from 1st internal state to "Done" if the checking condition is false	Integrity
SP-6	RSA-128	Process Completion Integrity	The "Done" state must be reached exclusively from FSM internal state which checks for encryption completion ensuring that only valid transitions lead to completion	Integrity, Availability
SP-7	NEORV32 SoC	Access Control	A user space application should never have access to the "Machine mode"	Confidentiality, Integrity
SP-8	NEORV32 SoC	Program Flow Control	The program counter (PC) must remain within the valid instruction memory (IMEM) address range	Confidentiality, Integrity

TABLE II
EMULATION RESULTS FOR DIFFERENT SECURITY PROPERTIES (SPS)

SP	HW (LUT)	Emul. Cycles	Violation Cycle
AES-128 Core			
SP-1a	23139	100	56
SP-2	23139	100	56
RSA-128 Core			
SP-1b	4974	200	3
SP-3	4974	200	139
SP-4	4974	200	10
SP-5	4974	200	N/A
SP-6	4974	200	8
NEORV32 SoC			
SP-7	32172	1000	202
SP-8	32172	1000	153

TABLE III
FAULT INJECTION SUMMARY FOR SP-2

Category	Signals (Examples)	Fault Injection Outcome
Final-round control path	valid_shift2key, valid_sub2shift	9 of 11 caused SP-2 violation
Intermediate counters	valid_round_key[...]	Mixed results; some masked
Key schedule enables	cipherkey_valid_in, data_valid_in	Partial effect; influenced other SPs
Two-site interaction	(valid_sub2shift, valid_shift2key)	$(0,1) \rightarrow$ SP-2 fail; $(1,0) \rightarrow$ no effect

corner-case vulnerabilities. We stopped at two bits because in reality each extra FIB edit demands a new align–mill–verify cycle, driving beam time and rental cost up almost linearly, so larger multi-site studies are not practical [8], [16].

C. Emulation Runtime, Efficiency and Comparative Analysis

Each design was emulated under thousands of cycles per test with runtime ranging from <1s to approximately 6s, significantly outperforming simulation-based campaigns. EmFIA's hardware-accelerated cycle injection and assertion tracking enabled fast and comprehensive fault exploration without requiring waveform dumping or repeated synthesis.

TABLE IV
COMPARISON OF FAULT INJECTION ASSESSMENT METHODS

Method	Fault	Approach	Stage	Scale/Success	Target
SoFI [8]	Laser	Sim.	Gate	Low/Low	ASIC
Spill [17]	Laser	Sim.	Layout	Low/Low	ASIC
ACME [18] [19]	Bitstream	FPGA	Bitstream	High/Low	FPGA
EmFIA	Laser	Emul.	RTL	High/High	ASIC, FPGA

Table IV summarizes EmFIA's strengths compared to prior simulation- and bitstream-level fault tools. While tools like SoFI [8] and Spill [17] focus on late-stage analysis using laser fault simulation, they are slow, limited in scope, and not easily adaptable to design-level verification. Bitstream-level fault frameworks (such as [18], [19]) operate on post-synthesis designs but offer limited semantic observability. Emulation overcomes these challenges by enabling real-time, cycle-accurate fault injection in full SoC environments. EmFIA builds on this by injecting faults into security-critical RTL signals and monitoring assertion violations without waveform overhead, enabling fast and accurate exploration of FIB-style attack scenarios.

V. CONCLUSION

We presented EmFIA, an emulation-driven fault injection framework for pre-silicon security assessment at the RTL level. EmFIA models realistic stuck-at fault scenarios inspired by FIB attacks and monitors the system using property-aware SystemVerilog assertions. By leveraging hardware-assisted emulation, EmFIA achieves high-speed fault evaluation across SoC and crypto IPs, uncovering security violations that are often missed by traditional simulation. Our results demonstrate that EmFIA provides fast, scalable, and accurate detection of pre-silicon vulnerabilities, enabling designers to harden systems early in the hardware design lifecycle. Future work will explore support for transient and multi-bit fault models and integration with automated signal classification tools.

979-8-3315-9813-6/25 $31.00 © 2025 IEEE

REFERENCES

[1] K. Murdock, D. Oswald, F. D. Garcia, J. Van Bulck, D. Gruss, and F. Piessens, "Plundervolt: Software-based fault injection attacks against intel sgx," in *2020 IEEE Symposium on Security and Privacy (SP)*. IEEE, 2020, pp. 1466–1482.

[2] A. Barenghi, L. Breveglieri, I. Koren, and D. Naccache, "Fault injection attacks on cryptographic devices: Theory, practice, and countermeasures," *Proceedings of the IEEE*, vol. 100, no. 11, pp. 3056–3076, 2012.

[3] A. Vasselle, H. Thiebeauld, Q. Maouhoub, A. Morisset, and S. Ermeneux, "Laser-induced fault injection on smartphone bypassing the secure boot-extended version," *IEEE Transactions on Computers*, vol. 69, no. 10, pp. 1449–1459, 2018.

[4] U. Farooq and H. Mehrez, "Pre-silicon verification using multi-fpga platforms: A review," *Journal of Electronic Testing*, vol. 37, no. 1, pp. 7–24, 2021.

[5] P. R. Maier, U. Sharif, D. Mueller-Gritschneder, and U. Schlichtmann, "Efficient fault injection for embedded systems: as fast as possible but as accurate as necessary," in *2018 IEEE 24th International Symposium on On-Line Testing And Robust System Design (IOLTS)*. IEEE, 2018, pp. 119–122.

[6] X. Meng, Q. Tan, Z. Shao, N. Zhang, J. Xu, and H. Zhang, "Seinjector: A dynamic fault injection tool for soft errors on x86," in *2017 International Conference on Computer Systems, Electronics and Control (ICCSEC)*. IEEE, 2017, pp. 1492–1495.

[7] I. Tuzov, D. de Andrés, and J.-C. Ruiz, "Accurate robustness assessment of hdl models through iterative statistical fault injection," in *2018 14th European Dependable Computing Conference (EDCC)*. IEEE, 2018, pp. 1–8.

[8] H. Wang, H. Li, F. Rahman, M. M. Tehranipoor, and F. Farahmandi, "Sofi: Security property-driven vulnerability assessments of ics against fault-injection attacks," *IEEE Transactions on Computer-Aided Design of Integrated Circuits and Systems*, vol. 41, no. 3, pp. 452–465, 2021.

[9] M. Eslami, B. Ghavami, M. Raji, and A. Mahani, "A survey on fault injection methods of digital integrated circuits," *Integration*, vol. 71, pp. 154–163, 2020.

[10] S. Mitra, S. A. Seshia, and N. Nicolici, "Post-silicon validation opportunities, challenges and recent advances," in *Proceedings of the 47th Design Automation Conference*, 2010, pp. 12–17.

[11] Synopsys, "Emulation systems," https://www.synopsys.com/verification/emulation.html, 2024, accessed: 2024-11-11.

[12] S. Nölting, "neorv32: A small and open-source risc-v processor core," https://github.com/stnolting/neorv32-verilog, 2024, accessed: 2024-11-11.

[13] OpenCores Project Contributors, "AES-128 Pipelined Encryption Core," https://opencores.org/projects/aes-128_pipelined_encryption, 2025, accessed: May 16, 2025.

[14] T. Rahman, M. K. Bepary, M. S. U. Haque, M. Tehranipoor, and F. Rahman, "Design and security-mitigation of custom and configurable hardware cryptosystems," in *2023 IEEE 16th Dallas Circuits and Systems Conference (DCAS)*. IEEE, 2023, pp. 1–6.

[15] M. Joye and S.-M. Yen, "The montgomery powering ladder," in *International workshop on cryptographic hardware and embedded systems*. Springer, 2002, pp. 291–302.

[16] University of Missouri Electron Microscopy Core, "*Scios FIB Cheat-Sheet: Estimating Milling Time*," https://docs.research.missouri.edu/emc/emc_MU-EMC_Scios-FIB-cheat-sheet-v1.pdf, 2024, accessed 30 May 2025.

[17] N. Pundir, L. Lin, H. Li, N. Chang, F. Farahmandi, and M. Tehranipoor, "Spill—security properties and machine-learning assisted pre-silicon laser fault injection assessment," in *International Symposium for Testing and Failure Analysis*, vol. 84437. ASM International, 2022, pp. 225–236.

[18] L. A. Aranda, A. Sánchez-Macián, and J. A. Maestro, "Acme: A tool to improve configuration memory fault injection in sram-based fpgas," *IEEE Access*, vol. 7, pp. 128 153–128 161, 2019.

[19] L. A. Aranda, O. Ruano, F. Garcia-Herrero, and J. A. Maestro, "Acme-2: improving the extraction of essential bits in xilinx sram-based fpgas," *IEEE Transactions on Circuits and Systems II: Express Briefs*, vol. 69, no. 3, pp. 1577–1581, 2021.

Work in Progress: Exploring Ferroelectric Oscillators for Solving NP-Hard Problems Using Mallick's Coupling Mechanism

Joaquin Welch[a], Jorge Gomez[a], Jaime Cisternas[a]

[a]*Universidad de los Andes, Chile, Facultad de Ingeniería y Ciencias Aplicadas,*
Monseñor Álvaro del Portillo 12455, Santiago, Chile

Abstract—**NP-hard problems pose a significant computational challenge for both classical and quantum computers, with no known efficient solution strategies. Recent research has explored networks of coupled oscillators as a promising analog alternative for solving such problems. While theoretically powerful, these systems remain difficult to implement in hardware and to map efficiently onto combinatorial problem structures. Recent models have provided a theoretical foundation for encoding optimization problems such as the Traveling Salesman Problem (TSP) into coupled oscillator dynamics. However, the lack of scalable, practical hardware platforms has limited experimental progress. In this work, we propose the use of silicon-compatible ferroelectric oscillators as a viable pathway toward physical implementation. We simulate the dynamics of coupled ferroelectric oscillators and validate their behavior by applying Mallick's coupling algorithm to TSP instances. Our preliminary results confirm the limit-cycle behavior predicted by Mallick's dynamic model and demonstrate the potential of implementing a system of ferroelectric coupled oscillators for solving NP-hard problems.**

Index Terms—**NP-hard problems, Combinatorial optimization, Travelling Salesman Problem (TSP), Coupled oscillators, Ferroelectric oscillators, Silicon-compatible hardware, Quantum and classical computation alternatives**

I. INTRODUCTION

Analog networks of coupled oscillators offer a hardware-friendly route to tackling NP-hard combinatorial problems by directly embedding problem structure into physical circuitry. In practice, however, building scalable oscillator arrays and mapping discrete problem constraints onto continuous device dynamics remains an open challenge. Prior theoretical work has shown how the Traveling Salesman Problem (TSP) and related optimization tasks can be formulated as phase-coupling patterns in idealized oscillator models ([1]–[4]), yet few studies have bridged the gap between these abstractions and silicon-compatible hardware.

In this paper, we propose a practical implementation strategy using ferroelectric-based relaxation oscillators, whose positive-feedback Landau–Khalatnikov dynamics can be implemented in SPICE-compatible circuits ([5], [6]). We adapt Mallick *et al.*'s coupling algorithm to generate resistor–capacitor networks that encode TSP distance matrices as weighted interactions between oscillator phases. By instantiating each city as a dedicated ferroelectric oscillator node,

Corresponding author: Joaquín Welch (e-mail: jnwelch@miuandes.cl)

and tuning the RC links according to Mallick's f_{TSP} coupling function, we form a complete circuit model that can be laid out and simulated using standard SPICE tools.

Our contribution is two-fold. First, we validate that the coupled oscillator array reproduces the limit-cycle synchronization patterns predicted by the Mallick model when solving small TSP benchmarks. Second, we automated a netlist generator for the TSP coupling network, enabling flexible exploration of problem instances without custom hardware. These preliminary SPICE results establish ferroelectric oscillators as a promising, silicon-compatible platform for analog combinatorial optimization, paving the way toward future FPGA or ASIC prototypes.

This paper is organized as follows. Section II presents the Traveling Salesman Problem (TSP) and its formulation using coupled oscillator dynamics. Section III defines the system's energy function and establishes its Lyapunov stability. Section IV describes the design of the ferroelectric oscillator circuit, the implementation of Mallick's coupling model in SPICE, and the generation of netlists based on TSP instances. Section V outlines the simulation setup and preliminary results. Finally, Section VI summarizes the main contributions and discusses directions for future work.

II. TRAVELING SALESMAN PROBLEM FORMULATION

The Traveling–Salesman Problem (TSP) asks for the shortest possible closed route that visits each of N cities exactly once and returns to its starting point. In the oscillator-based formulation, we assign one phase oscillator to each city, letting its phase ϕ_i vary continuously on the interval $[0, 2\pi)$. A candidate tour is then read off by sorting the oscillators' phases in ascending order: the oscillator with the smallest phase corresponds to the first city, the next-smallest to the second, and so on until the tour closes back at the origin.

To enforce that adjacent cities in the tour correspond to phases separated by exactly $2\pi/N$, Mallick [1] introduce a shaping function f_{TSP}. Whenever the phase difference (Eq. 1) deviates from $2\pi/N$, this function imposes a penalty, thereby biasing the system toward phase configurations that represent valid tours.

$$\Delta\phi_{ij} = \phi_i - \phi_j \tag{1}$$

Eq. 2 defines an energy function where $J_{ij} = -D_{ij}$ and D_{ij} is the distance between cities i and j. By choosing

$J_{ij} = -D_{ij}$, we ensure that pairs of cities that are adjacent in a short tour contribute the most negative energy, making those configurations energetically favorable. Each oscillator's phase then evolves according to Eq. 3 [1].

$$H_{\text{TSP}} = -\sum_{i<j} J_{ij} \cos\left(\Delta\phi_{ij} + f_{\text{TSP}}(\Delta\phi_{ij})\right) \quad (2)$$

$$\frac{d\phi_i}{dt} = -C_1 \sum_{j\neq i} J_{ij} \sin\left(\Delta\phi_{ij} + f_{\text{TSP}}(\Delta\phi_{ij})\right) - C_{\text{sync}} \sin(N\phi_i) \quad (3)$$

In Eq. 3, the first term drives the network toward lower H_{TSP} by adjusting phases according to the distance-weighted couplings and the TSP shaping function. The second term, $\sin(N\phi_i)$, acts as a synchronization force that "locks" each oscillator onto one of the N equally spaced phases $0, 2\pi/N, 4\pi/N, \ldots$, guaranteeing a one-to-one mapping between oscillators and tour positions. As the system relaxes, the oscillators naturally settle into a phase ordering that corresponds to a valid solution of the TSP. Thus, the coupled dynamics of phase oscillators provide an analog method for approximating solutions to this classic combinatorial problem.

III. ENERGY FUNCTION AND LYAPUNOV STABILITY

To steer the oscillator network toward valid TSP tours, we introduce the energy function shown in Eq. 4 [1]. This expression combines two terms: the first encourages phase alignments corresponding to short inter-city distances (via $J_{ij} = -D_{ij}$), and the second enforces discretization of each phase into one of the N evenly spaced positions.

$$E(\phi) = -\frac{N C_1}{2} \sum_{i\neq j} J_{ij} \cos\left(\Delta\phi_{ij} + f_{\text{TSP}}(\Delta\phi_{ij})\right)$$
$$- C_{\text{sync}} \sum_{i=1}^{N} \cos(N\phi_i) \quad (4)$$

By construction, $E(\phi)$ serves as a Lyapunov candidate for the system. In particular, one can show that its time derivative satisfies Eq. 5 so the energy never increases as the oscillators evolve and directly ties energy dissipation to the squared phase velocities. This guarantees that the system will converge to a stationary point of E, but it does not by itself ensure the resulting fixed point is dynamically stable (i.e., a local minimum rather than a saddle or maximum).

$$\frac{dE}{dt} = -N \sum_{i=1}^{N} \left(\frac{d\phi_i}{dt}\right)^2 \leq 0, \quad (5)$$

To investigate stability in detail, we linearize the phase dynamics around a candidate solution ϕ^* and examine the eigenvalues of the Jacobian matrix J, as described in [2]. If all eigenvalues have negative real parts, the fixed point is locally asymptotically stable, confirming that small perturbations will decay and the network will remain in a valid TSP configuration.

A. Coupling Function f_{TSP}

The coupling function $f_{\text{TSP}}(\Delta\phi_{ij})$ is the mechanism that marks certain phase differences between two oscillators i and j as more or less desirable. Its primary purpose is to reward those configurations in which the phase difference $\Delta\phi_{ij}$ corresponds exactly to two cities being adjacent in a valid tour (i.e. $|\Delta\phi_{ij}| = 2\pi/N$) and to heavily penalize all other phase differences.

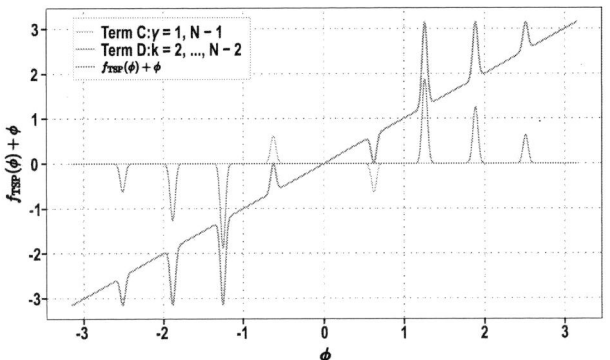

Fig. 1. The f_{TSP} function for $N = 10$, taking $N - 1$ discrete values in $\{0, \pm\pi\}$. The contributions of Term C ($\gamma = 1, N - 1$) and Term D ($k = 2, \ldots, N - 2$) are shown separately.

In this work, we utilize the multi-peak Gaussian energy profile introduced by Mallick et al. [1], as depicted in Fig. 1. This profile consists of a series of narrowly localized Gaussian functions centered at phase offsets corresponding to permissible city-to-city transitions, i.e., integer multiples of $2\pi/N$. In the limiting case where the standard deviation $\sigma \to 0$, the energy profile converges to the expression given in Eq. 6, with the amplitude of each peak governed by the coefficients defined in Eqs. 7 and 8. The function $f_{\text{TSP}}(\Delta\phi_{ij})$ assigns minimal energy values to phase differences $\Delta\phi_{ij} = 2\pi m/N$, where m is an integer, thereby corresponding to valid transitions in the traveling salesman problem. All other, non-permissible phase differences are associated with higher energy values. This energy shaping mechanism biases the dynamics of the coupled oscillator network toward phase-locking configurations that represent valid and near-optimal TSP solutions, analogous to how an electronic filter selectively passes desired frequency components.

$$f_{\text{TSP}}(\Delta\phi_{ij}) = -\sum_{\gamma=1,N-1} C\left(\gamma, \Delta\phi_{ij}\right) + \sum_{\substack{k=2 \\ k\neq N-1}}^{N} D\left(k, \Delta\phi_{ij}\right) \quad (6)$$

$$C\left(\gamma, \Delta\phi_{ij}\right) = \frac{2\gamma\pi}{N}\left[\exp\left(-\frac{(\Delta\phi_{ij} - 2\gamma\pi/N)^2}{2\sigma^2}\right) - \exp\left(-\frac{(\Delta\phi_{ij} + 2\gamma\pi/N)^2}{2\sigma^2}\right)\right] \quad (7)$$

979-8-3315-9813-6/25 $31.00 © 2025 IEEE

$$D\left(k, \Delta\phi_{ij}\right) = \left(\pi - \frac{2k\pi}{N}\right)\exp\left(-\frac{(\Delta\phi_{ij} - 2k\pi/N)^2}{2\sigma^2}\right) + \left(\frac{2k\pi}{N} - \pi\right)\exp\left(-\frac{(\Delta\phi_{ij} + 2k\pi/N)^2}{2\sigma^2}\right) \quad (8)$$

IV. Hardware Platform

We simulate the system in SPICE to ensure circuit compatibility. To implement the oscillator, we chose to use ferroelectric devices, which are promising candidates due to their scalability and silicon compatibility. The coupling is implemented through current sources governed by behavioral equations. We are currently working on a circuit-level implementation of the coupling.

A. Circuit-Level Oscillator Design

Each node in the oscillator network corresponds to a single city in the TSP and is implemented as a ferroelectric relaxation oscillator ([6] and [7]), as shown in Fig. 2. The design is based on the Landau-Khalatnikov (LK) framework, which models ferroelectric dynamics using a nonlinear double-well potential [8]. The oscillator operates by periodically charging and discharging the load capacitance (C_{load}) through a ferroelectric FET ($FeFET$) and a discharge transistor (M_{dis}). When the internal voltage reaches the coercive threshold (V_c), the ferroelectric switches, triggering rapid charging of C_{load}. The gate-source capacitance (C_{gs}) then inverts the polarization, turning the $FeFET$ OFF and restarting the cycle.

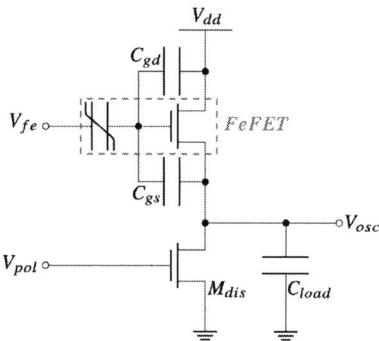

Fig. 2. Implementation of the ferroelectric-based relaxation oscillator.

B. System of coupled oscillators

Fig. 3 shows two oscillators in a system of coupled oscillators. Each oscillator (representing a city) is connected to the others through an RC coupling that encodes the distances between the cities. The system is orchestrated by the shaping function f_{TSP}, which rewards configurations with shorter total distances.

C. TSP Encoding and Netlist Generation

Each TSP instance is defined by a distance matrix D_{ij}, with resistor values in the coupling network inversely related to D_{ij}. A Python script automates this process, generating a complete SPICE netlist for a fully connected oscillator graph. Each pair of oscillators is linked through RC branches that encode the

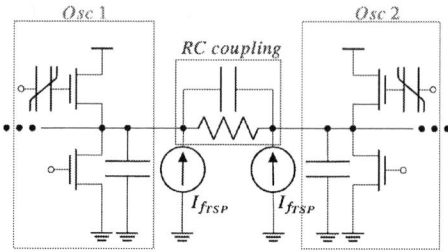

Fig. 3. Schematic of two coupled ferroelectric oscillators representing cities in a TSP instance. The oscillators are connected through an RC coupling network that encodes the distance between cities. The shaping function f_{TSP} drives the coupling current $I_{f_{\text{TSP}}}$, favoring configurations that minimize the total distance.

TSP cost structure, enabling easy reconfiguration of problem instances. To map physical distances into the electrical network we define the complex admittance of each branch as Eq. 9

$$Y_{ij}(j\omega) = \frac{1}{R_{ij}} + j\omega\,C_{ij} \quad (9)$$

where R_{ij} and C_{ij} are the selected resistance and capacitance between nodes i and j. As an example we can consider a Toy-model of 4 cities and the distances are summarized in Table I. The resistor and capacitor values were selected to maintain a consistent frequency of operation across the network, targeting a nominal oscillation frequency of approximately 25 kHz. To ensure practical implementation constraints, the minimum resistance was limited to $R_{\text{min}} = 21\,\text{k}\Omega$ and the maximum capacitance to $C_{\text{max}} = 1\,\text{nF}$. These bounds were respected while still encoding the relative distances between cities through the RC time constants. In the Toy model, the path with the lowest total distance is given by the sequence 1324. Due to the circular symmetry of the TSP, any cyclic permutation or reversal of this sequence (e.g., 2314, 4231) represents the same valid solution.

TABLE I
PAIRWISE COUPLING PARAMETERS

Pair	Distance	R (Ω)	C (F)
2–3	25	82 k	270 pF
2–4	31	62 k	330 pF
1–3	59	33 k	680 pF
3–4	66	30 k	720 pF
1–2	73	27 k	820 pF
1–4	92	21 k	1 nF

D. Implementation of the f_{TSP} Coupling Function in SPICE

The shaping function f_{TSP}, which biases the oscillator network toward valid tours, is implemented using SPICE's behavioral modeling capabilities. Eq. 10 shows how circular phase variables are obtained in the simulation, where V_{mid} is the average between the maximum and minimum voltage and V_{span} is the difference and to ensure correct wrapping of phase differences on the unit circle we use Eq. 11.

$$\theta(v) = \frac{2\pi \cdot (v(t) - V_{\text{mid}})}{V_{\text{span}}} \quad (10)$$

$$\phi(\theta) = \theta - 2\pi \cdot \left\lfloor \frac{\theta + \pi}{2\pi} \right\rfloor \quad (11)$$

Finally the f_{TSP} function is implemented as shown in Eq. 12, where k_{TSP} is a scalar to scale the resulting current. This expression approximates the behavior of the theoretical shaping function by penalizing non-adjacent phase differences and energetically favoring those that correspond to valid city adjacencies (i.e., angular separations near $\pm 2\pi/N$).

$$I\left(\Delta\phi_{ij}\right) = -k_{TSP} \cdot f_{\text{TSP}}\left(\Delta\phi_{ij}\right) \quad (12)$$

This analog implementation enables enforcement of TSP constraints directly through oscillator dynamics using voltage-to-phase transformation and current-based coupling.

E. Simulation Environment

All simulations were conducted in LTspice XVII using transient analysis over 100 milliseconds with a 0.1 microsecond maximum time step. Each oscillator is powered by a 3.3 V source and initialized with randomized voltage conditions to simulate diverse starting phases. The oscillator output voltage range is normalized between $V_{\min} = 0$ and $V_{\max} = 0.85$ volts to match the expected span of valid phase values.

V. RESULTS

We simulated the dynamics of the proposed Toy-model ferroelectric oscillator network for TSP instances with four cities. Each city is represented by a dedicated relaxation oscillator, coupled via RC networks encoding the TSP distance matrix. Fig. 4 (a) shows the oscillator voltages over time. Initially, all nodes begin from near-identical conditions. However, around $t \approx 2.5\,\text{ms}$, the system undergoes a phase separation: each oscillator stabilizes into a distinct peak-timing pattern.

A zoomed view highlight this transition (Fig. 4 (a) inset) in which the separation is sufficiently distinct to identify the relative order of peaks. This confirms that the system moves from a nearly synchronous initial state into a steady-state configuration where each oscillator maintains a consistent phase offset. The ordering of the peak voltages remains readable and consistent, allowing the encoded TSP tour to be directly extracted from the final voltage sequence.

Fig. 4 (b) shows the phase-space projections of voltage differences between oscillator pairs. The system settles onto a limit cycle—a closed trajectory that indicates stable, periodic behavior in the voltage-difference space. This behavior is consistent with the theoretical predictions of Mallick's model.

In particular, the order of oscillator phases in the final steady state corresponds directly to the best tour found for the TSP instance. That is, by reading the sequence of peak voltages, one can reconstruct the path through the cities that minimizes total distance. This confirms the analog circuit's ability to encode and converge toward valid TSP solutions purely through its natural dynamics.

Together, these results demonstrate that the ferroelectric oscillator array not only synchronizes as expected but also self-organizes into phase orderings that encode optimal or near-optimal solutions to combinatorial problems.

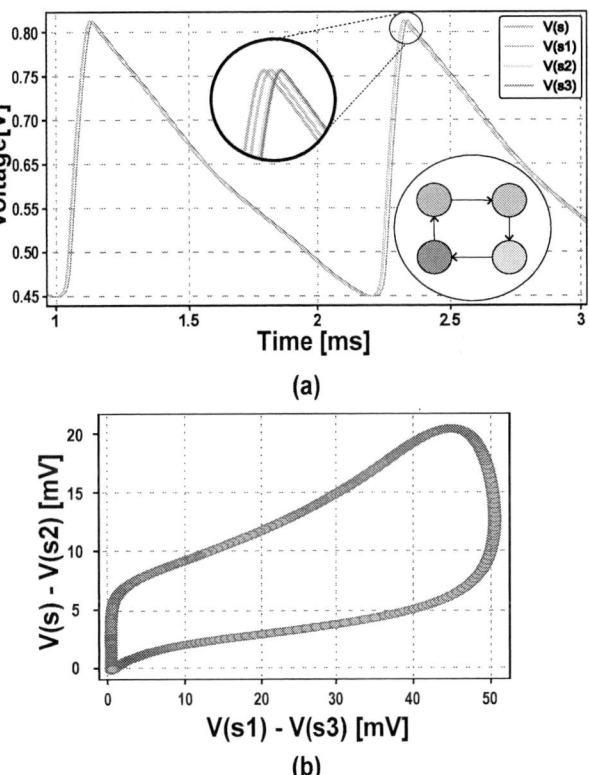

Fig. 4. Time evolution and phase behavior of the proposed ferroelectric oscillator network for a 4-city TSP instance. (a) Oscillator voltages over time, where each oscillator evolves to a distinct phase-separated configuration. (b) Phase-space trajectory of voltage differences between oscillator pairs. The system converges to a stable limit cycle, demonstrating consistent periodic behavior. The resulting city order is indicated by the circular nodes with arrows, representing the phase-based tour solution.

VI. CONCLUSIONS

In this work in progress, we demonstrated that the Mallick coupling model can be successfully applied to networks of ferroelectric relaxation oscillators. Using a behavioral SPICE implementation, we verified that the oscillator array converges to stable phase configurations that correspond to valid solutions of the Traveling Salesman Problem. This confirms that the dynamic principles behind the coupling are compatible with ferroelectric oscillator dynamics and can guide analog computation toward low-energy, combinatorial optima.

Although the coupling was implemented through behavioral functions in this initial study, the results lay the groundwork for future work focused on circuit-level realizations of the coupling mechanism. A hardware-accurate implementation of the coupling interaction would enable scalable, silicon-compatible architectures for analog optimization using ferroelectric oscillator arrays.

VII. ACKNOWLEDGMENT

This work was supported by the Chilean National Research Agency (ANID) through Fondecyt Regular under Grant No. 1250681 and Fondecyt de Iniciación under Grant No. 11250340.

REFERENCES

[1] A. Mallick, M. K. Bashar, Z. Lin, and N. Shukla, "Computational models based on synchronized oscillators for solving combinatorial optimization problems," *Phys. Rev. Appl.*, vol. 17, p. 064064, Jun 2022. [Online]. Available: https://link.aps.org/doi/10.1103/PhysRevApplied.17.064064

[2] M. K. Bashar, Z. Lin, and N. Shukla, "Stability of oscillator Ising machines: Not all solutions are created equal," *Journal of Applied Physics*, vol. 134, no. 14, p. 144901, 10 2023. [Online]. Available: https://doi.org/10.1063/5.0157107

[3] J. Hopfield and D. Tank, "Neural computation of decisions in optimization problems," *Biol. Cybern.*, vol. 52, pp. 141–152, 1985. [Online]. Available: https://doi.org/10.1007/BF00339943

[4] A. Lucas, "Ising formulations of many np problems," *Frontiers in physics*, vol. 2, p. 5, 2014.

[5] Z. Wang, B. Crafton, J. Gomez, R. Xu, A. Luo, Z. Krivokapic, L. Martin, S. Datta, A. Raychowdhury, and A. I. Khan, "Experimental demonstration of ferroelectric spiking neurons for unsupervised clustering," in *2018 IEEE International Electron Devices Meeting (IEDM)*, 2018, pp. 13.3.1–13.3.4.

[6] Y. Fang, J. Gomez, Z. Wang, S. Datta, A. I. Khan, and A. Raychowdhury, "Neuro-mimetic dynamics of a ferroelectric fet-based spiking neuron," *IEEE Electron Device Letters*, vol. 40, no. 7, pp. 1213–1216, 2019.

[7] Y. Fang, Z. Wang, J. Gomez, S. Datta, A. I. Khan, and A. Raychowdhury, "A swarm optimization solver based on ferroelectric spiking neural networks," *Frontiers in Neuroscience*, vol. Volume 13 - 2019, 2019. [Online]. Available: https://www.frontiersin.org/journals/neuroscience/articles/10.3389/fnins.2019.00855

[8] P. Chandra and P. B. Littlewood, "A landau primer for ferroelectrics," 2006. [Online]. Available: https://arxiv.org/abs/cond-mat/0609347

Lightweight Congruence Profiling for Early Design Exploration of Heterogeneous FPGAs

Allen Boston*, Biruk Seyoum†, Luca Carloni†, Pierre-Emmanuel Gaillardon*

*Department of Electrical and Computer Engineering, University of Utah, Salt Lake City, UT, 84112
{allen.boston, pierre-emmanuel.gaillardon}@utah.edu
†Department of Computer Science, Columbia University in the City of New York, New York, NY 10027
{biruk, luca}@cs.columbia.edu

Abstract—**Field-Programmable Gate Arrays (FPGAs) have evolved from uniform logic arrays into heterogeneous fabrics integrating digital signal processors (DSPs), memories, and specialized accelerators to support emerging workloads such as machine learning. While these enhancements improve power, performance, and area (PPA), they complicate design space exploration and application optimization due to complex resource interactions.**

To address these challenges, we propose a lightweight profiling methodology inspired by the Roofline model. It introduces three congruence scores that quickly identify bottlenecks related to heterogeneous resources, fabric, and application logic. Evaluated on the Koios and VPR benchmark suites using a Stratix 10–like FPGA, this approach enables efficient FPGA architecture co-design to improve heterogeneous FPGA performance.

Index Terms—**FPGA modeling, FPGA timing analysis, VTR/VPR, Heterogeneous FPGA**

I. INTRODUCTION

The development of new Field-Programmable Gate Array (FPGA) architectures typically requires extensive design space exploration (DSE), involving iterative performance evaluation and fine-tuning using a range of benchmark applications [1]. Over the years, the architecture of FPGAs has undergone significant evolution, transitioning from homogeneous arrays of general-purpose logic to sophisticated fabrics that incorporate heterogeneous compute blocks and memory resources [2]–[5]. This architectural shift sacrifices generality in favor of higher performance, aiming to narrow the gap between reconfigurable platforms and application-specific integrated circuits (ASICs) [6], [7]. Although these enhancements yield substantial gains in power, performance, and area (PPA), they also increase the complexity of balancing application characteristics with architectural resources, adding to the complexity of evaluating candidate FPGA architectures during DSE.

The performance of traditional FPGA architectures, characterized by limited resource heterogeneity, is largely determined by the complex interplay of spatial layout, resource delay, and the routing locality. As FPGAs evolve to include a broader range of *heterogeneous components* (H-blocks), the interaction between general-purpose logic and specialized resources becomes increasingly critical to overall performance. This intricate relationship is illustrated in Figure 1, which models

This material is based upon work supported by the NSF PPoSS Award No. 2217154. Any opinions, findings, and conclusions or recommendations expressed in this material are those of the author(s) and do not necessarily reflect the views of the National Science Foundation.

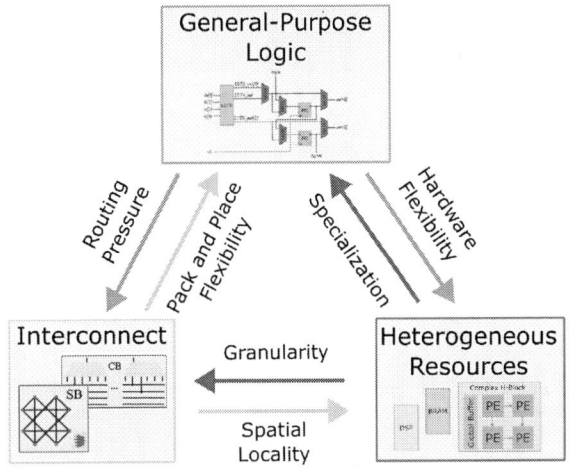

Fig. 1: Interplay among logic, heterogeneous resources, and interconnect in FPGA architectures. Arrows show directional influences, while labels indicate pressures such as specialization, flexibility, mapping feasibility that shape congruence between applications and hardware.

the interdependence among three fundamental elements: the general-purpose logic blocks (CLBs), the interconnect, and the specialized resources (H-blocks) embedded within the fabric. Each edge in the triangle represents a direction of influence. The architecture affects both the spatial organization of resources and the class of applications that can be efficiently supported. The application imposes specific demands on specialized blocks such as digital signal processing (DSP) blocks and memory, which may or may not be well-suited for its computation pattern; and finally, the design and distribution of H-blocks within the fabric feeds back into the architecture considerations, introducing placement complexities and increasing routing pressure. Therefore, the increasing heterogeneity of the fabric requires a fundamental rethinking of how FPGA architectures are evaluated and optimized.

Traditional design space exploration is often manual and ad hoc, overlooking how applications align with FPGA resources or how specialized components affect general-purpose reconfigurability (Figure 1). Existing CAD tools focus on low-level exploration but provide limited guidance for high-level

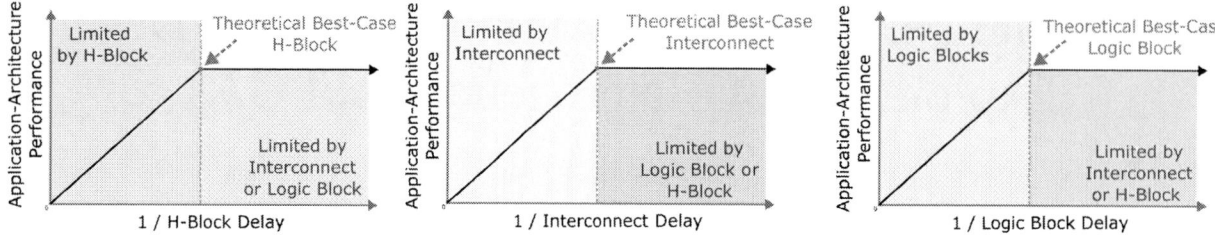

Fig. 2: Roofline-inspired model of FPGA subsystem bottleneck contribution. Arrows indicate shifting performance bottlenecks as specific components (e.g., H-blocks or interconnect) are improved, emphasizing the need for balanced architectural optimization.

architectural decisions. As FPGA fabrics grow more complex, fast and interpretable metrics are needed to guide co-design without exhaustive trial-and-error, motivating the development of a lightweight and straightforward profiling methodology.

Building on this insight, we introduce a *congruence score* that quantitatively evaluates how well a heterogeneous FPGA architecture aligns with a given application, enabling efficient and lightweight fabric optimization. Inspired by the Roofline model [8], our approach systematically analyzes delays and interconnect characteristics. By focusing on delay, we isolate application–architecture alignment, since specialized compute blocks naturally improve area and power. From this analysis, three congruence scores are derived to highlight key performance bottlenecks in heterogeneous FPGA designs.

- **Interconnect Congruence Score (ICS)** — measures how much the physical FPGA fabric limits application performance, revealing whether architectural layout, routing topology, or block placement creates bottlenecks.
- **Heterogeneous Resource Congruence Score (HRCS)** — quantifies how well the application utilizes specialized compute blocks, such as DSPs and memory elements, by evaluating their contribution to critical path delays or routing pressure.
- **Logic Block Congruence Score (LBCS)** — assesses the alignment between the application's logic structure and the FPGA's general-purpose logic resources, indicating how effectively the application utilizes the available fabric.

Applying this methodology using a Stratix 10–like FPGA to a subset of the Koios and VPR benchmark suites [9], [10], we demonstrate how these scores assist designers in optimizing FPGA layouts and facilitating faster and more informed co-design decisions. Finally, we pair each evaluated application with its best-fit architecture using the proposed scoring method.

The remainder of this paper is organized as follows. Section II outlines our proposed scoring methodology. Section III demonstrates the application of the congruence scoring system on well-known benchmark suites and presents the results. Section IV concludes the paper.

II. DEFINING APPLICATION-ARCHITECTURE CONGRUENCE

In this section, we present details about the congruence score used to assess how effectively an application leverages the various subsystems within a heterogeneous FPGA architecture. Building on a Roofline-inspired model, our approach isolates the contributions of key FPGA subsystems, represents their theoretical performance limits, and quantifies their influence on overall system behavior.

Figure 2 visualizes our conceptual model, showing the interplay between heterogeneous resources, FPGA architecture, and application behavior. For example, when an application relies heavily on heterogeneous resources, improving the timing of those resources initially reduces the critical path. However, beyond a certain point, the critical path shifts to other regions of the fabric, such as routing or general logic, requiring further optimization there. Similarly, applications constrained by routing may benefit from reduced congestion or shorter interconnects up to a limit, after which the logic structure becomes the primary constraint. Finally, if the logic fabric itself dominates timing, further improvements must come from better pipelining, restructuring, or architectural enhancements.

This evaluation uses VPR [11], a versatile architecture modeling tool for exploring reconfigurable platforms. VPR supports diverse configurations, including complex components, custom routing, and heterogeneous compute and memory resources. Its integrated placement and routing engine maps applications onto candidate architectures, enabling accurate performance evaluation. Using this framework, we can modify delays in the architecture description to isolate the theoretical limits of individual FPGA subsystems. We set these modified delays near-zero to emulate the Roofline ideal for each subsystem. This highlights how each component would contribute to overall performance if unconstrained, revealing the true bottlenecks in the application–architecture mapping.

$$\forall i \in \{ICS, HRCS, LBCS, \} \; Score_i = 1 - \frac{\alpha_i - \beta_i}{\gamma_i - \beta_i} \quad (1)$$

To quantify the application-architecture congruence score, we used Equation 1. Here, α_i is the delay after modifying one architectural subsystem, γ_i is the original timing result with no modifications, and β_i is a user-defined target delay. This formulation enables interpretable profiling of how each architectural component contributes to performance limitations. A score approaching zero indicates that the subsystem has minimal impact on the critical path, suggesting limited opportunity for further improvement. In contrast, a score approaching one suggests that the subsystem is a dominant performance bottleneck and a prime target for co-design optimization.

979-8-3315-9813-6/25 $31.00 © 2025 IEEE

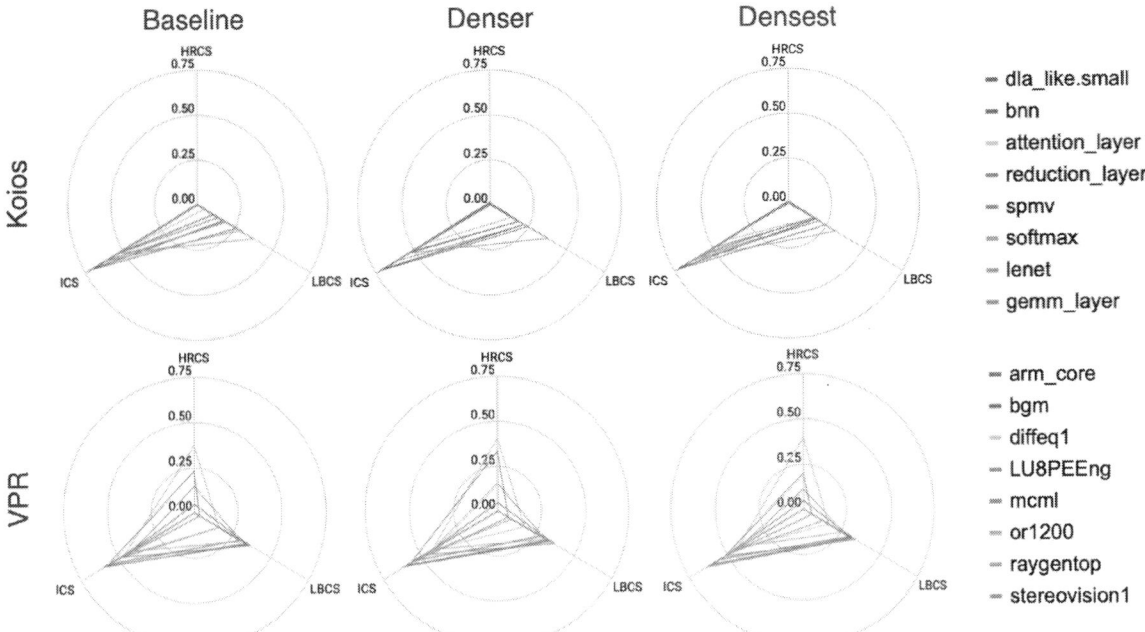

Fig. 3: Radar plots illustrating the congruence scores for a subset of Koios and VPR benchmarks across three architecture variants. The plots highlight how performance bottlenecks shift between heterogeneous resources, interconnect, and logic for varying benchmark sets.

A. Interconnect Congruence Score

The Interconnect Congruence Score (ICS) measures how much an application's performance is limited by the routing infrastructure. To compute ICS, all interconnect delays are set to near-zero, while logic and heterogeneous block delays remain unchanged. ICS is quantified using Equation 1. A high ICS indicates that the routing network significantly contributes to performance degradation. In such cases, architectural strategies like adding direct interconnects or hierarchical routing may help. Similarly, enhancements such as pipelining and buffering can reduce routing delays, although they may require application-level scheduling changes.

B. Heterogeneous Resource Congruence Score

The Heterogeneous Resource Congruence Score (HRCS) evaluates how much an application relies on specialized compute and memory elements such as DSPs, BRAMs, or custom accelerators [7], [12], [13]. To compute HRCS, we modify the architecture to assign near-zero delay values to all heterogeneous blocks, simulating ideal performance for these components.

By isolating blocks like DSPs or BRAMs, we can observe whether they form a significant portion of the critical path. If performance improves substantially under idealized heterogeneous block conditions, it indicates that the application is dependent on those resources and that they may be performance bottlenecks. While we group all specialized blocks into a single HRCS value for simplicity, the methodology can be extended to separately evaluate each component type. For example, one could isolate BRAMs or DSPs individually to determine their relative contributions to overall performance limitations.

C. Logic Block Congruence Score

The Logic Block Congruence Score (LBCS) quantifies the extent to which an application's internal logic structure constrains performance. Many FPGA applications achieve performance gains by using deep pipelining, inserting memory elements between stages of combinational logic to shorten the critical path. However, if the underlying logic fabric remains a bottleneck, further optimization may be needed at the application or architectural level.

To compute LBCS, the delays of all logic elements such as LUTs and multiplexers are set to zero, while delays for routing and heterogeneous resources remain unchanged. A high LBCS means logic is the dominant limiter and may require architectural changes (e.g., larger LUTs or more local interconnect) or software-level improvements like better pipelining and scheduling. A low LBCS, on the other hand, suggests that the logic fabric is not a critical contributor to performance constraints.

III. EXPERIMENTAL SETUP AND RESULTS

A. Benchmark Selection and Target Architectures

We evaluate our profiling methodology using a subset of benchmarks from the Koios and VPR suites [9], [10], [14], including compute-intensive machine learning tasks such as an Intel DLA–like accelerator and convolution layers, as well as general-purpose compute designs from VPR like an ARM core. The benchmarks target an FPGA architecture provided by the Koios framework, modeled after a Stratix 10 routing structure, with complex DSPs inspired by Agilex, configurable 20Kb BRAMs, and CLBs composed of 10 6-input LUTs.

To examine the impact of heterogeneous resource density, we study three architecture variants. The first is a *baseline* with a balanced resource mix similar to commercial FPGAs. The second, *denser*, increases the number of DSPs and BRAMs. The third, *densest*, further increases the ratio of specialized blocks to general-purpose logic. These variants form the foundation of our analysis.

This work builds on the original Koios study by extending the evaluation beyond traditional metrics like wire length and critical path delay [14]. Instead, we focus on the architectural interplay among general-purpose logic, routing fabric, and heterogeneous blocks, providing a multidimensional view of how well architectures align with application demands.

B. Experimental Procedure

For each architecture variant, we compute HRCS, ICS, and LBCS across selected Koios and VPR benchmarks using the VPR flow. This is achieved by modifying the architecture description files to isolate subsystem delays, as described in Section II, so that each score reflects the impact of a specific architectural component while preserving realistic placement and routing behavior. We first run VPR with full delay annotations to obtain baseline packing, placement, routing, and critical path delays. The resulting packing, placement, and routing information are then reused in subsequent runs with zeroed subsystem delays to prevent VPR from executing the full flow again. In these runs, only the final timing analysis is performed to extract the subsystem-specific impact.

C. Results

Figure 3 shows the results of our architectural exploration as radar plots, where each benchmark from a subset of Koios and VPR is evaluated using the three congruence scores from Section II, computed with Equation 1 using an optimistic ideal delay of 0.2 ns. As shown by the radar plots, Koios applications are primarily limited by routing, followed by logic, with minimal dependence on heterogeneous blocks. This suggests that the H-blocks are well-aligned with the application's needs, effectively supporting computation without becoming bottlenecks, as evidenced by the fact that critical paths seldom pass through them [15]. As H-block density increases, we observe variations in the influence of general-purpose logic, while routing consistently remains the dominant limiting factor. In contrast, the VPR benchmarks exhibit a more balanced distribution of bottlenecks across the evaluated resources, with greater reliance on heterogeneous components when compared to Koios benchmarks, indicating that their critical paths often include H-blocks. Additionally, the smaller radar plot areas for Koios benchmarks suggest that they are likely to perform better on this architecture set compared to the VPR designs.

While the radar plots provide intuitive visual insight into the independent contributions of heterogeneous and general-purpose resources, we introduce an aggregate scoring methodology to quantitatively assess overall application–architecture

Koios Benchmarks	Baseline	Denser	Densest
dla_like.small	0.545	0.565	0.573
bnn	0.722	0.737	0.740
attention_layer	0.485	0.462	0.448
reduction_layer	0.765	0.766	0.766
spmv	0.643	0.600	0.626
softmax	0.644	0.653	0.678
lenet	0.684	0.668	0.682
gemm_layer	0.562	0.534	0.522
Koios Mean	0.561	0.554	0.600
VPR Benchmarks	**Baseline**	**Denser**	**Densest**
arm_core	0.636	0.626	0.630
bgm	0.686	0.684	0.685
diffeq1	0.621	0.614	0.613
LU8PEEng	0.692	0.700	0.706
mcml	0.608	0.606	0.618
or1200	0.543	0.545	0.499
raygentop	0.512	0.493	0.497
stereovision1	0.612	0.665	0.555
VPR Mean	0.614	0.617	0.601
VPR/Koios Aggregate	1.175	1.170	1.160

TABLE I: Application–architecture congruence scores for each benchmark across three architecture variants. The scores highlight best-fit architectures for individual applications and overall trends across the Koios and VPR benchmark sets.

alignment. The application–architecture congruence score is calculated as the magnitude of an n-D vector (HRCS, LBCS, ICS), which can be extended to additional dimensions. These scores are summarized in Table I. Although the best-fit architecture varies across individual applications, averaging across each benchmark set reveals clear trends: the *Denser* architecture best matches Koios benchmarks, while the *Densest* architecture aligns best with VPR designs. When both benchmark sets are combined, the aggregate score indicates that the *Densest* architecture provides the best overall fit. This is likely due to the strong compatibility of Koios benchmarks with H-block–rich architectures and the relatively minor performance shifts in VPR designs as H-block density increases.

IV. CONCLUSION

We present a lightweight profiling methodology that quantifies the alignment between applications and heterogeneous FPGA architectures using congruence scores. These scores isolate bottlenecks in specialized compute blocks, routing, and general-purpose logic based on the roofline model. Applied to well adopted benchmark suites and FPGA architectures, our method quickly identifies architectural constraints and supports effective co-design decisions. It enables faster design space exploration and can be integrated into automated optimization tools, thereby advancing the understanding of application–architecture interactions in FPGAs.

REFERENCES

[1] S. Yazdanshenas and V. Betz, "Coffe 2: Automatic modelling and optimization of complex and heterogeneous fpga architectures," *ACM Trans. Reconfigurable Technol. Syst.*, vol. 12, no. 1, jan 2019. [Online]. Available: https://doi.org/10.1145/3301298

[2] "Versal ai engine," https://www.xilinx.com/products/intellectual-property/versal-ai-engine.htmloverview, 2023.

[3] "Flexlogix inferx ai," https://flex-logix.com/inferx-ai/inferx-ai-hardware/, 2023.

[4] M. Langhammer, E. Nurvitadhi, B. Pasca, and S. Gribok, "Stratix 10 nx architecture and applications," in *The 2021 ACM/SIGDA International Symposium on Field-Programmable Gate Arrays*, 2021, pp. 57–67.

[5] "Achronix machine learning processor," https://www.achronix.com/machine-learning-processor, 2023.

[6] A. Boutros and V. Betz, "Fpga architecture: Principles and progression," *IEEE Circuits and Systems Magazine*, vol. 21, no. 2, pp. 4–29, 2021.

[7] A. Arora, Z. Wei, and L. John, "The case for hard matrix multiplier blocks in an fpga," in *Proceedings of the 2020 ACM/SIGDA International Symposium on Field-Programmable Gate Arrays*, ser. FPGA '20. New York, NY, USA: Association for Computing Machinery, 2020, p. 323. [Online]. Available: https://doi.org/10.1145/3373087.3375360

[8] S. Williams, A. Waterman, and D. Patterson, "Roofline: an insightful visual performance model for multicore architectures," *Communications of the ACM*, vol. 52, no. 4, pp. 65–76, 2009.

[9] A. Arora, A. Boutros, D. Rauch, A. Rajen, A. Borda, S. A. Damghani, S. Mehta, S. Kate, P. Patel, K. B. Kent *et al.*, "Koios: A deep learning benchmark suite for fpga architecture and cad research," in *2021 31st International Conference on Field-Programmable Logic and Applications (FPL)*. IEEE, 2021, pp. 355–362.

[10] V. Betz and J. Rose, "Vpr: A new packing, placement and routing tool for fpga research," in *Proceedings of the 7th International Workshop on Field-Programmable Logic and Applications*, ser. FPL '97. Berlin, Heidelberg: Springer-Verlag, 1997, p. 213–222.

[11] K. E. Murray, S. Whitty, S. Liu, J. Luu, and V. Betz, "Timing-driven titan: Enabling large benchmarks and exploring the gap between academic and commercial cad," *ACM Trans. Reconfigurable Technol. Syst.*, vol. 8, no. 2, Mar. 2015. [Online]. Available: https://doi.org/10.1145/2629579

[12] A. Arora, S. Mehta, V. Betz, and L. K. John, "Tensor slices to the rescue: Supercharging ml acceleration on fpgas," in *The 2021 ACM/SIGDA International Symposium on Field-Programmable Gate Arrays*, ser. FPGA '21. New York, NY, USA: Association for Computing Machinery, 2021, p. 23–33. [Online]. Available: https://doi.org/10.1145/3431920.3439282

[13] A. Arora, A. Bhamburkar, A. Borda, T. Anand, R. Sehgal, B. Hanindhito, P.-E. Gaillardon, J. Kulkarni, and L. K. John, "Comefa: Deploying compute-in-memory on fpgas for deep learning acceleration," *ACM Trans. Reconfigurable Technol. Syst.*, vol. 16, no. 3, jul 2023. [Online]. Available: https://doi.org/10.1145/3603504

[14] A. Arora, A. Boutros, S. A. Damghani, K. Mathur, V. Mohanty, T. Anand, M. A. Elgammal, K. B. Kent, V. Betz, and L. K. John, "Koios 2.0: Open-source deep learning benchmarks for fpga architecture and cad research," *IEEE Transactions on Computer-Aided Design of Integrated Circuits and Systems*, vol. 42, no. 11, pp. 3895–3909, 2023.

[15] A. Mishra, N. Rao, G. Gore, and X. Tang, "Architectural exploration of heterogeneous fpgas for performance enhancement of ml benchmarks," in *2023 IEEE Asia Pacific Conference on Circuits and Systems (APCCAS)*, 2023, pp. 232–235.

ML4FPGA: An LLM Framework for Electronic Design Automation and Verification on FPGA

Uchechukwu Leo Udeji
Electrical and Computer Engineering department
University of Massachusetts Lowell
Lowell, USA
Leo_Udeji@student.uml.edu

Martin Margala
Computing and Informatics department
University of Louisiana at Lafayette
Lafayette, USA
martin.margala@louisiana.edu

Abstract— The application of Large Language Models (LLMs) to various machine learning tasks, including natural language processing (NLP), text, and code generation, has improved efficiency in these tasks. Despite their increasing prevalence, a comprehensive framework capable of generating and verifying Verilog code via a combination of LLMs and timing diagrams is lacking. This study introduces an LLM-based transformer framework that accelerates the process of reverse engineering electronic designs by analyzing the timing diagram via image recognition and generating and verifying Verilog design via pre-trained LLMs. We also compare the performance of our trained custom LLM model with much larger base models like Llama 4, GPT-4, and DeepSeek V3, Claude 3. The training of our model is done with datasets from HDLBits and publicly available Verilog code scrubbed from the web. Our custom model is trained/finetuned on Google A100 GPU accelerator and profiled. Our results indicate improvements compared to existing research in this domain and prospects of broader applications.

Keywords— Large Language Models, Natural Language Processing, Transformer, FPGA, Machine Learning.

1. INTRODUCTION

The process of translating electronic design specifications into appropriate designs on Field Programmable Gate Arrays (FPGA) hardware via Hardware Description Languages (HDL) like Verilog has mostly been done by hardware engineers which is not only time-consuming but also error-prone [1]. Large Language Models (LLMs) are transformer models that when given some input prompt predict outputs that most likely are a continuation of that prompt [2]. It does this by choosing the best token from a distribution of natural or structured language corpus [3]. LLMs can complete partial code, and, in this case, Verilog code for hardware design. This can be applied to electronic design and automation (EDA), design verification and reverse engineering efforts. Advances in EDA algorithms

and tools over the years have also provided significant improvement in design productivity [4].

The design of some logic circuits, sequential circuits, or application-specific integrated circuits (ASICs) on FPGAs can be time-consuming and, could require profound experience in the use of hardware description languages (HDLs) in writing Register Transfer Level (RTL) codes. Numerous studies have delved into this area, including recent works such as [5], which introduce the ChatGPT framework by OpenAI. This framework employs a prompt-based tool trained on a very large dataset, which comprises over 1 trillion tokens. Other state-of-the-art models like Claude3, Llama4, and DeepSeekV3 also exhibit robust generalization capabilities across various data distributions. Due to the size of these models and the computing power required to achieve the performance exhibited by these models, finetuning pre-trained models can be used to train these models to exhibit high performance on specific tasks, and these finetuned models have shown competitive performance against state-of-the-art models [3], trained on general datasets; performing better in certain scenarios.

In the study, we focus on the generation and verification of Verilog EDA scripts from natural language using LLMs via the ML4FPGA framework. We also attempt to incorporate reverse engineering capabilities into our framework, by using the analysis of circuit timing diagrams, which hasn't been achieved in any previous related study. In addition to these, we use the latest pre-trained models, in this case, Llama4 scout open source model, which has multi-modal data processing capabilities compared to earlier versions like Llama2 [6], to build our custom model. Fig. 1, shows a high-level description of the ML4FPGA framework.

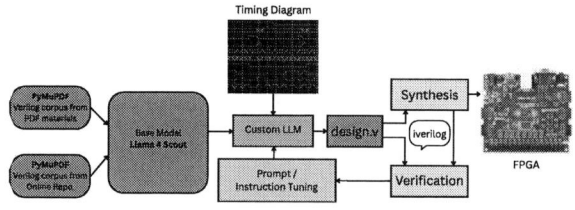

Fig. 1. ML4FPGA Framework.

The manuscript's structure is outlined as follows: Section 2 provides the background of this study and related work. Section 3 elucidates the various datasets used in this study. Section 4 describes the data-cleaning process. Section 5 expounds on training the LLM model for EDA. Section 6 provides the preliminary simulation results. Section 7 discusses the simulation results and, finally, Section 8 concludes the study.

2. BACKGROUND AND RELATED WORK

Attempts to improve HDL design time and quality using High-Level Synthesis (HLS) allow developers to quickly create HDL designs in languages like C, but this comes at the expense of hardware efficiency [3]. LLMs' self-attention capabilities enhance the memory capacity of transformer models, allowing them to make informed decisions based on prior data.

While traditional neural networks primarily focus on learning weights, transformers are designed to prioritize attention scores or maps, which facilitate the alignment of encoded inputs with decoder outputs. Generative Pre-trained Transformers (GPT) and Bidirectional Encoder Representation from Transformers (BERT) are examples of these networks [7], and the parameters used to develop these networks include the equations (1) and (2).

$$Attention(Q,K,V) = soft\max(\frac{QK^T}{\sqrt{d_k}})V \qquad (1)$$

$$Multihead(Q,K,V) = Concat(head_1,...,head_h)W^O \qquad (2)$$

$$\text{where } head_i = Attention(QW_i^Q, KW_i^k, VW_i^V)$$

Attention layers in transformers are also called heads [2], and multiple heads are usually used in training. Q, K, V, and d, in (1) which represent the query, key, value, and dimension variables. These variables are used to create key-value pairs of the inputs to the network which can be queried. The parameter d_k represents the dimension of the key variable K, and K^T represents the transpose of K. Together these parameters are used to create an attention matrix which can be concatenated to create the multi-head attention as in (2), and described in Fig. 2 in [2]. One of the first work exploring LLMs for use in hardware domain was by H. Pearce et. al [8]. It involved the fine-tuning of a GPT-2 model over synthetically generated Verilog code snippets and evaluated the model outputs lexically for an 'undergraduate level' task. However, due to limited training data, the model does not generalize to unfamiliar tasks [1]. This model was called DAVE. Other studies on the subject of EDA via LLM have included: LLM4EDA [9] which compares various studies addressing the application of LLMs in the EDA field categorizing them into: Assistant chatbot, HDL and script generation, and HDL verification and analysis. Another is AutoChip, the parent study that produced VeriGEN [3]. The framework attempts to iteratively improve Verilog designs without human feedback. ChatEDA [10] presents an autonomous agent for EDA empowered by an LLM, AutoMage, complemented by EDA tools serving as executors, to streamline the design flow from RTL to the Graphic Data

System Version II (GDSII) by managing task decomposition, script generation, and task execution. VerilogEval [11][12] develops a benchmarking framework tailored for evaluating LLM performance in the context of Verilog code generation for hardware design and verification. ChipNeMo [4] provides an engineering assistant chatbot, EDA script generation, and bug summarization and analysis framework. Chip-Chat [1] develops a novel 8-bit accumulator-based microprocessor architecture with LLM according to real-world hardware constraints. Despite all these studies on the subject, the challenges of translating specifications into efficient Verilog code persist as several of these studies were performed using less-powerful LLMs without multimodal capabilities. In this study, we create a more powerful framework to address some of the challenges encountered in previous studies such as code generation for problems that include time-diagrams.

3. DATASET

A large dataset of Verilog is required for fine-tuning a given LLM which was previously lacking but is currently growing [3]. The dataset used in this study includes Verilog Corpus from Github repositories which has over 2 million repositories gathered using BigQuery, and from Verilog text source using a Python script and pymuPDF tool. The pymuPDF is an optical character recognition tool that extracts text from PDF. Datasets from Github and Verilog textbooks are augmented to create the extensive dataset used for training which has a size of ~1.5GB.

4. DATA CLEANING

Even though Verilog code can be found online, the quality and coding technique of these programs such as those in the popular dataflow format or gate format varies. There is also a variation in the code complexity such as those for combinational logic and those for sequential logic. Hence, dataset preparation is vital to reduce the misclassification of signals, which for example, led to error in [8]. Duplicate files are removed using MinHash and Jaccard similarity metrics to remove duplicate Verilog code as applied in [3]. The files were also filtered further to only retain .v extension files containing at least one pair of "module" and "endmodule" statements. The resulting corpus contains about 53,000 Verilog files with a size of 1.3GB. To improve our dataset further we filter out large files with over 20,000 characters as applied in [3]. This restriction in file length helps narrow the context window for LLMs during training. By keeping file sizes smaller than the LLM maximum context window, the models can effectively encode entire files allowing better generalization. The final corpus after filtering is about 800MB.

5. TRAINING LLMS FOR EDA

Training LLMs from scratch can be resource-intensive [3] and requires a massive dataset and numerous parameters. Hence, we fine-tune existing models. General purpose LLMs like ChatGPT, Llama, and Claude trained on a vast amount of data have shown remarkable generative AI capabilities in many tasks across various domains. When fine-tuning LLMs for optimal results two key principles are usually followed.

Principle 1, entails providing clear and specific instructions. This is achieved via four tactics. The first entails using delimiters. This helps avoid prompt injections. The second entails asking for a structured output. The third entails asking the model if the conditions are satisfied, and the fourth, is the use of few-shot-prompting, where we provide useful examples of the task. In principle 2, we give the model time to think. This is achieved via two tactics. Firstly, we specify the required text to complete the task via a prompt. Next, we instruct the model to determine its solutions before jumping to conclusions. When fine-tuning for Verilog, multiple existing LLM models are fed with task descriptions [English] or a combination of task descriptions and images and are expected to produce results [Verilog]. The availability of relevant information for the task such as variable names, number of inputs and outputs, and the number of operators are expected to aid in the overall performance of the model. We use PyVerilog to extract syntax trees from Verilog code [11] because the quality of data used in training is of high importance. Model alignment for domain-specific tasks, Retrieval Augmented Generation (RAG), global batch size, and context window are techniques applied here to reduce hallucination in our LLM model [4] where inaccurate data is generated. We look out for signs of overfitting/hallucination which include repeating irrelevant answers as plausible results, after we have performed multiple rounds of domain adaptive pretraining as shown in Fig. 2. The RAG technique [4] helps ground the LLM to generate accurate information and to extract up-to-date information to improve knowledge-intensive NLP tasks. It was observed that smaller models with RAG can outperform larger models without RAG. We use the default LLM temperature parameter of zero (0) to ensure stable responses.

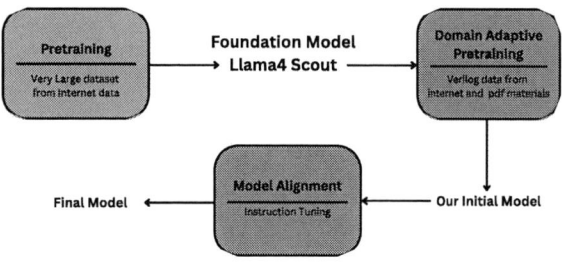

Fig. 2. LLM Model Training Flow.

ICARUS Verilog [13] is another tool used to compile generated code in this study. Training transformers can be very compute intensive, and one approach to speed up the model without a notable loss in prediction and training accuracy is via a mixed-precision algorithm [14]. The PyTorch framework and NVIDIA A100 GPU on Google Colab are also employed in some of our training. Mixed-precision is easily activated with this setup. Some training is also done on our workstation, which comprises a Dell Precision dual Intel Xeon E5- 2680 V3 CPU with 128GB DDR4 memory 1TB-SSD. We fine-tune a range of code-based LLMs on the Verilog training corpus, which we put together on these systems. Next, we evaluate these models on a range of 140 Verilog problems from HDLBits [15].

6. SIMULATION RESULTS

Functional correctness of the generated results is tested using the pass@k metric where k is the number of problems in a scenario time, and n is the number of suggestions per problem. A higher pass@k indicates a relatively better result [3] as it reflects the proportion of completions that compile. In [11], the pass@k metric is defined as in (3), where we generate $n \geq k$ samples per task in which $c \leq n$ samples pass testing.

$$pass@k := \mathop{E}_{Problems}\left[1 - \frac{\binom{n-c}{k}}{\binom{n}{k}}\right] \tag{3}$$

Result comparison shows differing performance across various problem levels such as in Fig.3. We also partially perform verification manually. Another important metric is the timing constraints of our design or the critical or maximum clock frequency, which could affect power consumption [1]. Table 1 shows the performance of other models according to the pass@k metric as compared to our model. We also report resource usage in one of our problems in Table 2. We check the functional correctness of generated code using a test suite featuring a custom problem set and testbenches as in [3].

Our results show that our fine-tuned LLMs can generate functioning code even for challenging problems including problems involving timing diagrams.

Problem	Difficulty	Description
1	Basic	A not gate
2	Basic	A 2-input and gate
3	Basic	A 2-input multiplexer
4	Basic	A 3-bit priority encoder
5	Basic	A NAND gate
6	Intermediate	A half adder
7	Intermediate	A counter
8	Intermediate	A Shift left and rotate register
9	Intermediate	Random Access Memory
10	Intermediate	Truth table
11	Advanced	FSM to recognize a given number
12	Advanced	An up-counter with enable signal
13	Advanced	A down-counter with enable signal
14	Advanced	An ALU
15	Advanced	A simple microprocessor

Fig. 3. Selected problems and difficulty levels.

We also test our results using HDLBits online judge system where each problem is scored on its ability to pass the extensive testbenches provided by HDLBits, which include rigorous functional tests to ensure the correctness of generated code. This system features an "upload and compile" feature that accepts the completed Verilog as a solution, synthesizes the Verilog code using Quartus, and verifies its correctness through simulations run in Modelsim. After execution, the system returns a detailed summary of the execution. A status of "success" is returned only when the Verilog code passes all the tests within the test-bench for the problem. Other responses such as "simulation error", "compiler error", and "incorrect" can also be reported, indicating different stages of failure in the evaluation process.

TABLE 1. PERFORMANCE COMPARISON OF LLM MODELS

Study	Dataset	Training	Testing
ChatGPT	Our Custom Dataset	-	80%
Llama4	Our Custom Dataset	-	79%
Claude	Our Custom Dataset	-	~
DeepSeekV3	Our Custom Dataset	-	79%
ML4FPGA	Our Custom Dataset	-	85%

TABLE 2. FPGA RESOURCE USAGE

Study	Problem #14	LUT	FF	I/O	DSP
ChatGPT4	ALU - 32bits	208	0	101	0
Llama4	ALU - 32bits	277	0	100	0
Claude3	ALU - 32bits	-	-	-	-
DeepSeekV3	ALU - 32bits	262	0	100	0
ML4FPGA	ALU - 32bits	257	0	100	0

Our generated code is tested on the Nexys4 board [16] and can used to implement designs in the Python-based Pynq z1 board or Alveo board [17]. The results are used to generate values in Table 2. We implemented the generated code in Vivado software 2019.2. Some scripts used to implement sections of this study can be found in [18].

7. DISCUSSION

Several tools were used to implement our custom LLM, and several parameters were taken into account during the implementation of this study. LMstudio [19] and Code llama or Ollama [20] were used to download these models and implement the finetuned versions locally on our workstation. These tools keep the training process simple.

Some of the factors that must be taken into account when implementing domain-specific models include the context length of the original model. The context length is the maximum amount of text that the model can process and remember at a given time, which can affect its overall performance in various tasks in terms of accuracy and latency. In several previous studies, models with very different context lengths were used to implement the studies which isn't a fair comparison. Llama 3 and Llama 4 for example have a very steep disparity in context length and size, with Llama 3 having a context length of ~128,000 (128K) and ~2GB size, while Llama 4 has a context length of between 1M and 10M and is ~67GB in size. In addition to this Llama 4 has multimodal capabilities which aren't found in Llama3. Comparing the performance of GPT-4 with that of Llama3 won't be a fair comparison also, because GPT-4 has multi-modal capabilities. The token structure is another factor related to context length.

For OpenAI models each token represents about 4 characters, and their context windows range up to 8000 tokens in size, which implies that they can support about 16000 characters of I/O [1]. Knowledge of such parameters can help to understand performance disparity in models.

Another factor to consider is the main tasks that the LLM performs best at, such as the various levels of knowledge: graduate, undergraduate, and grade school. We also need to look into the performance in tasks such as Math, Multilingual Math, Computer Code, reasoning, and general knowledge. For example, the Claude 3 model [21] is a multimodal model that comes in 3 versions, the Opus, Haiku, and Sonnet variants which all show high performance and latency across the aforementioned tasks. The Opus variant of the model is the most sophisticated and can handle complex problems. Another factor, that's easily overlooked is the prompt engineering side of LLM where various roles exist or are assigned to ensure optimal performance during query. Llama4 for example has four different (4) roles [22], namely: system, user, assistant, and tool. This is similar to other LLMs. System prompt sets the context in which to interact with the AI model and typically includes rules or necessary information to help the model. The user represents the human interacting with the model, which includes the inputs, commands, and questions to the model. Assistant, represents the model generating a response to the user, while tool represents the output of a tool call when sent back to the model from the executor. A richer and more detailed prompt greatly enhances the model's capability to generate high-quality Verilog code [3].Some models such as the Llama 4 Maverick variant can only be run on multiple GPUs [22], highlighting the profound computing power required for training or finetuning. Despite providing better performance larger models sometimes tend to have longer time to response or inference time. Inference time is the total time taken from the time a prompt is fed into the model to the time the complete response is generated by the respective LLMs.

8. CONCLUSION

Developing Instruction-tuned LLMs from base LLMs ideally entails reinforcement learning with human feedback (RLHF). Achieving functionally correct code, especially in complex problems, is still a challenge as the generated code tends to need finetuning to handle edge cases and meet specifications. Another problem is design generation via truth tables or timing diagrams. This paper addresses these problems and presents the outcomes of our investigation. Using a custom dataset trained both on our workstation and in Colab. Our findings showcase the ability of LLMs with multi-modal data processing capabilities to handle more complex HDL design tasks. Our Llama4-based custom model had the highest testing accuracy and has a resource usage similar to that of Llama4 Scout. ChatGPT-4 produced the most resource-efficient design on the first attempt, Claude 3 had the slowest inference speed and produced non-synthesizable results. DeepseekV3 provided the most detailed working result but was only faster than Claude in terms of inference. Moving forward, we hope to apply this to industrial chip design and rapid reverse engineering tasks.

REFERENCES

[1] J. Blocklove, S. Garg, R. Karri and H. Pearce, "Chip-Chat: Challenges and Opportunities in Conversational Hardware Design," 2023 ACM/IEEE 5th Workshop on Machine Learning for CAD (MLCAD), Snowbird, UT, USA, 2023, pp. 1-6, doi: 10.1109/MLCAD58807.2023.10299874.

[2] Ashish Vaswani, Noam Shazeer, Niki Parmar, Jakob Uszkoreit, Llion Jones, Aidan N. Gomez, Łukasz Kaiser, and Illia Polosukhin. 2017. "Attention is all you need". In Proceedings of the 31st International Conference on Neural Information Processing Systems (NIPS'17). Curran Associates Inc., Red Hook, NY, USA, 6000–6010.

[3] Shailja Thakur, Baleegh Ahmad, Hammond Pearce, Benjamin Tan, Brendan Dolan-Gavitt, Ramesh Karri, and Siddharth Garg. 2024. VeriGen: A Large Language Model for Verilog Code Generation. ACM Trans. Des. Autom. Electron. Syst. 29, 3, Article 46 (May 2024), 31 pages. https://doi.org/10.1145/3643681.

[4] Liu, M., Ene, T., Kirby, R., Cheng, C., Pinckney, N.R., Liang, R., Alben, J., Anand, H., Banerjee, S., Bayraktaroglu, I., Bhaskaran, B., Catanzaro, B., Chaudhuri, A., Clay, S., Dally, B., Dang, L., Deshpande, P., Dhodhi, S., Halepete, S., Hill, E., Hu, J., Jain, S., Khailany, B., Kunal, K., Li, X., Liu, H., Oberman, S.F., Omar, S., Pratty, S., Raiman, J., Sarkar, A., Shao, Z., Sun, H., Suthar, P.P., Tej, V., Xu, K., & Ren, H. (2023). ChipNeMo: Domain-Adapted LLMs for Chip Design. *ArXiv, abs/2311.00176*.

[5] GPT-4 Technical Report, OpenAI, March, 2023: (https://arxiv.org/abs/2303.08774)

[6] Touvron, Hugo et al. "Llama 2: Open Foundation and Fine-Tuned Chat Models." ArXiv abs/2307.09288 (2023): n. pag.

[7] Topal, M. Onat, Anil Bas and Imke van Heerden. "Exploring Transformers in Natural Language Generation: GPT, BERT, and XLNet." ArXiv abs/2102.08036 (2021): n. pag.

[8] Hammond Pearce, Benjamin Tan, and Ramesh Karri. 2020. DAVE: Deriving Automatically Verilog from English. In Proceedings of the 2020 ACM/IEEE Workshop on Machine Learning for CAD (MLCAD '20). Association for Computing Machinery, New York, NY, USA, 27–32. https://doi.org/10.1145/3380446.3430634.

[9] Llm4eda: Emerging progress in large language models for electronic design automation R Zhong, X Du, S Kai, Z Tang, S Xu, HL Zhen, J Hao... - arXiv preprint arXiv:2401.12224, 2023.

[10] H. Wu et al., "ChatEDA: A Large Language Model Powered Autonomous Agent for EDA," in IEEE Transactions on Computer-Aided Design of Integrated Circuits and Systems, vol. 43, no. 10, pp. 3184-3197, Oct. 2024, doi: 10.1109/TCAD.2024.3383347.

[11] Liu, M., Pinckney, N., Khailany, B., & Ren, H. (2023). VerilogEval: Evaluating Large Language Models for Verilog Code Generation. *ArXiv*. https://arxiv.org/abs/2309.07544

[12] Nathaniel Pinckney, Christopher Batten, Mingjie Liu, Haoxing Ren, and Brucek Khailany. 2025. Revisiting VerilogEval: A Year of Improvements in Large-Language Models for Hardware Code Generation. ACM Trans. Des. Autom. Electron. Syst. Just Accepted (February 2025). https://doi.org/10.1145/3718088.

[13] ICARUS Verilog (https://bleyer.org/icarus/)

[14] S. Narang, G. Diamos, E. Elsen, P. Micikeviciuss, J. Alben, D. Garcia, B. Ginsburg, M. Houston, O. Kuchaiev, G. Venkatesh, H. Wu, "Mixed Precision Training". Baidu Research and NVIDIA. February 2018.

[15] HDLBits (https://hdlbits.01xz.net/wiki/Main_Page)

[16] Nexys 4 Board (https://digilent.com/reference/programmable-logic/nexys-4/start?srsltid=AfmBOoq1W6BFVXw0jcN1w5OvsI40Cty-OBeem3cySOiDBqimsIZiDB6H)

[17] Pynq Z1 (https://digilent.com/shop/pynq-z1-python-productivity-for-zynq-7000-arm-fpga-soc/)

[18] Repo: (https://github.com/Leoudeji/ML4FPGA)

[19] LMStudio (https://lmstudio.ai/)

[20] Code llama (https://ollama.com/library/codellama)

[21] Claude 3 Model (https://www.anthropic.com/news/claude-3-family)

[22] Llama 4 (https://www.llama.com/docs/model-cards-and-prompt-formats/llama4_omni/)

Scalability analysis of multi-bank near-memory computing in low-power SoCs

Luigi Giuffrida* Pasquale Davide Schiavone† Michele Caon† Guido Masera* Maurizio Martina* David Atienza†

*VLSI Lab
Politecnico di Torino, Italy
{luigi.giuffrida,guido.masera,maurizio.martina}@polito.it

†Embedded Systems Laboratory
EPFL, Switzerland
{davide.schiavone,michele.caon,david.atienza}@epfl.ch

Abstract—Machine learning and artificial intelligence are moving towards the edge, where the need for high throughput with a constrained energy budget is more urgent than ever. During the last few years, near-memory computing has emerged as a promising solution to address the memory bandwidth and energy efficiency limitations of conventional von Neumann systems. The recently proposed NM-Carus architecture combines vector-oriented computing capabilities within a RISC-V programmable, configurable, and autonomous memory macro, addressing the usability of near-memory computing from a software deployment standpoint. In this paper, we explore the scalability of NM-Carus in terms of computation parallelism, memory size and energy consumption, As a benchmarking platform, we rely on a low-power microcontroller that features multiple instances of NM-Carus that target the execution of biomedical applications. This exploration was performed on 16nm TSMC NM-Carus implmentation, and we highlighted the benefits of technology scaling for a previous implementation on 65nm with respect to the overhead of replacing conventional on-chip data SRAMs with near-memory computing banks. Overall, the paper presents a solid baseline regarding the trade-offs in terms of area, performance, and energy efficiency of integrating programmable near-memory computing in an existing edge-oriented system on chip towards efficient edge AI architectures at the system level.

Index Terms—Near-memory computing, Machine Learning, RISC-V

I. INTRODUCTION

The increasing prevalence of data-driven workloads executed close to sensors necessitates a paradigm shift in computing toward enhanced energy efficiency and performance. The conventional von Neumann architecture, the foundation of computing for decades, exhibits inherent inefficiencies for data-intensive tasks due to constant data and instruction movement between the memory hierarchy and processing units [1]. This issue is compounded by the "memory wall", where the advancements of SRAM integration technology lag behind logic scaling, resulting in SRAM accesses consuming significantly more energy than arithmetic operations [2].

To address this challenge, Compute-In-Memory (CIM) paradigm, which includes In-Memory Computing (IMC) and Near-Memory Computing (NMC), has been proposed to reduce data movement and optimize utilization of the available memory bandwidth [3]. nmcNMC does not require the memory-array layout to be redesigned, but relies on commercially available memories to store data tightly connected to specialized data paths. However, controlling operations on IMC and NMC IPs presents difficulties. While a simple approach involves streaming micro-operations to the CIM blocks, complex runtime control demands either significant CPU overhead for encoding operations or reliance on predefined command sequences. This limitation has been addressed or partially mitigated by Single Instruction Multiple Data (SIMD) vector capabilities as in [4], [5], [6], [7], [8]. In particular, the NM-Carus NMC IP [7], integrates a CPU-based controller supporting a custom, vector-inspired CIM-oriented RISC-V ISA extension (xmnmc) to handle and autonomously execute virtually any data processing kernel, while the host system is only responsible for storing the data in the NMC memory, adhering to the kernel data placement requirements. Depending on the target physical and performance constraints, selecting the best combination of memory size and amount of computing resources of the NMC blocks requires exploration of the trade-off of wiring costs, performance, and energy [6]. In this paper, we provide an exploration of the scalability of NM-Carus in terms of throughput, area, and energy consumption. Unlike [7], where the performance assessment was limited to a single 32kB instance with 4 lanes, this work proposes a broader design space exploration, whose major contributions are: (1) scalability analysis of NM-Carus single-instance throughput and energy efficiency with varying memory size and amount of computing resources; (2) Analysis of multi-instance performance scalability; (3) Assessment of the area overhead at the instance and system level; (4) evaluation of the technological benefits of a 16nm NM-Carus implementation compared to its original 65nm variant. Section II analyses the timing, area, and energy scalability of different configurations of NM-Carus, while Section III evaluates the performance improvement enabled by a system featuring multiple instances of NM-Carus. Finally, concluding remarks and a summary of the findings are presented in Section IV.

II. EXPLORATION

In this Section, different configurations in terms of memory size and number of lanes of the NM-Carus IP [7] are explored to find the best trade-offs and scalability figures.

979-8-3315-9813-6/25 $31.00 © 2025 IEEE

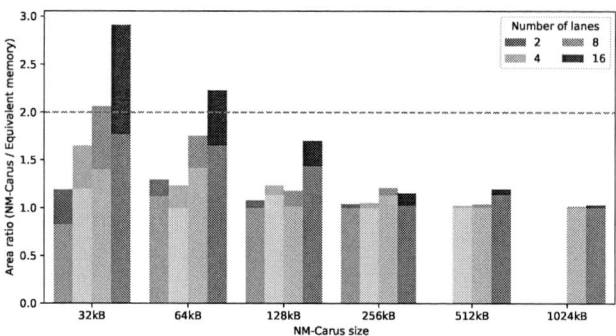

Figure 1: Ratio between the area of differently configured NM-Carus instances and conventional single-bank SRAM memories with equivalent sizes. The dashed line represents the NMC overhead threshold defined in [7] as equal to 2. Bars are split into internal SRAM area (lighter color) and the logic cell area (darker color).

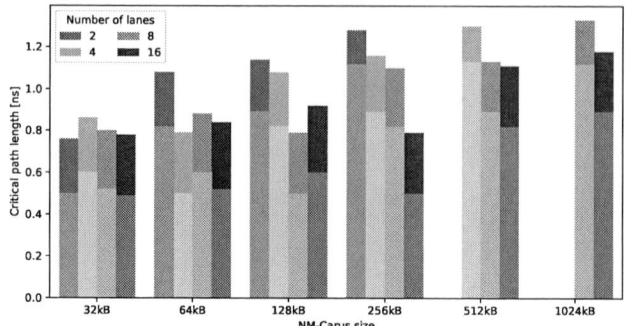

Figure 2: Critical path of the NM-Carus configurations. Bars are split into two: the lighter part represents the portion of the timing portion taken by the SRAM banks, while the darker part represents the delay introduced by logic cells.

The NM-Carus architecture can be configured with two main parameters: the number of processing lanes and the total memory size. The architecture is built such that the number of lanes is equal to the number of internal SRAM banks. Increasing the number of lanes not only determines a linear replication of the computational datapath (i.e., ALUs) but also results in an increase in total area due to the inherent overhead associated with memory sub-banking (i.e., the combined area of multiple smaller memory banks exceeds that of a single larger bank with equivalent size). Conversely, for kernels exhibiting optimal mapping onto SIMD architectures, like matrix multiplications, performance demonstrates a near-linear scaling with the number of processing lanes, directly increasing the throughput in terms of operations per cycle. For instance, when running a matrix multiplication kernel, 4-, 8-, and 16-lane configurations are $1.93\times$, $3.61\times$, and $6.38\times$ faster than a configuration with 2 lanes, respectively. The size of each sub-bank, M_{LANE} is determined by:

$$M_{LANE} = M_{TOT}/N_{LANES}, \qquad (1)$$

where M_{TOT} is the total NM-Carus memory size and N_{LANES} is the number of lanes. The area overhead of NM-Carus, relative to a standard SRAM bank of equivalent capacity, is the ratio between the area of an NM-Carus instance and that of a conventional single-bank SRAM with the same capacity. Adopting the same threshold established in [7], a ratio exceeding 2 indicates excessive area overhead, making the NM-Carus IP instance impractical as a direct replacement for a conventional SRAM bank. Figure 1 depicts the area overhead of NM-Carus instances with different combinations of memory capacity and lane count, subdivided in SRAM area and logic cells area. To keep the total NM-Carus area under constraints, both the internal sub-banks and the reference single-bank SRAM memories were generated with the PDK compiler targeting the minimal area. As shown in Figure 1, the 32 16-lane NM-Carus configuration exhibits the highest area overhead, at $2.9\times$. This substantial overhead can be attributed to two primary factors: the area increase resulting

from partitioning the 32 memory into 16 individual 2 sub-banks ($2.08\times$) and the area contribution of the 16 processing elements, including the integrated CPU controller and its instruction memory, which constitutes 39.7% of the total area. In general, the area overhead of the considered NM-Carus variants demonstrates an inverse relationship with their memory capacity. This trend arises from the diminishing relative overhead associated with sub-bank splitting as the total memory size increases. Furthermore, the area of the computing logic is solely determined by the number of lanes and remains constant regardless of the total memory size, thus contributing to a smaller relative overhead for larger configurations. Consequently, for memory capacities of 128 and greater, the 16-lane configurations exhibit lower area overhead compared to their counterparts with smaller total memory sizes.

Increasing the complexity of the near-memory processing logic not only affects the silicon area but also the maximum operating frequency, which is still primarily influenced by both the number and the timing characteristics of the sub-banks. As illustrated in Figure 2, the 128, 16-lane configuration, comprising 16 8 sub-banks, exhibits a $1.14\times$ reduction in operating frequency compared to a 128 variant configured with 8 lanes and thus 8 larger 16 sub-banks. This disparity arises from a combination of factors: the higher speed of the larger 16 sub-banks (to minimize the area, banks are generated with wider rows, thus slower decoders and increased depth of multiplexers) and the increased latency introduced by the more complex vector slide logic required to move data across different lanes. In contrast, for a 32 instance, the configuration with 2 banks of 16 achieves the highest operating frequency, running $1.02\times$ faster than the 32 variant with 16 smaller 2 banks. In this scenario, while the smaller 2 sub-banks exhibit better timing characteristics, the deeper vector slide logic associated with the larger number of banks ultimately limits the overall performance. Figure 2 highlights how the critical path is dominated by the clock-to-output delay of the memories (lighter color). Configurations that share the same internal memory bank sizes have the same critical path, as the 32 4-lane and the 64 8-lane variants, both using 8 internal

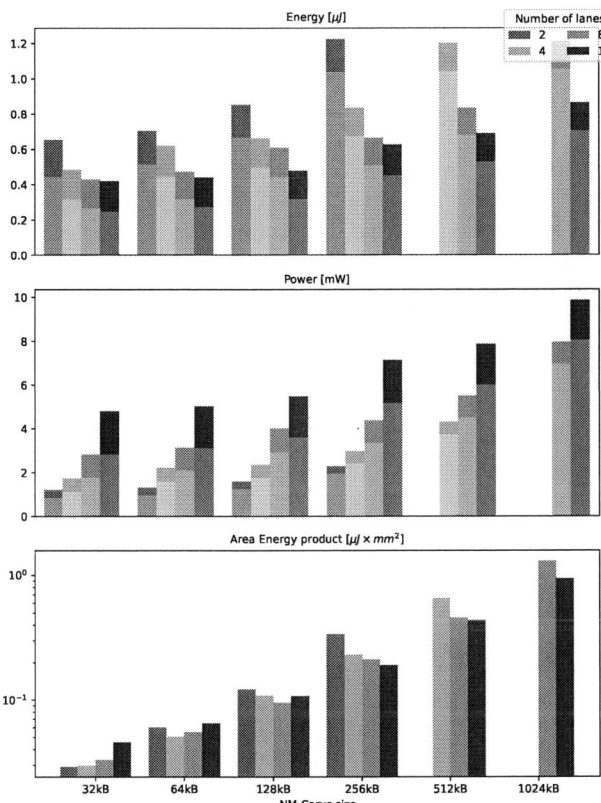

Figure 3: Power, energy and area energy product of NM-Carus for different configurations when running at 100 MHz an 8-bit matrix multiplication ($A[10 \times 10] \times B[10 \times 1024]$)). Power and energy bars are split into two: the lighter part represents the portion of the power dissipated by SRAM banks, while the darker part represents the logic cells.

memory sub-banks. Figure 3 shows the power consumption of NM-Carus configurations when running an 8-bit matrix multiplication at 100 MHz under typical conditions (0.8 V, 25 °C). Power consumption is dominated by the internal memory sub-banks (lighter color). Given a fixed memory capacity, a higher number of lanes is associated with a near-linear increase in computational logic power, mostly due to its dynamic power contribution, whereas the embedded CPU controller's consumption remains negligible, accounting for only 5% to 7% of the total power depending on the configuration. Regarding energy consumption, Figure 3 illustrates how the increase in power consumption associated with a higher number of processing lanes is compensated by the improved performance of the matrix multiplication kernel executions, showing that a higher number of lanes improves performance more than the power overhead in all configurations. However, when taking into account the area overhead, which is crucial for considering NM-Carus an SRAM replacement, the best compromise depends on the configuration. Figure 3 shows the area-energy product on a logarithmic scale, revealing that 2, 4, and 8 lanes are optimal for 32, 64, and 128 respectively. While for 256 and beyond, 16 lanes offer the best trade-off.

Comparing the 32 4-lane 16 nm NM-Carus configuration from Figures 1 to 3 with its 65 nm counterpart from [7], it is possible to appreciate the benefits of technological scaling on such memory-oriented architectures. In the former case, the area overhead is 27%, significantly lower than the 58% overhead in 65 nm. The lower scaling trend in SRAM manufacturing compared to logic scaling ensures more headroom for cost-performance optimization. From a timing perspective, the 16 nm implementation features a much higher operating frequency at 1.16 GHz compared to the 330 MHz in 65 nm. Such timing characteristics support the exploitation of NM-Carus in integrated last-level data cache, as proposed in [8]. Similarly, technological scaling allows for an estimated 2.1 × reduction in power consumption at 100 MHz.

III. SYSTEM INTEGRATION USE-CASE

To demonstrate the performance improvement of the NM-Carus instances, two distinct configurations were integrated into the memory subsystem of the X-HEEP [9] RISC-V microcontroller. The X-HEEP system has been configured to include one CV32E40P [10] core, four DMA channels, standard peripherals, and a 352 on-chip SRAM memory subsystem partitioned into 64 for instruction storage, 32 for private data, two 64 banks for extra data, and 128 with interleaving addressing scheme on four banks. The two 64 banks for extra data have been replaced with the NM-Carus instances, configured as one with 4 processing lanes, and one with 8 processing lanes. Unlike Arcane [8], where several homogeneous instances of NM-Carus are used within a cache and exploited by means of host CPU custom ISA extensions, in this work the memories are scratch-pad memories which require explicit data tiling and layout at application level, as well as loading specific near-memory computing kernels for each instance, which allows running heterogeneous kernels in parallel. The two NMC instances can operate either in memory mode or as computational acceleration engines, depending on the operational mode as detailed in [7]. The X-HEEP [9] system was synthesized using the TSMC 16nm LVT process node, targeting a worst-case operating frequency of 330 MHz (0.72 V, 125 °C, slow process conditions). Figure 4 shows the area breakdown of the X-HEEP system, including the NMC IPs. The area overhead of the 4-lane and 8-lane NM-Carus instances relative to equivalent SRAM implementations is 1.23× and 1.74×, respectively, contributing with an increase of 10% of the overall system area. To showcase the speedup that NM-Carus can introduce in an actual system, a matrix multiplication task has been executed on the X-HEEP system with the two NM-Carus instances as a benchmarking application. The size of the two matrix operands are $A[30 \times 15]$ and $B[15 \times 2048]$. Both are stored in an off-chip flash chip and copied with quad SPI to the on-chip SRAN memory. We analyzed three different scenarios: execution of the matrix multiplication on the CPU, execution on a single instance of NM-Carus, and execution on both the NMC instances. For the reference execution on the CPU, first, the DMA copies both matrices from the external flash to the interleaved memory banks, then the CPU executes the kernel, storing the result in another portion of the same memory region. For the single

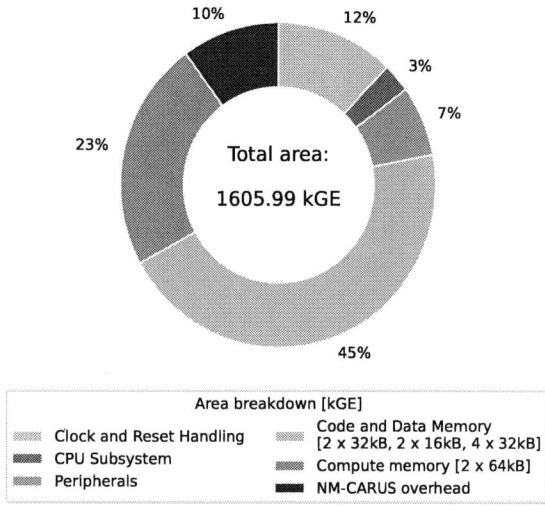

Figure 4: Area breakdown of X-HEEP with two NM-Carus instances. Area reported on the pie-chart is in equivalent kilogates (kGE).

NM-Carus execution, tiling is performed as the output of the matrix multiplication would not fit the NMC memory. After the DMA has copied the 15 rows of A and the complete matrix B, NM-Carus executes the kernel in place. When done, the results are copied to the system memory to leave space for the next tile. This solution requires an extra data-movement operation as the NM-Carus size is smaller than the problem size. While the deployment using both instances of NM-Carus splits the matrix A into two and replicates the matrix B, allowing for the execution of the matrix multiplication using two parallel NM-Carus banks. The multi-channel DMA is used to hide latencies in a double-buffering scheme. Employing configurations with 4 and 8 lanes balances execution. Due to the sequential nature of data transfers from flash memory, the higher speed of the configuration with 8 lanes, is exploited to complete concurrently, despite starting later. The execution of the application with NM-Carus shows a speedup of $19.85\times$ and $29.34\times$ faster than the CPU for the single and double NM-Carus configurations, respectively.

IV. CONCLUSIONS

This study has presented a scalability analysis of NM-Carus IP [7], investigating the trade-offs between memory capacity and the number of parallel processing lanes. We quantitatively compared the area, maximum frequency, power, and energy characteristics of various NM-Carus configurations against equivalent standard SRAM implementations in the TSMC 16nm LVT technology node. The results indicate that the most significant contributor to area overhead is the partitioning of the total memory into smaller sub-banks. In contrast, the impact on the area of the integrated computational logic decreases for larger total memory capacities. The maximum frequency exhibits a nonlinear relationship with the number of lanes and memory organization. In some configurations, increasing the number of sub-banks (resulting from a higher

lane count) leads to a reduction in maximum frequency due to the increased complexity of the addressing. Conversely, in other configurations, the utilization of larger capacity sub-banks with faster access due to area-optimized sub-bank configurations can mitigate or even reverse this trend. In addition, while the power is dominated by the internal sub-banks' consumption, energy shows that more parallelism is always beneficial. However, when combining area and energy, the best configurations depend on the total memory size. Finally, we integrated two heterogeneous configurations into the X-HEEP system, showing that at the system level, using two instances of NM-Carus increases only by 10% the total area, while improving speed up to $29.34\times$.

ACKNOWLEDGMENTS

This work was supported in part by the Swiss State Secretariat for Education, Research, and Innovation (SERI) through the SwissChips research project. This work is part of the SERICS project (PE00000014), under the MUR National Recovery and Resilience Plan funded by the European Union - NextGenerationEU. In addition, we thank Mr. Francesco Poluzzi for supporting the deployment of end-to-end applications.

REFERENCES

[1] J. Backus, "Can programming be liberated from the von neumann style? A functional style and its algebra of programs," *Commun. ACM*, vol. 21, no. 8, pp. 613–641, aug 1978.

[2] S. A. McKee, "Reflections on the memory wall," in *Proceedings of the 1st Conference on Computing Frontiers*. Association for Computing Machinery, 2004, p. 162.

[3] S. Srinivasa, A. K. Ramanathan, J. Sundaram, D. Kurian, S. Gopal, N. Jain, A. Srinivasan, R. Iyer, V. Narayanan, and T. Karnik, "Trends and opportunities for SRAM based in-memory and near-memory computation," in *2021 22nd International Symposium on Quality Electronic Design (ISQED)*, 2021, pp. 547–552.

[4] Y. Wang, M. Yang, C.-P. Lo, and J. P. Kulkarni, "30.6 Vecim: A 289.13GOPS/W RISC-V vector co-processor with compute-in-memory vector register file for efficient high-performance computing," in *2024 IEEE International Solid-State Circuits Conference (ISSCC)*, vol. 67, 2024, pp. 492–494.

[5] M. Kooli, A. Heraud, H.-P. Charles, B. Giraud, R. Gauchi, M. Ezzadeen, K. Mambu, V. Egloff, and J.-P. Noel, "Towards a truly integrated vector processing unit for memory-bound applications based on a cost-competitive computational SRAM design solution," *J. Emerg. Technol. Comput. Syst.*, vol. 18, no. 2, apr 2022.

[6] R. Gauchi, V. Egloff, M. Kooli, J.-P. Noel, B. Giraud, P. Vivet, S. Mitra, and H.-P. Charles, "Reconfigurable tiles of computing-in-memory SRAM architecture for scalable vectorization," in *Proceedings of the ACM/IEEE International Symposium on Low Power Electronics and Design*, ser. ISLPED '20. New York, NY, USA: Association for Computing Machinery, 2020, pp. 121–126.

[7] M. Caon, C. Choné, P. D. Schiavone, A. Levisse, G. Masera, M. Martina, and D. Atienza, "Scalable and risc-v programmable near-memory computing architectures for edge nodes," *IEEE Transactions on Emerging Topics in Computing*, pp. 1–15, 2025.

[8] V. Petrolo, F. Guella, M. Caon, P. D. Schiavone, G. Masera, and M. Martina, "Arcane: Adaptive risc-v cache architecture for near-memory extensions," *arXiv preprint arXiv:2504.02533*, 2025.

[9] S. Machetti, P. D. Schiavone, T. C. Müller, M. Peón-Quirós, and D. Atienza, "X-heep: An open-source, configurable and extendible risc-v microcontroller for the exploration of ultra-low-power edge accelerators," *arXiv preprint arXiv:2401.05548*, 2024.

[10] M. Gautschi, P. D. Schiavone, A. Traber, I. Loi, A. Pullini, D. Rossi, E. Flamand, F. K. Gürkaynak, and L. Benini, "Near-threshold risc-v core with dsp extensions for scalable iot endpoint devices," *IEEE transactions on very large scale integration (VLSI) systems*, vol. 25, no. 10, pp. 2700–2713, 2017.

979-8-3315-9813-6/25 $31.00 © 2025 IEEE

LiC: Low-Cost Cache Replacement Algorithm for All Cache Levels

Varun Venkitaraman[1*], Tejeshwar Thorawade[1**], Mitul Tandon[1], Keerthisagar Kokkiligadda[1],
Virendra Singh[1***] and Janak Patel[2]

[1]*Department of Electrical Engineering, Indian Institute of Technology Bombay*, India
[2]*Department of Electrical & Computer Engineering, University of Illinois Urbana-Champaign*, United States
*varunvbel@gmail.com, **tejeshwar@iitb.ac.in, ***singhv@iitb.ac.in

Abstract—Modern processors use caches to reduce memory access time. However, their limited size leads to frequent misses, requiring an efficient replacement policy. The Least Recently Used (LRU) policy is widely adopted for its effectiveness but becomes impractical in highly associative caches due to its high area and power costs. To address inefficiencies in the last-level cache (LLC), researchers have proposed sophisticated replacement policies. However, their complexity and hardware overhead make them unsuitable for level-one (L1) and level-two (L2) caches, which require fast and lightweight decision-making. Additionally, low-cost microcontrollers demand simple and efficient replacement mechanisms. This paper introduces the Lightweight Cache Replacement Policy (LiC) as a low-cost, power-efficient alternative to LRU. Unlike conventional policies that focus on eviction decisions, LiC prioritizes protecting the last accessed block. This approach significantly reduces hardware complexity and power consumption while maintaining performance. We evaluate LiC through simulations in both single-core and multi-core environments. Results show that LiC matches LRU's performance while drastically reducing storage overhead. Compared to sophisticated LLC replacement policies, it reduces storage costs by up to 28×. Against low-cost policies like NRU and PLRU, it achieves 4× and 3.75× lower storage demands. Additionally, LiC reduces area overhead by 16× compared to LRU. With its low hardware overhead and strong performance, LiC emerges as an efficient and scalable solution across all cache levels.

Index Terms—Cache replacement policy, Low-cost, Least recently used(LRU), Cache memory, Low complexity

I. INTRODUCTION

Modern processors use caches to reduce memory access latency. However, their limited size leads to frequent misses, requiring an efficient replacement policy. Caches are typically set-associative, where each set contains multiple blocks. A block can be placed in any way within its set. When the CPU accesses a block, its address is compared with stored addresses. A match results in a cache hit; otherwise, it causes a miss, requiring the block to be fetched from a higher memory level. Before inserting a new block, the replacement policy decides which block to evict. LRU evicts the least recently used block, assuming it is less likely to be reused soon.

While LRU performs well in L1 and L2 caches, its efficiency drops in large last-level caches (LLCs). Researchers have proposed advanced replacement policies to optimize LLCs [1]–[9]. However, their complexity and high hardware overhead make them impractical for L1 and L2, where quick

and simple decisions are essential. Despite LRU's effectiveness, its high cost has led to various approximations. Approximate LRUs, such as those in IBM mainframes since the 1970s, perform worse than ideal LRU, yet most research focuses on the ideal case. Belady's Optimal Algorithm [3], [10] theoretically achieves the best replacement by evicting the block with the longest future reuse distance. However, it is impractical due to the lack of future knowledge. Additionally, cache power consumption contributes 12% to 45% of total core power [11], making power-efficient policies essential for modern processors.

To address these challenges, we propose the **Lightweight Cache Replacement Policy (LiC)**. Unlike conventional policies that focus on eviction, LiC prioritizes protecting the last accessed block. This approach reduces hardware complexity and power consumption while maintaining performance close to LRU. Extensive simulations confirm that LiC matches LRU's performance with significantly lower cost. It requires less area and power while employing a simpler algorithm, making it effective across all cache levels. In summary, **LiC** offers a low-cost, power-efficient alternative to LRU with the following key advantages:

1) **Low-Cost Implementation:** LiC minimizes area requirement, reduces controller complexity and simplifies implementation challenges.
2) **Comparable Performance:** LiC achieves performance on par with LRU across all cache levels.
3) **Scalability:** LiC works efficiently in L1, L2, and LLC, integrating well with existing policies.

The paper is structured as follows. Section II discusses LRU's complexity. Section III details LiC's algorithm and implementation. Section IV describes our simulation framework. Section V presents performance evaluations in single-core and multicore setups. Finally, Section VI concludes the paper.

II. MOTIVATION

In this section, we explore the objectives of cache replacement policies and the challenges of implementing Least Recently Used (LRU). An ideal benchmark for cache replacement is Belady's optimal policy [3], [10], which evicts the block accessed farthest in the future. However, since future accesses are unknown, practical policies attempt to approximate this decision [1]–[9]. Among them, LRU follows a straightforward

979-8-3315-9813-6/25 $31.00 © 2025 IEEE

heuristic: recently accessed blocks are likely to be reused soon, so the least recently accessed block is chosen for eviction. While effective, LRU is costly to implement because it requires tracking access recency for all blocks, demanding a complex state machine.

The complexity of LRU increases with associativity. In an n-way set-associative cache, the number of states grows factorially with n. A 4-way cache manages 24 states, an 8-way cache requires 40,320, and a 16-way cache demands 21 trillion states. This exponential growth makes LRU impractical for highly associative caches. Additionally, each set requires its own controller, further increasing hardware costs. Given these challenges, a more efficient and cost-effective replacement policy is needed. In the next sections, we introduce an alternative that overcomes LRU's limitations while remaining practical for modern cache architectures.

III. PROPOSAL

In this section, we present the algorithm and implementation details of our proposed cache replacement policy, LiC. LiC serves as an efficient alternative to LRU across all cache levels by prioritizing the retention of the last accessed cache block. This approach significantly reduces complexity while maintaining performance comparable to LRU.

To achieve this, LiC employs a simple pointer-based mechanism. Each cache set maintains a pointer R, which identifies the block to be replaced on a miss. The pointer size depends on the cache associativity: a 4-way set-associative cache requires a 2-bit pointer, an 8-way cache requires 3 bits, and a 16-way cache requires 4 bits. More generally, a cache with n-way associativity requires $\log_2(n)$ bits. This compact design ensures minimal hardware overhead. The LiC algorithm operates as follows:

1) **Initialization:** At startup, caches are empty (as shown in ❶ in Fig. 1). The pointer R is in an arbitrary state and does not require initialization (as shown in ❷ in Fig. 1).

2) **First Memory Reference:** The CPU accesses a cache set (as shown in ❸ in Fig. 1), resulting in a miss (as shown in ❹ in Fig. 1). The requested block is fetched from memory and placed at the position indicated by R (as shown in ❽ in Fig. 1). The pointer is then incremented modulo the set size n, i.e., $R = R + 1$ (as shown in ❾ in Fig. 1).

3) **Cache Operations:** Once the cache is filled, memory accesses continue. If a reference hits a block other than the one pointed to by R, the pointer remains unchanged. The cache is accessed, and execution proceeds without modification.

4) **Cache Hit and Miss Handling:** If a reference hits the block pointed to by R, the pointer is incremented modulo n, i.e., $R = R + 1$ (as shown in ❻ in Fig. 1). On a miss, the block at R is evicted (as shown in ❼ in Fig. 1), and the new block is inserted at the same location (as shown in ❽ in Fig. 1). The pointer is then updated as $R = R + 1$ (as shown in ❾ in Fig. 1).

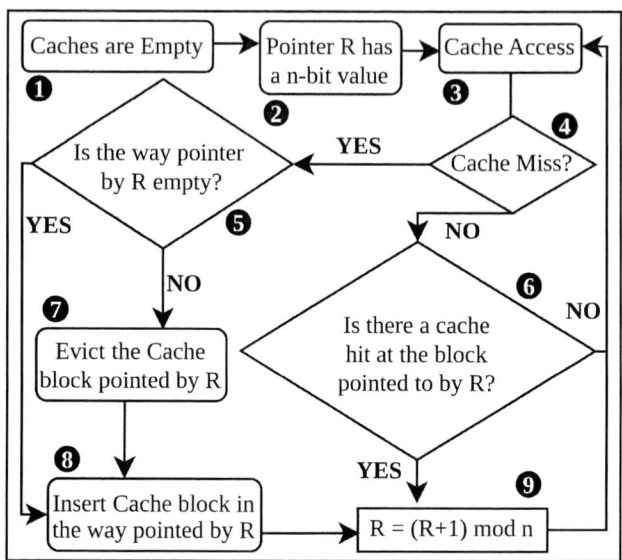

Fig. 1: **LiC Algorithm Flowchart**

TABLE I: Baseline Configuration

OoO Core	2.667 GHz, 6-way fetch, 4-way commit, ROB = 256, LQ = 128, SQ =72
L1-I Cache	32KB, 8-way, 4-cycle access latency, private
L1-D Cache	32KB, 8-way, 4-cycle access latency, private
L2 Cache	256KB, 8-way, 8-cycle access latency, private
L3 Cache (LLC)	2MB, 16-way, 20-cycle access latency, shared
DRAM	800 MHz, 4KB Page size

This section detailed the LiC algorithm and its lightweight implementation. In the next section, we describe the simulation environment and workloads used to evaluate LiC. We also compare its performance with LRU and other advanced cache replacement policies designed to optimize LLC efficiency.

IV. EVALUATION METHODOLOGY

This section describes the experimental setup for evaluating LiC. Performance and cache miss rates are measured using Champsim [12]. Core and memory configurations are in Table I. Cache parameters, including latencies, sizes, and associativities, are modeled using Cacti [13] with a 45nm technology node and a 64B block size. Each cache level has one read and one write port. LiC is evaluated across all cache levels, with LRU as the baseline. For single-core evaluation, we use 49 applications from SPEC CPU2006 and SPEC CPU2017: 29 from SPEC CPU2006 and 20 from SPEC CPU2017. Each application runs with reference inputs. Using SimPoint, we generate 2 billion instruction traces per application, selecting the most representative one. Caches warm up for 500 million instructions. Performance is measured over the next 1 billion instructions. We present results for 15 randomly selected applications (Table II). The *mean* value is the geometric mean across all 49 applications.

TABLE II: Single-core Application List

App. Name	Abbr.	App. Name	Abbr.
410.bwaves-945B	4.bwa	470.lbm-1274B	4.lbm
433.milc-127B	4.mil	473.astar-153B	4.ast
447.dealII-3B	4.dea	603.bwaves_s-3699B	6.bwa
454.calculix-104B	4.cal	627.cam4_s-573B	6.cam
458.sjeng-1088B	4.sje	638.imagick_s-10316B	6.ima
459.GemsFDTD-1491B	4.Gem	641.leela_s-800B	6.lee
462.libquantum-1343B	4.lib	648.exchange2_s-1699B	6.exc
geomean	mean	654.roms_s-842B	6.rom

TABLE III: Multi-core Application List

App. Name	Abbr.	Homogeneous Mix	Abbr.
cassandra_phase3	cas	410.bwaves-945B	hm_1
classification_phase4	cla	454.calculix-104B	hm_2
cloud9_phase5	clo	458.sjeng-1088B	hm_3
nutch_phase5	nut	603.bwaves_s-3699B	hm_4
streaming_phase5	str	648.exchange2_s-1699B	hm_5
mean_cloud	mn_cl	mean_homo	mn_hm
Heterogeneous Mix			**Abbr.**
454.calculix / 401.bzip2 / 447.deal / 625.x264_s			ht_1
458.sjeng / 657.xz_s / 416.gamess / 627.cam4_s			ht_2
600.perlbench_s / 416.gamess / 401.bzip2 / 416.gamess			ht_3
603.bwaves_s / 623.xalancbmk / 458.sjeng / 603.bwaves_s			ht_4
641.leela_s / 648.exchange2_s / 600.perlbench_s / 648.exchange2_s			ht_5
mean_hetero			mn_ht

TABLE IV: Storage Req. for Low-cost Cache Replacement Policy

Policy	LRU	NRU [6]	PLRU [15]	LiC	LRU w.r.t. LiC	NRU w.r.t. LiC	PLRU w.r.t. LiC
L1-I	192 B	64 B	56 B	24 B	8 x	2.67 x	2.33 x
L1-D	192 B	64 B	56 B	24 B	8 x	2.67 x	2.33 x
L2	1536 B	512 B	448 B	192 B	8 x	2.67 x	2.33 x
LLC	16384 B	4096 B	3840 B	1024 B	16 x	4 x	3.75 x

TABLE V: Storage Req. at LLC Replacement Policy

Policy	Storage Req.	Storage w.r.t. LiC	Policy	Storage Req.	Storage w.r.t. LiC
LRU	16 KB	16 x	SRRIP [6]	16 KB	16 x
NRU [6]	4 KB	4 x	DRRIP [6]	16 KB	16 x
PLRU [15]	3.75 KB	3.75 x	SHiP [9]	14 KB	14 x
LiC	1 KB	1 x	Hawkeye [4]	28 KB	28 x

For multi-core evaluation, we use five applications from Cloud Suite (multi-threaded) from the 2nd Cache Replacement Championship [14]. Each application has six traces, each with four threads. One trace per application is randomly selected for graph representation (Table III). These workloads run in a quad-core setup for 100 million instructions. Caches warm up for 10 million instructions. Performance is collected over the next 90 million instructions. We also evaluate LiC in a quad-core setup using homogeneous and heterogeneous workload mixes from the 49 SPEC traces. Homogeneous mixes run the same application on all four cores. Heterogeneous mixes run different applications on each core. Caches warm up for 500 million instructions. Performance is measured over the next 1 billion instructions. We present results for five randomly selected homogeneous and heterogeneous mixes (Table III). Mean values for multi-threaded, homogeneous, and heterogeneous mixes are reported, along with an overall mean for all quad-core workloads. In all multi-core experiments, the LLC is 2MB per core. The next section analyzes LiC's performance against LRU and other advanced cache replacement policies.

V. RESULTS & ANALYSIS

This section evaluates the performance of **LiC** in both single-core and multi-core systems. We consider four scenarios: (i) LiC deployed across all cache levels, (ii) LiC applied only at the LLC with LRU at L1 and L2, (iii) LiC used at L1 and L2 with various LLC policies, and (iv) LiC at all levels with increased L1-D capacity using the area overhead of complex LLC policies.

A. Performance Analysis

LiC at all cache levels: Figure 2a shows single-core IPC normalized to the LRU baseline. The configurations

all_NRU, all_PLRU, and all_LiC use NRU, PLRU, and LiC at all cache levels. LiC matches LRU closely, with a deviation of only $\approx 0.001\%$. It slightly improves *6.ima* and shows a minor drop for *4.Gem*. Figure 3a reports MPKI across L1-D, L2, and LLC, showing minimal variation. In the multi-core setting (Figure 2b), LiC improves average performance by 0.05%. The gain is consistent across workload types. Tables IV and V show that LiC incurs minimal storage overhead. It is an area-efficient alternative to LRU. For the remaining analysis, we report mean values across all applications for each configuration due to space constraints.

LiC only at LLC (LRU at L1 and L2): We evaluate LiC against NRU [6], PLRU [15], SRRIP [6], DRRIP [6], SHiP [9], and Hawkeye [4], when applied only at the LLC. Figure 4a presents speedup for single-core and multi-core environments, respectively normalized to baseline LRU at LiC. LiC exhibits comparable performance to other policies, with a maximum gain of 2.1% (Hawkeye on *6.rom*). In multi-core workloads, LiC offers a marginal 0.06% improvement over LRU. Cache-level breakdowns are omitted due to negligible differences. Table V highlights the significant area cost of advanced LLC policies. LiC achieves similar performance with substantially lower hardware overhead.

LiC at L1 and L2 (varying LLC policy): We compare **LiC** and LRU at L1 and L2 cache levels while varying LLC replacement policies. Figure 4b shows single-core and multi-core results, respectively, with speedup of LiC at L1/L2 normalized to LRU at L1/L2 for each LLC replacement policy. Replacing LRU with LiC at upper levels yields negligible performance impact. Slight gains are observed for *6.ima* (single-core) and *ht_1*, *ht_2* (multi-core). LiC's primary advantage is the elimination of LRU control logic, reducing area and design complexity at latency-sensitive cache levels.

LiC with increased L1-D capacity: We repurpose the storage overhead of advanced LLC policies to enlarge the L1-D cache, while running LiC at all levels. L1-D latency is modeled using CACTI for the increased size. Figure 5a compares performance with this configuration against advanced LLC policies with LRU at L1 and L2. In both single- and multi-core scenarios, the performance benefits of increased L1-D

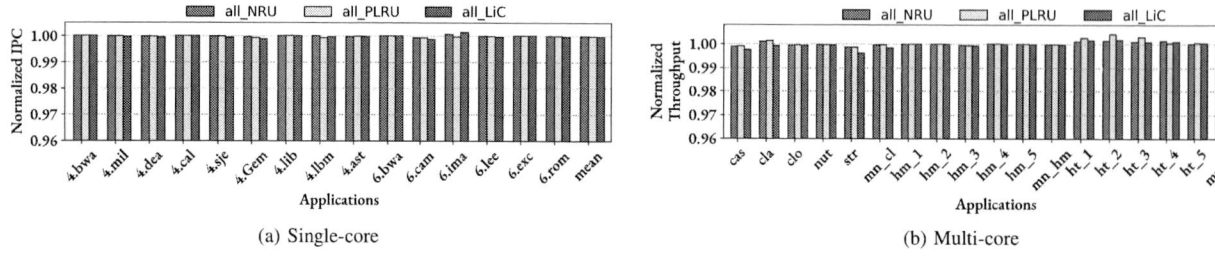

(a) Single-core

(b) Multi-core

Fig. 2: System Performance Analysis

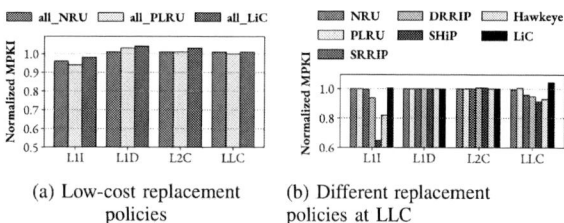

(a) Low-cost replacement policies

(b) Different replacement policies at LLC

Fig. 3: Single-core Cache Performance Analysis

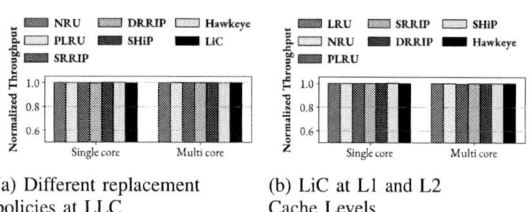

(a) Different replacement policies at LLC

(b) LiC at L1 and L2 Cache Levels

Fig. 4: System Performance Analysis

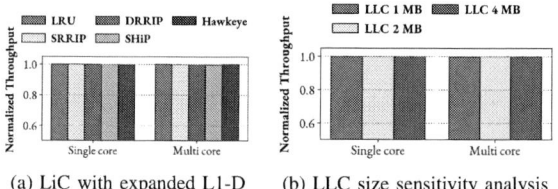

(a) LiC with expanded L1-D

(b) LLC size sensitivity analysis

Fig. 5: Impact of LiC on System's Performance

capacity are limited and often offset by higher access latency. Nonetheless, eliminating complex LLC policies reduces power and design complexity. LiC thus provides an area- and energy-efficient replacement strategy with competitive performance across cache levels.

B. LLC Size Sensitivity Analysis

We perform an LLC size sensitivity analysis to assess LiC's effectiveness across different LLC sizes. Figure 5b shows performance, normalized to LRU at all cache levels, in a single-core setup. LLC latencies for various sizes are modelled using CACTI. Results indicate minimal performance fluctuations, confirming LiC's stability across LLC sizes. A similar trend is observed in the quad-core setup (Fig. 5b), demonstrating LiC's robustness. Thus, LiC remains an efficient and resilient replacement policy across varying LLC configurations.

C. Hardware Overhead

Cache replacement policies require metadata storage for eviction decisions. For an n-way cache, LiC uses a $\log_2 n$-bit pointer per set, totaling $m \cdot \log_2 n$ bits across m sets. LRU requires $m \cdot n \cdot \log_2 n$ bits. Advanced LLC policies demand even more. Low-cost policies like NRU [6], PLRU [15], and LiC reduce LRU's area overhead. Table IV shows that LiC has

the lowest storage cost. NRU and PLRU require 2.6x and 2.3x more at L1/L2, and 4x and 3.75x more at LLC. LiC incurs only a 0.001% performance trade-off.

At the LLC, LiC significantly reduces storage compared to advanced policies [4], [6], [9], as shown in Table V. These policies improve performance but incur high area costs. LiC is ideal for low-power systems due to its minimal hardware and power efficiency. LiC's finite state machine (FSM) is simpler. LRU requires $n!$ states for an n-way cache, while advanced policies need more. LiC's lightweight FSM ensures an area-efficient, low-cost design with LRU-comparable performance across all cache levels.

VI. CONCLUSION

This paper has introduced LiC, an efficient cache replacement policy designed to minimize area, power consumption, and complexity. By focusing on protecting the last accessed cache block within a set, LiC ensures optimal performance with minimal overhead. Through comprehensive evaluations in both single-core and multi-core environments, we demonstrated that LiC achieves performance on par with LRU, while requiring substantially less hardware. In terms of storage efficiency, LiC reduces the storage requirements at the LLC (for a 2MB, 16-way configuration) by 16x, 14x, and 28x compared to LRU, SRRIP, DRRIP, SHiP, and Hawkeye, respectively. When compared to other low-cost replacement policies, LiC still offers significant reductions, achieving 4x and 3.75x lower storage requirements compared to NRU (a variant of NMRU) and PLRU (an approximate version of LRU). Additionally, the controller implementing LiC is remarkably simpler than those required for both low-cost and sophisticated LLC replacement policies. Ultimately, LiC achieves parity with LRU in terms of performance while drastically reducing hardware demands, positioning it as an power-efficient and scalable solution for all cache levels.

979-8-3315-9813-6/25 $31.00 © 2025 IEEE

REFERENCES

[1] M. Chaudhuri, J. Gaur, N. Bashyam, S. Subramoney, and J. Nuzman, "Introducing hierarchy-awareness in replacement and bypass algorithms for last-level caches," in *PACT*, 2012.

[2] Newton, S. K. Mahto, S. Pai, V. Singh *et al.*, "Daaip: Deadblock aware adaptive insertion policy for high performance caching," in *ICCD*, 2017.

[3] I. Shah, A. Jain, and C. Lin, "Effective mimicry of belady's min policy," in *HPCA*, 2022.

[4] A. Jain and C. Lin, "Back to the future: Leveraging belady's algorithm for improved cache replacement," in *ISCA*, 2016.

[5] A. Jaleel, E. Borch, M. Bhandaru, S. C. Steely Jr., and J. Emer, "Achieving non-inclusive cache performance with inclusive caches: Temporal locality aware (tla) cache management policies," in *MICRO*, 2010.

[6] A. Jaleel, K. B. Theobald, S. C. Steely, and J. Emer, "High performance cache replacement using re-reference interval prediction (rrip)," in *ISCA*, 2010.

[7] S. M. Khan, Y. Tian, and D. A. Jimenez, "Sampling dead block prediction for last-level caches," in *MICRO*, 2010.

[8] M. Yan, B. Gopireddy, T. Shull, and J. Torrellas, "Secure hierarchy-aware cache replacement policy (sharp): Defending against cache-based side channel attacks," in *ISCA*, 2017.

[9] C.-J. Wu, A. Jaleel, W. Hasenplaugh, M. Martonosi, S. C. Steely, and J. Emer, "Ship: Signature-based hit predictor for high performance caching," in *MICRO*, 2011.

[10] L. A. Belady, "A study of replacement algorithms for a virtual-storage computer," *IBM Systems Journal*, vol. 5, no. 2, pp. 78–101, 1966.

[11] A. Sodani and C. Processor, "Race to exascale: Opportunities and challenges," in *Keynote at MICRO*, 2011.

[12] ChampSim, "https://github.com/champsim/champsim."

[13] N. Muralimanohar, R. Balasubramonian, and N. P. Jouppi, "Cacti 6.0: A tool to model large caches," *HP laboratories*, 2009.

[14] 2nd Cache Replacement Championship, "http://crc2.ece.tamu.edu/," 2017.

[15] H. Al-Zoubi, A. Milenkovic, and M. Milenkovic, "Performance evaluation of cache replacement policies for the spec cpu2000 benchmark suite," in *Proceedings of the 42nd annual Southeast regional conference*, 2004, pp. 267–272.

979-8-3315-9813-6/25 $31.00 © 2025 IEEE

Automated Generation of Microfluidic Netlists using Large Language Models

Jasper Davidson*, Skylar Stockham*, Allen Boston*, Ashton Snelgrove*, Valerio Tenace†, Pierre-Emmanuel Gaillardon*

*Department of Electrical and Computer Engineering, University of Utah
{jasper.davidson, skylar.stockham, allen.boston, ashton.snelgrove, pierre-emmanuel.gaillardon}@utah.edu
† Primis AI, Inc.
valerio@primis.ai

Abstract—**Microfluidic devices have emerged as powerful tools in various laboratory applications, but the complexity of their design limits accessibility for many practitioners. While progress has been made in microfluidic design automation (MFDA), a practical and intuitive solution is still needed to connect microfluidic practitioners with MFDA techniques. This work introduces the first practical application of large language models (LLMs) in this context, providing a preliminary demonstration. Building on prior research in hardware description language (HDL) code generation with LLMs, we propose an initial methodology to convert natural language microfluidic device specifications into system-level structural Verilog netlists. We demonstrate the feasibility of our approach by generating structural netlists for practical benchmarks representative of typical microfluidic designs with correct functional flow and an average syntactical accuracy of 88%.**

Index Terms—**Microfluidics, Microfluidic Design Automation, Large Language Models**

I. INTRODUCTION

Microfluidic devices (MFDs) are small-scale systems that precisely manipulate tiny volumes of fluids for chemical, biological, or physical processes. Recent advancements in MFDs have resulted in promising improvements to laboratory procedures across various scientific disciplines, leading to the development of high-performance "lab-on-chip" devices [1]. While these advances progress, the microfluidic design automation (MFDA) field is still in its early stages [2]. MFDA tools demand both microfluidic expertise and additional proficiency in computer-aided design tools, algorithms, and manufacturing techniques. As a result, practitioners in microfluidic design have shown limited adoption of MFDA techniques [3]. Meanwhile, Large Language Models (LLMs) have emerged as powerful tools, serving as a bridge between non-experts and high-productivity design across various domains.

In this work, we propose a methodology to leverage the capabilities of LLMs to generate microfluidic device designs from natural-language prompts provided by practitioners. The resulting output is a structural netlist composed of assembled predefined microfluidic primitives that achieve the functions outlined in the prompt. To the best of our knowledge, this represents the first application of LLMs for MFDA netlist generation, offering a preliminary and basic demonstration of its potential.

We showcase the capabilities of our proposed introductorymethodology by creating structural netlists from benchmark prompts representative of standard microfluidic manipulations. Our results highlight that LLMs can accurately map the appropriate device primitives, establish correct connectivity between them, and adhere to the adopted netlist syntax. These findings mark an important first step in providing biologists and chemists with direct access to design tools for developing microfluidic devices. Using our proposed LLM prompting methodology, we produce microfluidic netlists with correct functional flow and average syntactical accuracy of 88%. The proposed methodology is illustrated in Figure 1.

The remainder of this paper is organized as follows: Section II reviews the state of the art in MFDA and LLM-based hardware design. Section III presents the proposed prompting methodology. Section IV evaluates its performance, and Section V concludes the paper.

II. BACKGROUND AND RELATED WORK

A. Microfluidic Design Automation

Traditionally, microfluidic designers have created devices using general-purpose computer-aided design (CAD) tools intended for mechanical engineering. These methods involve manually drawing channels and components as geometric shapes with dimensions, which are then used directly for manufacturing. This introduces two key challenges: a time-consuming, high-barrier design process for non-experts, and strong dependency on manufacturing limitations such as fabrication resolution, which constrains device and channel dimensions.

Microfluidic devices share many characteristics with digital electronics, and researchers have sought to apply EDA techniques to microfluidics [2]. At a high level, both can be described as a netlist of internal components, their interconnections, and input/output ports. A desired but still missing capability is specifying devices using libraries of domain-specific components [3]. As in EDA, providing manufacturing-dependent components introduces a crucial level of abstraction and portability.

This material is based upon work supported by the National Science Foundation under Grant No. 2245494. Any opinions, findings, and conclusions or recommendations expressed in this material are those of the author(s) and do not necessarily reflect the views of the National Science Foundation.

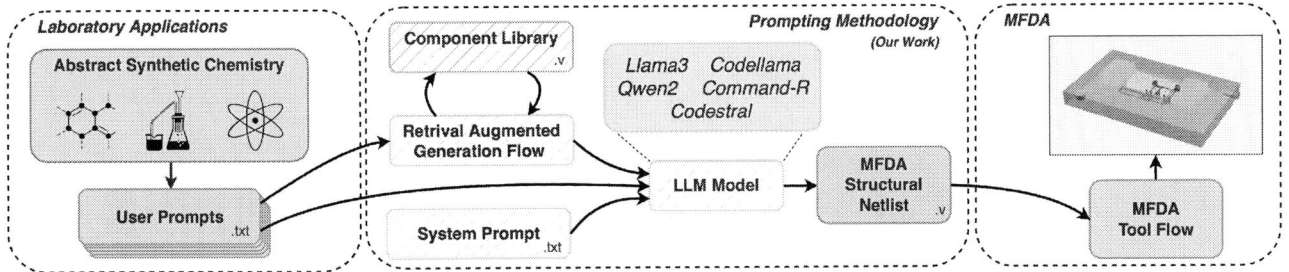

Fig. 1: Microfluidic netlist generation methodology using LLMs, spanning from laboratory applications to Verilog netlists that are compatible with existing MFDA techniques.

Current state-of-the-art microfluidic design automation (MFDA) tools include Columba [4], Fluigi [5], and Open-MFDA [6]. Columba and Fluigi use domain-specific languages to describe devices, while OpenMFDA leverages Verilog hardware description language (HDL) to tap into existing EDA tool flows. Nonetheless, all MFDA tools require designers to manually specify the syntax and low-level implementation details. While domain experts may be comfortable designing in structured textual formats, typical microfluidics practitioners are unlikely to adopt such approaches [3].

B. LLMs for Hardware Development

The use of LLMs for code generation is a growing, popular topic in current research. LLMs can potentially enhance engineers' productivity, including code completion [7], interactive design development [8], and generating test benches for validation from natural language specifications [9].

This technique is also applied to hardware design, where early models were able to generate simple Verilog code snippets from human prompts [10]. More recent works demonstrate a flow for optimizing LLMs to output Verilog HDL and define a standard for gauging the performance of a model, as well as proposals for improvement, achieving syntax-error-free code 65% of the time [11]. CodeGen [12] highlights the challenges of synthesizing code from natural language input.

While making user prompts more explicit can improve the output of LLMs, it may also increase prompt size and complexity, potentially hindering usability. Chip-Chat [8] describes an iterative process for hardware development via LLM rather than focusing on single prompt methodologies. To complement LLMs for code and HDL generation, recent works have established benchmark and evaluation data sets for LLMs in circuit hardware design [13][7].

These works highlight the utility of LLMs in the automated generation of hardware description languages within the EDA domain. However, at the time of writing, no published research has been found on the use of LLMs for generating microfluidic design specifications.

III. METHODOLOGY

A. Describing Microfluidic Systems with Verilog

We employ Verilog as a framework to represent components' structural composition and connectivity within the microfluidic device. Primitive elements such as reservoirs, mixers, valves, and pumps are defined as cells, which contain a behavioral description of the device function with various metadata, such as geometric dimensions and flow rate, represented using Verilog parameters. Predefined cells are interconnected in a structural netlist to construct a microfluidic system that executes the desired chemical manipulations.

B. LLM Software Structure

To develop our prompting methodology, we leverage existing open-source LLMs from Ollama [14]. Ollama provides a framework that consolidates various LLM models with diverse architectural characteristics, such as the number of parameters, layers, and context length, in one code base. Of the available models we select two models optimized for code generation (Codellama and Codestral) and two general purpose models (Llama3 and Qwen2). Additionally, we include the general-purpose Deepseek-R1 due to its state-of-the-art performance and efficiency in handling diverse workloads.

TABLE I: Architectural Features of the Selected LLMs

Model	Parameters	Layers	Context Length (Number of Tokens)
CodeLlama	7 billion	32	16,384
Codestral	22 billion	56	32,768
Llama3	70 billion	32	8,192
Qwen2	72 billion	80	32,768
DeepSeek-R1	32 billion	60	131,072

The number of parameters in a model reflects the weights in its layers, with more parameters enabling finer feature extraction. Network depth influences the ability to capture complex features, with deeper models improving performance but increasing overfitting risk. Context length determines how much of a prompt the model can process at once; prompts exceeding this length are split, causing each section to be considered separately. Tokenization defines the text units the model processes, with shorter context lengths reducing available contextual information. Table I summarizes the key characteristics of the LLM architectures used in this paper.

C. Prompt Methodology

We propose a prompting methodology with two primary components: retrieval-augmented generation (RAG) and a sys-

TABLE II: Microfluidic domain benchmark prompts utilized to develop and refine our LLM prompting methodology.

	Benchmarks used for generating Prompting Methodology	Tests for:
1	Take 2 solutions as input. Mix them together.	Basic interpretation of prompts
2	Take 5 solutions as input to the experiment module. Mix the 5 solutions together parallel.	Reasoning in parallel
3	Take 5 solutions as input to the experiment module. Mix the 5 solutions together sequentially.	Reasoning in sequence
4	Take 3 solutions as input. Dilute the first solution, then mix with the other two solutions.	Reasoning with dilution
5	Put an algae solution into a heater. Mix the heated algae solution with itself to create a stirring effect. Pass through a membrane filter	Abstract reasoning

TABLE III: Microfluidic domain benchmark prompts applied to assess and validate our LLM prompting methodology.

	Test Benchmarks	Tests for:
6	Take 2 solutions as input. Mix them together to create the output solution.	Basic interpretation of prompts
7	Take 4 solutions as input. Mix the 4 solutions sequentially to create the output.	Reasoning in sequence
8	Take 6 solutions as input. Mix the 6 solutions together in parallel to create the output.	Reasoning in parallel
9	Take two solutions as input. Dilute the first solution, then mix with the other solution.	Reasoning with dilution
10	Heat up a solution of water. Filter the water to purify it, then mix with a diluted solution of oil.	Abstract Reasoning

tem prompt. The user prompt, library context from RAG, and system prompt are combined and then given to the model as visualized in Figure 1. The RAG process compares the similarity of the user prompt to computes vector similarity to the library and retrieves the top-k relevant primitives as context. This process gives the model access to information outside its training through vector spaces, and bypasses the need to retrain a new model every time a module is added to the library. We created a library of Verilog modules describing microfluidic primitives for MFDA and used RAG to provide the model with relevant context about their functionality. At the same time, a system prompt is developed throughout this study to guide the model's behavior for the specific target application.

Using the system prompt, we limit the model to generating structural Verilog code that includes only user-defined primitives from the RAG library. This ensures the tool does not produce behavioral Verilog or functionality outside the microfluidic domain and prompts the user if certain requests are restricted based on the available set of primitives. In addition to giving the AI model further insight into the workings of Verilog, the system prompt is able to direct the model towards certain lines of thinking (such as the most effective ways to utilize different classes of Microfluidic devices e.g. heaters, diluters, mixers, etc.) as well as encourage chain-of-thought reasoning. By stepping through each of the logical steps the model takes, it's easier for the model to identify logical inconsistencies in its own reasoning.

D. Finishing the Workflow

We integrate LLM netlist generation with the open-source MFDA toolchain, OpenMFDA, to demonstrate a complete workflow [6]. OpenMFDA takes a structural Verilog netlist as input and generates a standardized microfluidic chip model ready for 3D printing. The OpenMFDA rendering of a mixer tree is shown in Figure 2. With this compatibility, we enable non-experts in the EDA or MFDA domains to generate ready-to-print devices using natural language as a frontend to an automated framework.

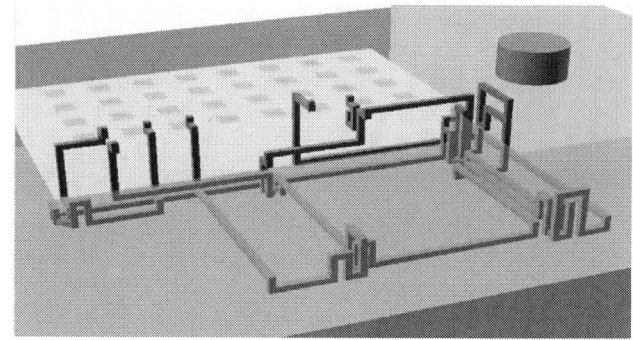

Fig. 2: A three-dimensional rendering of a microfluidic chip designed to mix six solutions in parallel, produced using OpenMFDA with a Qwen2-generated system of six mixers.

IV. PERFORMANCE EVALUATION AND ANALYSIS

A. Experimental Setup

We assume that end users of this methodology lack the technical expertise to modify the model's output, making it crucial to generate a valid Verilog netlist on the first attempt. Furthermore, if this work were to extend beyond a single prompt, each subsequent interaction with the model would yield varying results, adding complexity and hindering our ability to report consistent quantitative outcomes effectively. Therefore, for our experiment, we selected five fundamental chemical manipulations commonly used in various laboratory procedures as user prompts, as detailed in Table II. We then refined our prompt methodology to consistently generate representative, syntactically accurate structural netlists based on these foundational benchmarks. To assess the accuracy of the module generation, we utilized two key parameters. The "microfluidic function" parameter assesses whether the output demonstrates a logically correct flow of components and accurately infers the appropriate microfluidic primitives according to fundamental microfluidic design principles. The "Verilog syntax" parameter evaluates the output for accuracy and synthesizability as Verilog code.

TABLE IV: Pass@1 (single-short success rate) on MFD benchmarks testing for Verilog syntax and microfluidic functionality.

Specification	Verilog Syntax					Microfluidic Function				
Pass@1[%]	CodeLlama	Codestral	Llama3	Qwen2	DeepDeek-R1	CodeLlama	Codestral	Llama3	Qwen2	DeepDeek-R1
Benchmark 6	80%	100%	100%	100%	100%	80%	100%	100%	100%	100%
Benchmark 7	80%	80%	100%	100%	100%	100%	100%	100%	100%	100%
Benchmark 8	0%	20%	60%	80%	100%	0%	40%	80%	100%	100%
Benchmark 9	80%	100%	60%	100%	80%	100%	100%	100%	100%	100%
Benchmark 10	0%	100%	60%	60%	60%	0%	100%	80%	100%	100%

We identified the following as the most common points of error and subsequently structured our system prompt to specifically address these issues:

1) Complete Verilog Initialization: ensuring adherence to proper Verilog syntax, ultimately determining if the netlist is compatible with existing tools that accept Verilog.

2) Correctly Defined Primitives: ensuring the LLM property infers the correct microfluidic primitives, given the steps listed in the chemical manipulations.

3) Correct Connections: validating whether the LLM accurately connects the ports of each primitive to implement the final microfluidic system.

4) Correct Component Flow: determining whether the resulting netlist accurately emulates the specified benchmark.

B. Experimental Results and Discussion

1) Prompting Methodology Outcomes: To assess the developed prompting methodology, we evaluate a new set of test benchmarks (user prompts) detailed in Table III. Each benchmark was executed five times, and the generated netlists were evaluated based on the specified Verilog syntax and microfluidic function parameters. These results are compiled in Table IV, presented as a percentage of passing outputs. We observed that, in some cases, the correct microfluidic primitives were inferred, but the Verilog syntax was incorrect. Conversely, there were instances where the generated netlist was syntactically correct yet did not conform to the defined microfluidic function.

On average, the microfluidic correctness of each model's output was better than their syntactical correctness. This uncovers a limitation of the prompting methodology, which struggled to enable the LLM to apply correct syntax across similar benchmarks. The system prompt included sections aimed at teaching the LLM proper syntax, such as example prompts followed by correct netlists and specifications for calling modules and creating wires. One way to address these syntactical errors is to fine-tune the model on the specifics of Verilog syntax. Alternatively, including a more comprehensive library of Verilog syntax in the RAG procedure could also help improve output quality without further extending the system prompt.

2) Comparison between LLMs: Qwen2 and DeepSeek-R1 outperformed all other models, benefiting from the highest number of parameters and the largest context size. However,

Codestral, despite having fewer parameters than Llama3 and CodeLlama, performed better due to its context length being four times larger. This is crucial given the extensive system prompt, as a larger context size allows the LLM to connect different parts of the prompt and understand their relationships.

In Table IV, our most important metric, correct component flow, passes for all but one benchmark. This shows that large-language models that are *not* fine-tuned are able to perform extremely well in the design of microfluidic architecture. While this paper focuses on using currently open and available LLMs, building a model fine-tuned for Verilog syntax and microfluidic design automation could see syntactical accuracy improved beyond the 88% demonstrated by Qwen2 and Deepseek-R1.

C. Future Work

Our results demonstrate that the LLMs are capable of creating basic systems with user prompts; future work can be focused on creating more robust models for syntactical issues which will allow for more complex systems to be built with less detailed user prompts.

Recent works have proposed effective automatic verification techniques to produce near-perfect syntax for HDL [ ]. These works use a feedback system where each output is passed through an EDA tool, which checks it for syntactical errors. The prompt is automatically iterated with corrective prompts. This process is repeated until the final output is syntactically correct. Implementing a similar verification system could significantly increase the syntactical accuracy of models. Combined with the higher functionality of our RAG system which is tuned to allow greater functional accuracy, LLMs could become extremely accurate and consistent for MFDA netlist generation.

V. Conclusion

We have presented the initial and foundational demonstration for generating system-level descriptions of microfluidic devices by using a well-adopted hardware description language and advanced "language-to-hardware" capabilities of existing open-source LLMs. Our LLM prompting methodology achieved netlists with combined microfluidic functionality and Verilog syntactical accuracy of 88%. This approach offers a streamlined and accessible method for researchers across various fields to harness the power of microfluidic platforms without requiring specialized knowledge of MFDA tools and corresponding device descriptions.

979-8-3315-9813-6/25 $31.00 © 2025 IEEE

REFERENCES

[1] N. Convery and N. Gadegaard, "30 years of microfluidics," *Micro and Nano Engineering*, vol. 2, pp. 76–91, 2019.

[2] X. Huang, T.-Y. Ho, *et al.*, "Computer-aided design techniques for flow-based microfluidic lab-on-a-chip systems," *ACM Computing Surveys*, vol. 54, no. 5, pp. 1–29, 2022.

[3] J. McDaniel, W. H. Grover, and P. Brisk, "The case for semi-automated design of microfluidic very large scale integration (mVLSI) chips," in *Design, Automation & Test in Europe Conference & Exhibition (DATE), 2017*, 2017, pp. 1793–1798.

[4] T.-M. Tseng, M. Li, *et al.*, "Columba s: A scalable co-layout design automation tool for microfluidic large-scale integration," in *Proceedings of the 55th Annual Design Automation Conference*, ser. DAC '18, New York, NY, USA: Association for Computing Machinery, 2018, pp. 1–6.

[5] H. Huang and D. Densmore, "Fluigi: Microfluidic device synthesis for synthetic biology," *ACM Journal on Emerging Technologies in Computing Systems*, vol. 11, no. 3, 26:1–26:19, 2015.

[6] A. Snelgrove, D. Wakeham, *et al.*, "Openmfda: Microfluidic design automation in three dimensions," *Design Automation and Test in Europe*, 2025.

[7] M. Liu, N. Pinckney, *et al.*, "Invited paper: Verilogeval: Evaluating large language models for verilog code generation," in *2023 IEEE/ACM International Conference on Computer Aided Design (ICCAD)*, 2023, pp. 1–8.

[8] J. Blocklove, S. Garg, *et al.*, "Chip-chat: Challenges and opportunities in conversational hardware design," in *2023 ACM/IEEE 5th Workshop on Machine Learning for CAD (MLCAD)*, 2023, pp. 1–6.

[9] F. Aditi and M. S. Hsiao, "Validatable generation of system verilog assertions from natural language specifications," in *2023 Fifth International Conference on Transdisciplinary AI (TransAI)*, 2023, pp. 102–109.

[10] H. Pearce, B. Tan, and R. Karri, "Dave: Deriving automatically verilog from english," in *Proceedings of the 2020 ACM/IEEE Workshop on Machine Learning for CAD*, ser. MLCAD '20, Virtual Event, Iceland: Association for Computing Machinery, 2020, pp. 27–32.

[11] S. Thakur, B. Ahmad, *et al.*, "Benchmarking large language models for automated verilog rtl code generation," in *2023 Design, Automation Test in Europe Conference Exhibition (DATE)*, 2023, pp. 1–6.

[12] E. Nijkamp, B. Pang, *et al.*, "Codegen: An open large language model for code with multi-turn program synthesis," *arXiv preprint arXiv:2203.13474*, 2022.

[13] Y. Lu, S. Liu, *et al.*, *Rtllm: An open-source benchmark for design rtl generation with large language model*, 2023. arXiv: 2308.05345 [cs.LG].

[14] O. Developers, *Ollama Website*, https://ollama.com, 2024.

[15] M. ul Islam, H. Sami, *et al.*, *Eda-aware rtl generation with large language models*, 2024. arXiv: 2412.04485 [cs.AR].

PVT-Robust Analog Control Stage for Buck DC-DC Converters in Open-Source SKY130

Giordano de Moraes Rossa[1], Henrique Beque[1], Iuri Tinti[1],
Jorge Marín[2], Juan Pablo Martínez Brito[3]

[1]CI Inovador Program - Federal University of Rio Grande do Sul (UFRGS), Porto Alegre, Brazil
[2]Advanced Center for Electrical and Electronic Engineering (AC3E), Valparaiso, Chile
[3]CEITEC Semiconductores, Porto Alegre, Brazil

Abstract—**This paper presents a post-layout validated analog control loop for a 1.2 V Buck DC-DC converter tape-out ready using the SKY130 open-source 130 nm CMOS process. The system integrates five modular blocks: a Constant-g_m Current Reference, a Bandgap Voltage Reference, a Miller OTA-based Error Amplifier, a Non-overlapping Gate Driver, and a Sawtooth Ramp Generator. The design achieves a peak efficiency greater than 85%, output ripple below 10 mV, and stable closed-loop operation at 100 kHz from a 1.8 V supply, confirmed across 36 PVT (Process, Voltage, Temperature) corners. Each block was independently verified with DRC/LVS-clean layouts. The design occupies 0.06 mm^2 and consumes 98 µA. The system showcases the viability of robust analog IP design using open-source flows for SoC power management integration.**

Keywords—*Buck converter, Bandgap reference, Ramp generator, Current reference, Low-power analog design, SKY130 CMOS.*

I. INTRODUCTION

Advances in semiconductor technology call for compact, energy-efficient voltage regulators for SoC integration. Buck converters are widely adopted for their simplicity and efficiency [1], [2]. On-chip control integration reduces I/O and improves regulation [3]. This work presents a post-layout verified analog control system for a 1.2 V Buck converter operating from a 1.8 V supply, implemented in the open-source SKY130 130 nm CMOS process [4]. The system integrates five blocks: a constant-g_m current reference and gate driver [5], a bandgap voltage reference with OTA-based error amplifier [6], and a sawtooth ramp generator for PWM (Pulse Width Modulation) [7]. A proportional voltage-mode control loop is adopted to minimize quiescent power and area. The full architecture is shown in Fig. 1, highlighting the Power and Control stages, the latter being the focus of this paper.

The modular design achieves robust PVT (Process, Voltage, Temperature) operation with sub-500 µW power and compact area. Key specifications include: a 100 kHz ramp with 0.18–1.62 V swing; gate driver with ≥35 ns deadtime; current reference delivering 10 µA with >100 dB PSRR and <5 µs

The authors would like to thank the CI Inovador program, UFRGS, and CEITEC Semicondutores for infrastructure support, design tool access, and technical guidance. This research was partially funded by AC3E (ANID/BASAL/AFB240002), by ANID + Fondecyt de Exploración 2025 + 13250147, and by ANID + FONDECYT Initiation Research Project 11240947.

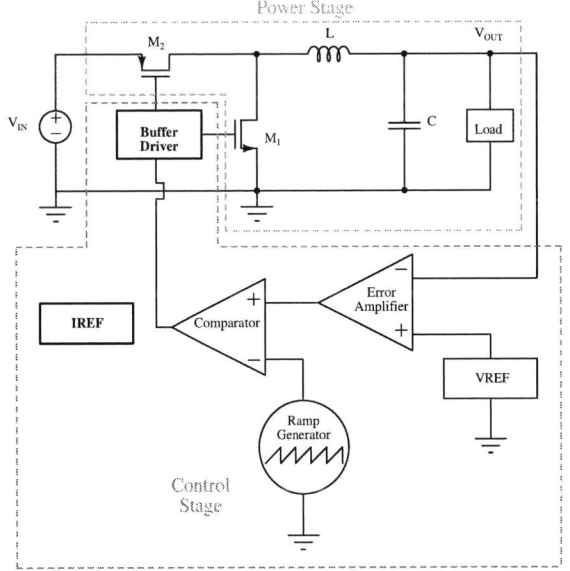

Fig. 1. Proposed Buck converter architecture with integrated analog control.

startup; voltage reference of 1.1 V with <60 ppm/°C of temperature coefficient (TC) ; and an error amplifier based on Operational transconductance amplifier (OTA) achieving ~80 dB gain, 70° phase margin, and 30 MHz unity-gain bandwidth. Sections II–IV describe the analog blocks. Section V presents PVT results; Section VI concludes.

II. CURRENT REFERENCE AND GATE DRIVER

A. Constant-g_m Current Reference (I_{REF})

The current reference adopts a modified Vittoz topology [8], generating a 10 µA output current using Proportional-To-Absolute-Temperature (PTAT) biasing and active source degeneration. Matched NMOS transistors develop a temperature-proportional voltage across a precision polysilicon resistor, mirrored by a cascode stage to improve PSRR. A start-up circuit ensures proper initialization, avoiding metastable zero-current states [9]. The core is scaled to a 10 µA output target and exhibits a gm variation of 1.911 µS ($\approx$ 2.26 %) over -40 °C

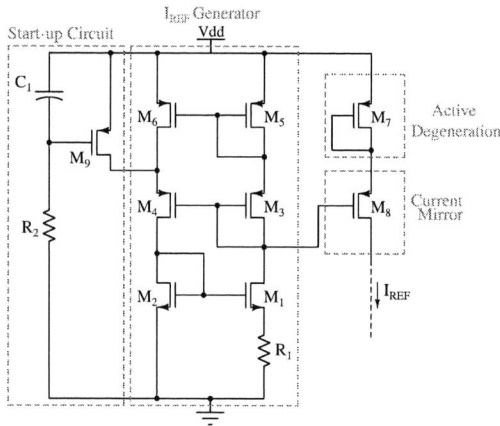

Fig. 2. Modified constant-g_m current reference with integrated start-up circuit and enhanced output stage.

to 85 °C; under ±10% supply variation, *gm* stayed below 4 µS. PSRR improves up to about 111 dB in the final topology. Temperature coefficient remained within 2.5 kppm/°C, and startup completed reliably within 3.6 µs under worst-case conditions. Monte Carlo simulations with 1000 samples verified resilience to mismatch and process variations. The final layout occupies 76.57 µm × 58.21 µm and consumption is 17 µA.

B. Gate Driver

The non-overlapping buffer driver ensures safe power-stage switching by inserting a controlled deadtime between transitions. A skewed inverter chain implements tri-state logic to prevent simultaneous PMOS/NMOS conduction. The schematic is shown in Fig. 3. Simulations were performed with a 100 pF

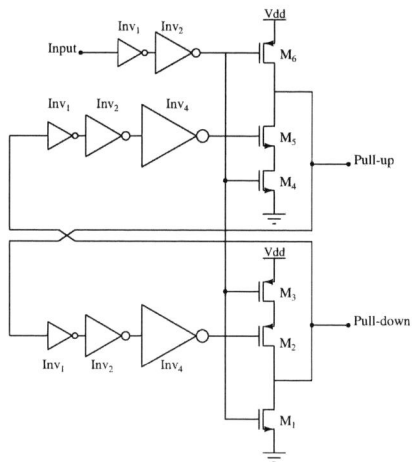

Fig. 3. Non-overlapping buffer driver with shoot-through suppression logic.

capacitive load, mimicking the power transistors. The nominal deadtime was chosen near 50 ns to place the design near the center of the simulated spread while preventing cross-conduction. Post-layout corner simulations produced 36–58 ns

deadtime across 36 PVT corners with no shoot-through observed, confirming the nominal choice meets protection and efficiency constraints. At 100 kHz switching, this is only 0.5% of the period and does not compromise peak efficiency (>85%) or ripple (<10 mV) as reported. This aligns with established guidelines that optimal deadtime minimizes diode conduction while preventing cross-conduction (e.g. [2], [10]). The layout area is 14.82 µm × 13.37 µm and the consumption is about 540 nA.

III. VOLTAGE REFERENCE AND ERROR AMPLIFIER

A. Bandgap Voltage Reference (V_{REF})

The voltage reference circuit implements a classic bandgap topology, combining Complementary-To-Absolute-Temperature (CTAT) and PTAT terms. As shown in Fig. 4, ΔV_{BE} (Q_1-Q_2) is used to generate a PTAT current, added with a CTAT base-emitter voltage (Q_3) to yield a temperature-stable output. Post-layout simulations show an output voltage

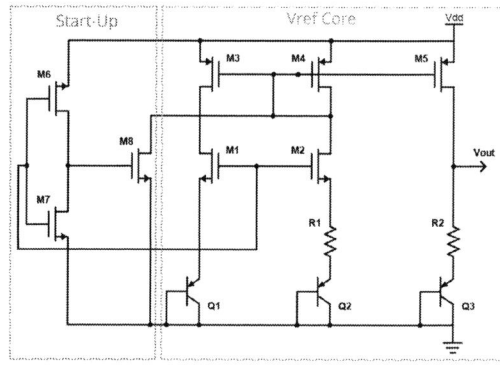

Fig. 4. Bandgap reference architecture showing Vref Core and Startup Circuit.

of 1.08 V with a temperature coefficient of 57 ppm/°C over -40 °C to 85 °C and supply variation under 2% across 1.62 V to 1.98 V. Monte Carlo (100 runs) showed a standard deviation of 2.1 mV. The layout area is 207.64 µm × 106.05 µm, with 3 µA of power consumption.

B. Miller OTA-based error amplifier

The OTA adopts a two-stage Miller-compensated architecture (Fig. 5b), with a differential pair, current mirror load, and compensation capacitor introducing the dominant pole. The error amplifier (Fig. 5a) is implemented as a simple inverter Op Amp configuration. Post-layout simulations report a DC gain of 60 dB, phase margin of 67°, and 2.1 MHz unity-gain bandwidth, with PSRR exceeding 80 dB. Monte Carlo simulations indicated low sensitivity to mismatch. The layout area is 101.24 µm × 66.2 µm with a consumption of 26 µA.

IV. RAMP GENERATOR

The ramp generator produces a linear sawtooth waveform serving as the PWM reference. It targets 100 kHz with a voltage swing from 10–90% of the 1.8 V supply (defined as $0.1 V_{DD}$ and $0.9 V_{DD}$, respectively). The circuit charges

979-8-3315-9813-6/25 $31.00 © 2025 IEEE

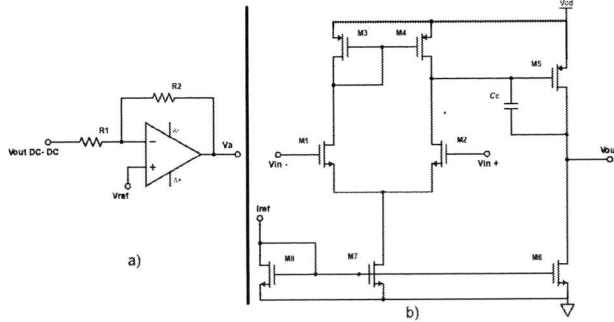

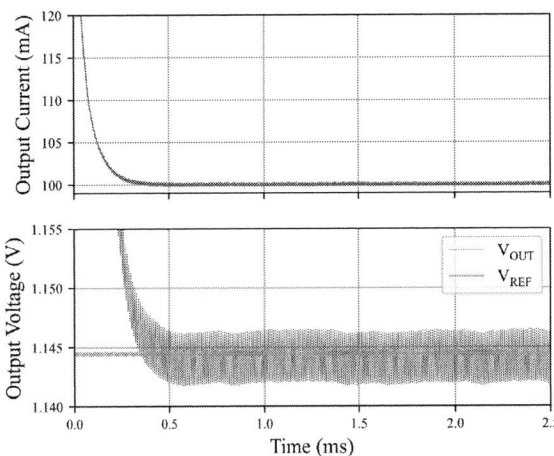

Fig. 5. a) Error amplifier schematic. b) Miller-compensated OTA topology.

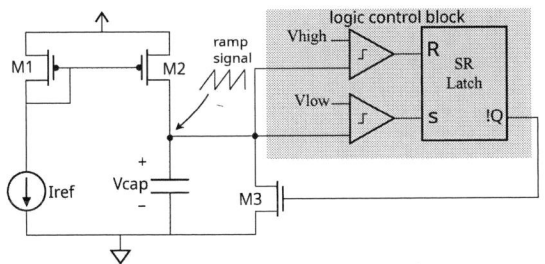

Fig. 6. Ramp generator architecture using current mirror, comparator, and SR latch logic.

Fig. 7. Top: Output current settling curve. Bottom: Output voltage ramp-up and regulation compared to V_{REF}.

an 8 pF capacitor using the current source. When the ramp reaches $0.9\,V_{DD}$, a comparator triggers a rapid discharge via a PMOS switch. An SR latch ensures stable toggling. Two rail-to-rail comparators (NMOS and PMOS input pairs) improve transition sharpness near the supply rails. Post-layout simulations across 36 PVT corners showed robust performance, with operating frequency ranging from 95.6 kHz to 102.1 kHz. The rising edge spans 9.78 μs at a slew rate of 160.7 kV/s, and the falling edge occurs in 143 ns at 11.4 MV/s. CMRR at 100 kHz was 36.0 dB and 75.1 dB for the high- and low-threshold comparators, respectively. The total layout area is 83.3 μm × 70.8 μm and it consumes 22 μA.

V. Top-Level Simulation Results and Layout

This section presents the post-layout top-level simulation results of the integrated Buck DC-DC converter, incorporating both the control and power stages (as shown in Fig. 1). M_2 and M_1 power-stage transistors are sized at $W/L = 4.38\,\mu m/500\,nm$ with multipliers of 1734 (PMOS) and 970 (NMOS). LC values for the Power Stage are L=21 μH and C=42 μF.

Fig. 7 shows the output current stabilization and output voltage ramp-up under typical conditions. The converter reaches steady state, with the output voltage converging to approximately 1.145 V, closely tracking the 1.1 V reference, within 10 μs. No overshoot is observed, confirming effective loop compensation. In steady state, the voltage ripple remains below 10 mV$_{pp}$.

Figure 8 presents the measured power conversion efficiency

versus output current for three PVT conditions: typical (V_{DD}, TT, 27 °C), worst power corner (1.1 V_{DD}, FF, 85 °C), and worst delay corner (0.9 V_{DD}, SS, –40 °C). Under typical conditions, the peak efficiency reaches 86.77% at a load of 41.3 mA (load 31 Ω). For the worst-power corner, the peak shifts rightward to 62.8 mA with an improved maximum efficiency of 89.2% (load 21 Ω). The worst-delay scenario exhibits significant degradation in regulation, yet maintains a peak efficiency of 85.5% near 31.7 mA (load 41 Ω). Beyond that, efficiency drops sharply, reflecting the limited voltage headroom at low supply and slow devices.

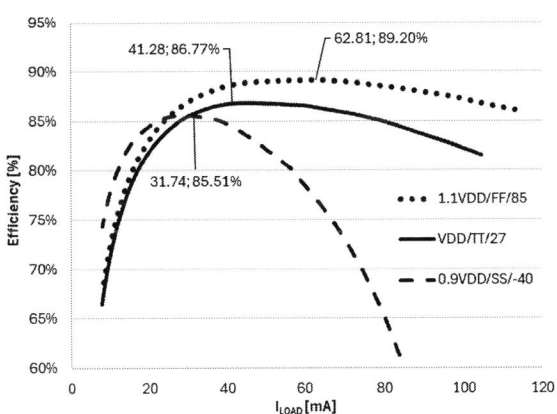

Fig. 8. Efficiency versus output current for three PVT scenarios.

The corresponding output voltage characteristics are shown in Figure 9. At high loads, the $0.9\,V_{DD}$/SS/–40 °C curve exhibits nonlinear behavior, with the output voltage dropping to below 0.9 V as I_{LOAD} increases, confirming the limitations of control headroom under slow, low-voltage-temperature conditions. In contrast, the other corners demonstrate improved linearity and regulation, with output voltage maintained above 1.15 V for most of the current range.

979-8-3315-9813-6/25 $31.00 © 2025 IEEE

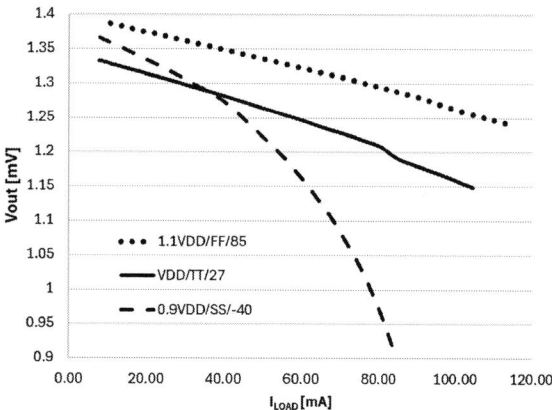

Fig. 9. Output voltage versus output current for different corners, illustrating regulation behavior and load dependence.

Figure 10 shows the measured output ripple (V_{pp}) across the same load sweep. In all three corners, the ripple voltage generally decreases with increasing I_{LOAD}, indicating effective feedback loop action. For the $0.9\,V_{DD}$/SS/–40 °C condition, the ripple initially trends downward, but deviates from linear behavior beyond 60 mA, coinciding with degraded voltage regulation and reduced loop gain at higher loads.

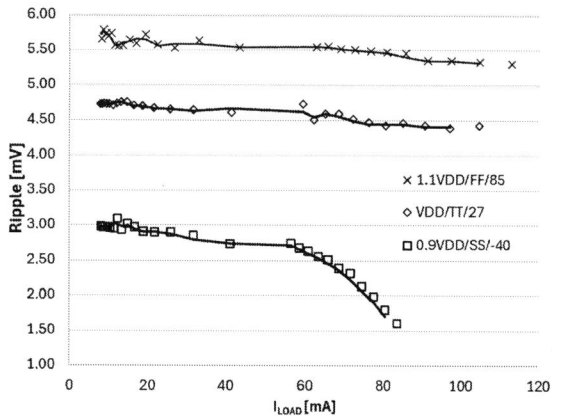

Fig. 10. Output voltage ripple (V_{pp}) versus load current. Decreasing ripple with higher current reflects effective feedback operation.

Figure 11 shows the preliminary top-level floorplan. While full hierarchical integration into a unified GDSII layout was not completed, each block was independently implemented, verified, and validated via DRC and LVS checks.

Analog layout techniques were used, such as common-centroid geometries, dummy structures, symmetric routing, and guard rings, with a total layout area of $207.64\,\mu$m x $276.10\,\mu$m. The whole control stage consumes about 98 µA typically.

Table I compares this work with recent literature. Our analog-controlled Buck converter, implemented in the open-source SKY130 process, achieves over 85% efficiency with sub-10 mV ripple at 100 kHz. A distinguishing feature is the thorough PVT analysis across 36 corners, including edge-case

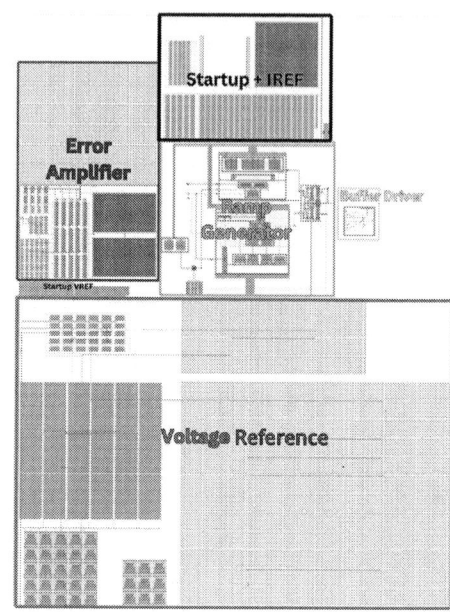

Fig. 11. Top-level floorplan showing the placement of the individual analog blocks. All blocks were verified for DRC and LVS compliance.

TABLE I.   COMPARISON WITH PRIOR WORKS. N/R MEANS NOT REPORTED.

	This Work	[11]	[12]	[13]
CMOS Technology [nm]	**130 (SKY130)**	130	180	180 BCD
Operating Frequency [MHz]	**0.1**	100	2.5	5
Input/output voltages [V]	**1.8/1.2**	1.2/0.9	3.3/0.6–2.0	5/1.2
I_{Load} Max [A]	**0.12**	1.2	N/R	5
Peak Efficiency [%]	**>85**	82.4	95.4	95.3
Ripple [mV]	**<10**	N/R	43.4	N/R
Area [mm²]	**0.06**	1.25	1.89	N/R
Control consumption	**98 µA**	N/R	N/R	35.6 µA
PVT Analysis	**36 corners**	N/R	N/R	N/R

conditions such as $0.9\,V_{DD}$/SS/–40 °C and $1.1\,V_{DD}$/FF/85 °C. None of the referenced works reports such an analysis. Furthermore, our area is an order of magnitude smaller than that of [11], with slightly better efficiency.

VI. CONCLUSION

A compact, modular analog control stage for a 1.2 V Buck DC-DC converter has been designed and post-layout verified in the SKY130 open-source CMOS process. The architecture integrates five core blocks, each independently implemented, DRC/LVS-verified, and evaluated over 36 PVT corners. The system achieved peak efficiency above 85%, sub-10 mV ripple, and robust regulation, even under worst-case conditions. The complete control stage occupies just $0.06\,\mathrm{mm}^2$ and consumes less than 100 µA. These results validate the potential of open-source analog IP development for low-cost, tapeout-ready SoC power management. Future work includes implementing a proportional-integral-derivative (PID) control loop and the dynamic analysis of the closed-loop system.

REFERENCES

[1] S. Chang and K. Lee, "A low-power cmos analog integrated circuit for dc-dc converter applications," in *Proceedings of the IEEE International Symposium on Circuits and Systems*, 2000, pp. 49–52.

[2] R. J. Didonet, "Conversor buck cmos com controle pwm de frequência fixa em modo de tensão," Universidade de Brasília, Faculdade de Tecnologia, Trabalho de conclusão de curso (TCC), Dec 2015, defesa: 27-Sep-2016. Available: https://bdm.unb.br/handle/10483/14786.

[3] L. Wei, L. Zheng, L. Gu, and Z. Zhai, "A high-efficiency wide-load-range dc-dc buck converter with adaptive ripple-based on-time control," *IEEE Transactions on Power Electronics*, vol. 27, no. 6, pp. 2868–2872, 2012.

[4] D. Lee, S. Yoo, and C. Kim, "A fully integrated low-power high-efficiency dc-dc converter for portable applications," in *Proceedings of the IEEE International Solid-State Circuits Conference*, 2004, pp. 526–527.

[5] G. de Moraes Rossa and J. P. M. Brito, "Design of buffer driver and current reference for dc-dc converter control on sky130 cmos process," *Phase 1 Final Project, CI Inovador Polo UFRGS (INOVA-ME)*, 2025.

[6] H. F. Beque and J. P. M. Brito, "Design and implementation of a closed-loop controller for dc-dc buck converter in sky130nm cmos process," *Phase 1 Final Project, CI Inovador Polo UFRGS (INOVA-ME)*, 2025.

[7] I. M. Tinti and J. P. M. Brito, "Low-power ramp generator design for buck dc-dc converter using sky130nm cmos process," *Phase 1 Final Project, CI Inovador Polo UFRGS (INOVA-ME)*, 2025.

[8] E. Vittoz and J. Fellrath, "Cmos analog integrated circuits based on weak inversion operations," *IEEE Journal of Solid-State Circuits*, vol. 12, no. 3, pp. 224–231, 1977.

[9] T. H. Lee, "Preface to the second edition," in *The Design of CMOS Radio-Frequency Integrated Circuits*. Cambridge: Cambridge University Press, Dec. 2003, pp. xiii–xiv.

[10] Texas Instruments, "Optimizing efficiency through dead time control," Texas Instruments, Tech. Rep. SNVA815, 2012, application Report. [Online]. Available: https://www.ti.com/lit/pdf/snva815

[11] C. Huang and P. K. T. Mok, "A 100 mhz 82.4% efficiency package-bondwire based four-phase fully-integrated buck converter with flying capacitor for area reduction," *IEEE Journal of Solid-State Circuits*, vol. 48, no. 12, pp. 2977–2988, 2013.

[12] H.-Y. Huang, H. M. Villaruz, and N. M. Mapula, "High-efficiency low-emi buck converter using multistep pwl and pvt insensitive oscillator," *IEEE Transactions on Power Electronics*, vol. 37, no. 8, pp. 9325–9332, 2022.

[13] Z. Tang, R. P. Martins, and M. Huang, "A fast-transient buck converter with one-cycle-balancing control for single and consecutive load steps," in *2025 IEEE Custom Integrated Circuits Conference (CICC)*, 2025, pp. 1–3.

979-8-3315-9813-6/25 $31.00 © 2025 IEEE

From Secure Storage to Compute-in-Memory: A Versatile Memory System using 1T-nC FeRAM

Rakesh Acharya*, Rudra Biswas*, Jiahui Duan†, Prapti Panigrahi*, Kai Ni†, and Vijaykrishnan Narayanan*
*The Pennsylvania State University, University Park, PA, USA
†University of Notre Dame, South Bend, IN, USA
{rqa5569, vxn9}@psu.edu

***Abstract*—This study explores the versatility of 1T-nC ferroelectric RAM (FeRAM) memory capable of operating in both secure storage and highly energy-efficient Compute-in-Memory (CiM) modes without incurring any extra design overhead. In contrast to conventional, volatile 1T-1C Dynamic Random Access Memory (DRAM), having significant energy cost due to the continuous refresh cycles, FeRAM presents a robust non-volatile memory solution, enabling efficient data retention with minimal power consumption. Bitwise logic operations (AND, OR) are efficiently implemented using a single, unmodified 1T-nC FeRAM cell and single-row activation (SRA), while NOT operation is realized by a modified sense amplifier. These facilitate bulk bitwise operations with significant energy benefits compared to multi-row based DRAM solutions. Furthermore, we demonstrate key-based (K) cryptographic operations which help in storing Plain Text (PT) in Cipher Text (CT) format tightly coupled with the memory write/read cycles in secure-mode operation with minimal overload while maintaining robust security. In addition, the advantages of vertical 3D integration of 1T-nC FeRAM are evaluated, and thermal profiling of the remnant polarization (P_R) is done. Furthermore, we demonstrate that 1T-nC FeRAM memory reduces energy consumption by 31% compared to DRAM by assessing 8 real-world data-intensive workloads utilizing bulk bitwise operations.**

***Index Terms*—Ferroelectric capacitor, Data encryption, Compute-in-Memory, 3D integration, Vertical stack.**

I. Introduction & Background

The exponential growth of data-intensive applications in the era of Artificial Intelligence (AI) has placed unmatched demands on computing architectures in terms of performance, energy efficiency, and data security. The conventional von Neumann architecture is proving inefficient in frequent transferring of data between memory and processing units and incurs notable energy and latency overheads. To overcome this, CiM architectures have surfaced which enable computation straight inside memory arrays, reducing data transfer, and enhancing general system efficiency. CiM has been thoroughly investigated not only using conventional volatile memory technologies like SRAM [1] [2], DRAM [3], but also emerging NVMs like RRAM [4], MRAM [5], PCM [6], Ferroelectric memories [7] [8].

Among many memory technologies, FeRAM has drawn great interest due to its non-volatility, DRAM like high density, low-energy switching properties, high endurance, and CMOS process compatibility. In addition to the traditional 1T-1C structure, the continuous drive to enhance memory density and 3D integration have led to 2T-nC, 1T-nC FeRAM cell structures. While 2T-nC architecture demonstrates in-memory logic capabilities along with its quasi non-destructive readout (QNRO) [9] [10], the two transistors increase area footprint. Compared to the 1T-1C structure, the 1T-nC FeRAM offers a more compact and scalable design with higher effective storage density by having multiple FE capacitors connected to a single access transistor. Although read operation remains destructive, recent developments can be utilized to enable efficient write-back mechanisms [11], alleviating performance penalties during read cycles.

	1T-1C DRAM	1T-1C FeRAM	1T-nC FeRAM (This Work)
Memory Cell Structure	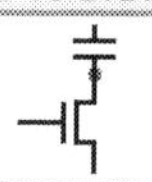	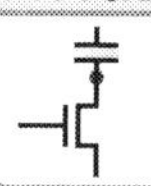	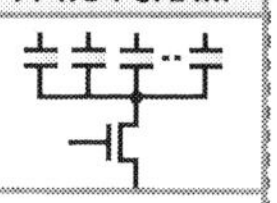
Data Retention	Volatile	Non-volatile	Non-volatile
Sensing Method	Destructive	Destructive	Destructive
Logic-in-Memory	✓	✓	✓
Memory Density	Limited	Limited	Enhanced
Bulk-Bitwise Operation Energy	High	Low	Lower

Fig. 1: Comparison of 1T-1C DRAM, 1T-1C FeRAM and 1T-nC FeRAM for CiM operation.

The non-volatility of new NVM technologies, including FeRAM, creates new security risks [12] [13] that previously did not exist in volatile counterparts such as SRAM and DRAM. Persistent data retention makes FeRAM vulnerable to "cold-boot" [14] and "stolen memory" [15] attacks, which pose risks to information-sensitive systems such as military applications, financial infrastructure, and consumer electronics. Conventional encryption techniques such as Advanced Encryption Standard (AES) create significant energy and computational overhead, particularly when implemented externally or near memory. Though they still struggle with granularity, programmability, or integration complexity, recent work in in-memory encryption has tried to close this gap [16].

This work presents a 1T-nC FeRAM-based memory architecture that simultaneously supports CiM and hardware-level encryption/decryption operations using a unified, area-efficient design. The approach enables AND/OR logic and secure key-based data transformation within the memory array using Fe-CAP polarization switching and the proposed SRA technique.

979-8-3315-9813-6/25 $31.00 © 2025 IEEE

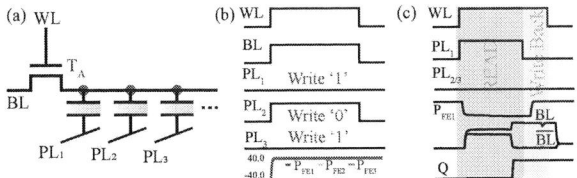

Fig. 2: (a) 1T-nC FeRAM cell structure (b) Write operation stimulus for data pattern '101' in 1T-3C FeRAM cell (c) Stimulus for read operation from cell X in 1T-3C FeRAM cell.

II. PROPOSED DESIGN & SIMULATION

In this work, we demonstrate bitwise logic operations using 1T-nC FeRAM cells, building on prior approaches in DRAM [11] and 2T-(n+1)C FeRAM [17]. Furthermore, we show that 1T-nC FeRAM cells can intrinsically convert PT into CT and successively decrypt them during readout using a specific key, without any additional circuitry.
For circuit simulation, the 1T-nC FeRAM cell has been modeled using the ASU 45 nm technology node [18] along with the metal-ferroelectric-metal (MFM) capacitor adopted from [19], which precisely mimics key ferroelectric properties like performance scaling, stochastic and domain switching, etc, and validated against Micron's experimental data [20]. The entire simulation setup has been built in the Cadence Virtuoso environment with SPECTRE netlists to ensure higher accuracy and compatibility within Cadence framework.

A. Structure and Basic Operation of 1T-nC FeRAM

A single 1T-nC FeRAM cell, as illustrated in Fig. 2(a), consists of an NMOS access transistor (T_A) connected to 'n' MFM FeCAPs, operates similarly to 1T-1C FeRAM but enables parallel storage of 'n' data per cycle, significantly enhancing data density.
During the write operation, Wordline (WL) voltage is raised, which connects Bitline (BL) and Plateline (PL) by enabling T_A. Once T_A is active, the BL is driven high and complementary data is applied to the PL. This assists in polarization switching in FeCAPs. Once data is written, BL is precharged for the next operation. For e.g., a '101' pattern can be stored by applying WL = 1, BL = 1, PL_1 = 0, PL_2 = 1, and PL_3 = 0, illustrated in Fig. 2(b).
During read operation, initially the BL is precharged to 0. T_A is enabled by increasing the WL voltage and target FeCAP PL voltage, leading to charge sharing between the MFM capacitor and the BL through destructive polarization switching. Successively, the built-up charge on the BL is fed into a sensing circuit against a complementary reference bitline ($\overline{BL}$), which resolves the differential voltage between the two by driving the BL to either '1' or '0', depicted in Fig. 2(c).

B. 1T-nC AND-OR Implementation

We leverage the inherent advantages of the 1T-nC FeRAM cell structure and explore plausible logical operations by introducing SRA technique. Triple-row activation (TRA) [11] can significantly increase power overhead and control logic

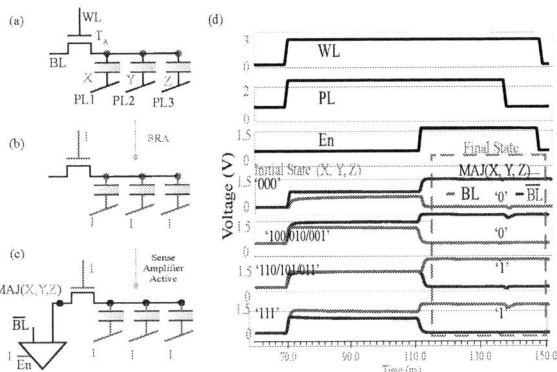

Fig. 3: (a) Initial state before SRA. (b) Cell under SRA, all three FeCAPs are activated by raising WL and PL voltages. (c) Sense Amplifier activated, Majority output available at the BL. (d) SPICE simulation of majority operation.

complexity as it necessitates three simultaneous WL to be activated. By modulating the reference voltage used for data sensing, the bitwise MAJORITY (MAJ) function can be implemented utilizing SRA and using only one 1T-nC cell where n is set to 3. MAJ function can be expressed by the following boolean logic -

$$MAJ(X, Y, Z) = Z.(X + Y) + \overline{Z}.(X.Y) \qquad (1)$$

As per (1), at least two positive polarization (P_{FE}) cells are needed to get a logical 1. Further, depending on the state of cell Z, 1T-nC enables bitwise AND or OR operations between cell X and cell Y, allowing logical computation within 1T-nC FeRAM cells. Fig. 3(a-c) describes the transition steps during the SRA mechanism using 1T-nC FeRAM cell. Initially, data is stored in cell X, Y, and Z, shown in Fig. 3a. BL is set to 0. Later, WL and PL of three cells are simultaneously activated, illustrated in Fig. 3b. Charge sharing starts due to destructive polarization switching and reflects on the BL. Successively, sense amplifier is activated and based on the reference voltage, read operation commences, Fig. 3c. Z = 0 produces bitwise AND output, while Z = 1 generates a bitwise OR.
The SPICE simulation results are presented in Fig. 3d. To validate the operation, three FeCAPs connected to the same T_A are initialized with all possible binary states ranging from '000' to '111'. When all three PLs are driven high simultaneously, the sense amplifier detects the resulting BL voltage by comparing it against a reference level. The final output captures most of the stored data, enabling AND-OR logic via a MAJ function.

C. Encryption-Decryption using 1T-nC

Polarization switching mechanism can be leveraged to convert stored PT to encrypted CT based on K in an 1T-nC FeRAM cell. In this encryption design, the entire memory is encrypted in blocks, utilizing one K (0/1) per block. Each block consists of two 1T-nC FeRAM cells, as illustrated in Fig. 4(a-c). The encryption mechanism is broken down in four steps. First, both rows (R1 & R2) are activated, and identical data is written to the corresponding cells, Fig. 4(a).

979-8-3315-9813-6/25 $31.00 © 2025 IEEE

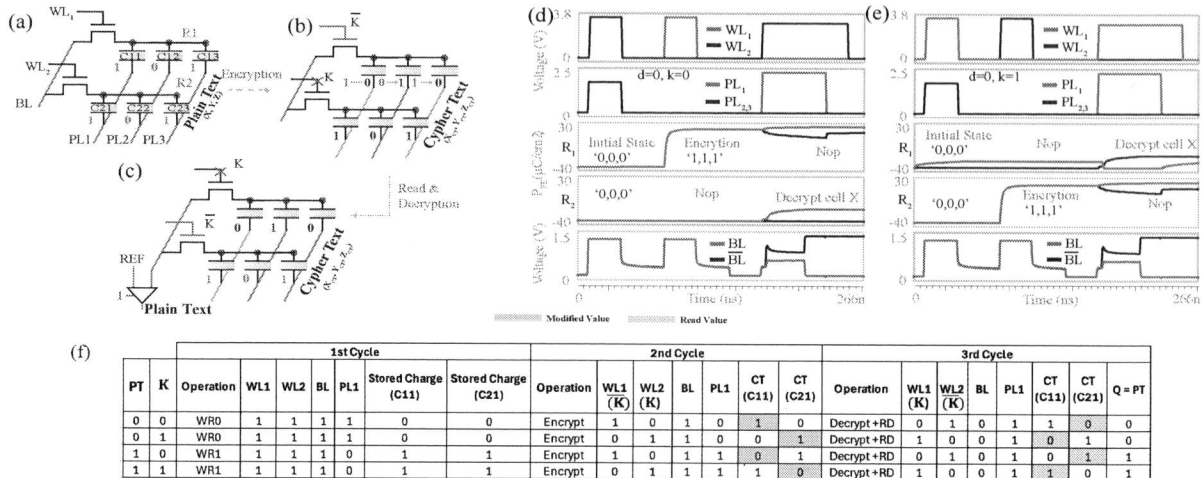

Fig. 4: 1T-nC FeRAM secure mode operation (a) Initial state (b) Encryption process based on K (c) Decryption using same K used in Encryption (d) SPICE simulation of Encryption-Decryption for Data=0 & K=0 (e) SPICE simulation of Encryption-Decryption for Data=0 & K=1 (f) Detailed logical table for proposed encryption-decryption scheme, shown with two FeCAPs.

PT	K	1st Cycle							2nd Cycle							3rd Cycle							Q = PT
		Operation	WL1	WL2	BL	PL1	Stored Charge (C11)	Stored Charge (C21)	Operation	WL1 ($\overline{K}$)	WL2 (K)	BL	PL1	CT (C11)	CT (C21)	Operation	WL1 (K)	WL2 ($\overline{K}$)	BL	PL1	CT (C11)	CT (C21)	
0	0	WR0	1	1	1	1	0	0	Encrypt	1	0	1	0	1	0	Decrypt +RD	0	1	0	1	1	0	0
0	1	WR0	1	1	1	1	0	0	Encrypt	0	1	1	0	0	1	Decrypt +RD	1	0	0	1	0	1	0
1	0	WR1	1	1	1	0	1	1	Encrypt	1	0	1	1	0	1	Decrypt +RD	0	1	0	1	0	1	1
1	1	WR1	1	1	1	0	1	1	Encrypt	0	1	1	1	1	0	Decrypt +RD	1	0	0	1	1	0	1

Second, BL is driven high, and depending on the K, R1 or R2 is selected. Third, PLs are programmed according to the PT: if PT is 0, PL is grounded; else PL is high. Fourth, for K = 0: WL1 is asserted, while WL2 is disabled, Fig. 4(b). Similarly, for K = 1: only WL2 is asserted. Thus, PT stored [X, Y, Z] in each FeCAP of 1T-nC cell undergo transformation $[X_{CT}, Y_{CT}, Z_{CT}]$ based on K, and cannot be retrieved without the knowledge of the corresponding K.

In the decryption process, Fig. 4(c), distinct logic values are applied to the gate of the T_A which differs from the encryption process and solely determined by the K used earlier.

More specifically, if K is '0', only WL2 is driven high, enabling access to R2 for the read operation, while WL1 remains inactive, Fig. 4(d). Conversely, if K is '1', WL1 is asserted high to access R1, and WL2 remains disabled. This selective K-dependent row activation aids in decrypting CT and read out the PT $[X_{CT} \rightarrow X]$, shown in Fig. 4(e). Fig. 4(f) shows an exhaustive logical table capturing all PT-K combinations for the proposed scheme, illustrated with two FeCAPs for simplicity.

D. Advantage in 3D Integration

The 1T-nC structure offers significant area reduction due to its compatibility with vertical 3D integration. Fig. 5(a) shows a vertically stacked 1T-nC FeRAM layout that enhances area efficiency. The planar 1T-3C array in Fig. 5(b) occupies $78\lambda \times 18\lambda$, while the vertical counterpart in Fig. 5(c) reduces the footprint to $23\lambda \times 16\lambda$, achieving a 3.81× area savings.

III. EXPERIMENTAL VERIFICATION

Encryption-decryption functionalities are verified with a FeRAM cell fabricated following the process flow in Fig. 6(a). Its schematic and cross-sectional TEM images are shown in Fig. 6(b-c). Transistor's width, length and FeCAP's area are 50 μm, 4 μm and 100 × 100 μm², respectively. The transfer curve of the transistor at V_d = 0.1V is shown in

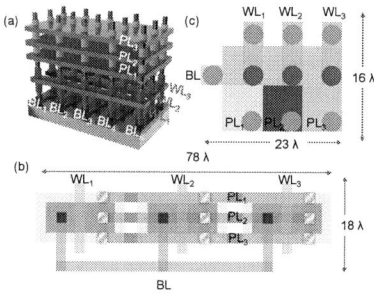

Fig. 5: (a) Vertical stacked 1TnC array. The layout size for three 1T-3C cells of (b) planar and (c) vertical stacked designs.

Fig. 6(d). The polarization-voltage loop in Fig. 6(e) confirms the ferroelectricity of the FeCAP, with P_R of 22.3 μC/cm². The impact of temperature on P_R is analyzed using our experimentally calibrated model from [22]. At 243K, 298K, and 360K, the P_R values are 18.66, 20.18, and 20.62 μC/cm², respectively, showing minor variation around the measured 22.3 μC/cm² at the applied voltage. The 1T-1C FeRAM cell's basic sensing operation is verified with write and read schemes for data '0' (D0) and '1' (D1) shown in Fig. 6(g). During a write, the access transistor is enabled by a 3V WL bias, while a 2.5V or –2.5V pulse is applied to the PL for storing '0' or '1', respectively. For reading, the transistor is again activated and a 2.5V pulse is applied to the PL, with the BL grounded at 0V. As shown in Fig. 6(i), the output current at BL distinctly differentiates between D0 and D1, confirming a successful readout. The encryption-decryption functionality was validated using a 2×1 FeRAM array. Fig. 6(h) illustrates the case with D0 and K = 0. Initially, PT = 0 is written by applying a 2.5V pulse to both PL$_1$ and PL$_2$. Encryption with K = 0 is performed by enabling WL$_1$, disabling WL$_2$, and applying –2.5V to both PL$_1$ and PL$_2$, inverting the data in cell 1. Decryption reverses the process by enabling WL$_2$, disabling WL$_1$, and applying 2.5V to both PLs. The data sensed from

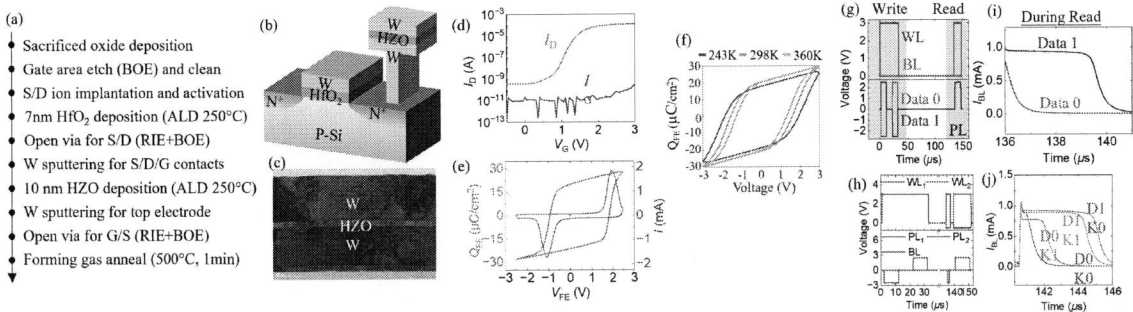

Fig. 6: (a) FeRAM fabrication process flow. (b) Fabricated FeRAM schematic (c) Cross-section TEM image of MFM capacitor (d) Transfer curve of fabricated access transistor. (e) PV loop of fabricated MFM capacitor. (f) Q-V loops under different temperatures. The basic operation (g) waveform and (i) readout results in FeRAM. (h) Encryption operation and (j) result after encryption and decryption in 2x1 FeRAM.

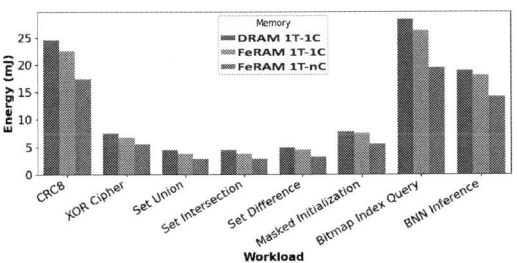

Fig. 7: Energy consumption for eight real-world, data-intensive workloads executed on DRAM, 1T-1C FeRAM, and 1T-nC FeRAM.

cell 2 remains '0'. Fig. 6(j) shows BL current for various data-key pairs, confirming correct encryption-decryption operation.

IV. WORKLOAD EVALUATION

The `pLUTo` simulator [23] has been used to model three memory configurations: 1T-1C DRAM/ FeRAM, and 1T-nC FeRAM with integrated refresh cycle analysis. Each memory type is simulated with an 8 GB capacity and a row size of 8192 bytes. Fig. 7 presents a comparative evaluation of energy consumption across eight data-intensive workloads—CRC8 [24], XOR Cipher [25], Set Union, Set Intersection, Set Difference, Masked Initialization [26], Bitmap Index Query [27], and BNN Inference [28]—each operating on a 1 GB dataset. The workloads are mapped to bulk bitwise operations executed using row-level primitives. DRAM and FeRAM 1T-1C employ the `AAP` (`ACTIVATE–ACTIVATE–PRECHARGE`) primitive [11], where the first `ACTIVATE (ACT)` triggers RowClone-based data transfer [29] , followed by a second `ACT` for a high energy TRA, and a `PRECHARGE (PRE)` to reset the bitlines. In contrast, FeRAM 1T-nC uses a sequence of `ACT`, `PLA` (Plate Line Activation), and `PRE`, with `ACT` addressing the target row and `PLA` toggling the FE capacitor plate lines to induce polarization switching, avoiding high-energy TRA operations with SRA instead. For example, an AND/OR operation in 1T-1C DRAM/ FeRAM needs 4 `AAP` operations, including high-energy TRA [11], consuming 28.4 nJ/28.65 nJ according to our simulation, while 1T-nC FeRAM can complete this with 3 `AAP` and 1 SRA (2 `ACT` + 3

`PLA` + 1 `PRE`) at 24.2 nJ, reducing energy overhead. Our analytical model uses energy parameters from simulations, estimating 2.82 nJ for DRAM, 2.85 nJ for FeRAM per `ACT`. For 1T-nC FeRAM the `ACT` energy includes 1.28 nJ for `PLA`. The `PRE` costs ∼ 0.038 nJ/row for all types. We assume that n-capacitors are stacked with negligible impact on area, however this maybe optimistic based on layout and needs to be revisited.

As shown in Fig. 7, the bitwise logic is extended to multiple workloads where FeRAM technologies substantially reduce energy consumption compared to DRAM, primarily by eliminating the need for periodic refresh cycles—a major contributor to DRAM's energy overhead. Specifically, 1T-1C FeRAM reduces energy by 9% and 1T-nC FeRAM by 31.46% compared to DRAM.

V. CONCLUSION

This work has demonstrated the potential of 1T-nC FeRAM that can serve as a compute-in-memory engine and a secure storage medium. Key-based encryption and decryption schemes ensure data confidentiality with almost no performance penalty. Concurrently, the novel SRA based CiM approach enables highly energy efficient computing using 1T-nC FeRAM and reduces energy consumption by 31% compared to DRAM configuration. The intrinsic non-volatility, and low-energy consumption of FeRAM position it as a clear and better alternative compared to its traditional DRAM counterpart. Notably, 1T-nC FeRAM's scalability is greatly increased by its compatibility with vertical 3D integration, making high-density, stacked memory-compute architectures perfect for environments with limited space. These qualities make 1T-nC FeRAM a promising option for low-power, secure computing in embedded platforms, IoT systems, and future edge devices.

ACKNOWLEDGMENT

This work is supported in part by PRISM under the JUMP 2.0 program (SRC) sponsored by DARPA, as well as by NSF Grants 2312886 and 2132918.

REFERENCES

[1] X. Si et al., "A Local Computing Cell and 6T SRAM-Based Computing-in-Memory Macro With 8-b MAC Operation for Edge AI Chips," in IEEE Journal of Solid-State Circuits, vol. 56, no. 9, pp. 2817-2831, Sept. 2021, doi: 10.1109/JSSC.2021.3073254.

[2] Chih, Yu-Der, et al. "16.4 An 89TOPS/W and 16.3 TOPS/mm 2 all-digital SRAM-based full-precision compute-in memory macro in 22nm for machine-learning edge applications." 2021 IEEE International Solid-State Circuits Conference (ISSCC). Vol. 64. IEEE, 2021.

[3] Sudarshan, Chirag, et al. "A novel DRAM-based process-in-memory architecture and its implementation for CNNs." Proceedings of the 26th Asia and South Pacific Design Automation Conference. 2021.

[4] Lu, Zhaojun, et al. "An RRAM-based computing-in-memory architecture and its application in accelerating transformer inference." IEEE Transactions on Very Large Scale Integration (VLSI) Systems 32.3 (2023): 485-496.

[5] Yusuf, Alaba, Tosiron Adegbija, and Dhruv Gajaria. "Domain-specific STT-MRAM-based in-memory computing: A survey." IEEE Access 12 (2024): 28036-28056.

[6] Sebastian, Abu, Manuel Le Gallo, and Evangelos Eleftheriou. "Computational phase-change memory: Beyond von Neumann computing." Journal of Physics D: Applied Physics 52.44 (2019): 443002.

[7] Soliman, Taha, et al. "First demonstration of in-memory computing crossbar using multi-level Cell FeFET." Nature Communications 14.1 (2023): 6348.

[8] Yoo, Jaewook, et al. "Recent research for HZO-based ferroelectric memory towards in-memory computing applications." Electronics 12.10 (2023): 2297.

[9] Y. Xiao, S. Deng, Z. Zhao, Z. Faris, Y. Xu, T.-J. Huang, V. Narayanan, and K. Ni, "Quasi-nondestructive read out of ferroelectric capacitor polarization by exploiting a 2tnc cell to relax the endurance requirement," IEEE Electron Device Letters, vol. 44, no. 9, pp. 1436–1439, 2023.

[10] .R. Biswas, J. Duan, S. Deng, X. Niu, Y. Qin, P. Panigrahi, V. D. Parekh, R. Joshi, K. Ni, and V. Narayanan, "Single-cell universal logic-in-memory using 2T-nC FeRAM: An area- and energy-efficient approach for bulk bitwise computation," in Proc. IEEE Int. System-on-Chip Conf. (SOCC), 2025.

[11] Seshadri, Vivek, et al. "Ambit: In-memory accelerator for bulk bitwise operations using commodity DRAM technology." Proceedings of the 50th Annual IEEE/ACM International Symposium on Microarchitecture. 2017.

[12] Khan, M. N. I. & Ghosh, S. Comprehensive study of security and privacy of emerging non-volatile memories. Journal of Low Power Electronics and Applications 11 (2021).

[13] Rajendran, Jeyavijayan, et al. "Nano Meets Security: Exploring Nano-electronic Devices for Security Applications." (2015

[14] Yitbarek, Salessawi Ferede, et al. "Cold boot attacks are still hot: Security analysis of memory scramblers in modern processors." 2017 IEEE International Symposium on High Performance Computer Architecture (HPCA). IEEE, 2017.

[15] Skorobogatov, Sergei. "Physical attacks and tamper resistance." Introduction to Hardware Security and Trust. New York, NY: Springer New York, 2011. 143-173.

[16] Zhang, Jingyao, Hoda Naghibijouybari, and Elaheh Sadredini. "Sealer: In-sram aes for high-performance and low-overhead memory encryption." Proceedings of the ACM/IEEE International Symposium on Low Power Electronics and Design. 2022.

[17] Xiao, Yi, et al. "A Compact Ferroelectric 2T-(n+ 1) C Cell to Implement AND-OR Logic in Memory." 2023 IEEE Computer Society Annual Symposium on VLSI (ISVLSI). IEEE, 2023.

[18] W. Zhao and Y. Cao, "New generation of predictive technology model for sub-45nm design exploration," in Proceedings of the 7th International Symposium on Quality Electronic Design, ser. ISQED '06. USA: IEEE Computer Society, 2006, p. 585–590.

[19] Alessandri, Cristobal, et al. "Monte Carlo simulation of switching dynamics in polycrystalline ferroelectric capacitors." IEEE Transactions on Electron Devices 66.8 (2019): 3527-3534.

[20] Ramaswamy, Nirmal, et al. "NVDRAM: A 32Gb dual layer 3D stacked non-volatile ferroelectric memory with near-DRAM performance for demanding AI workloads." 2023 International Electron Devices Meeting (IEDM). IEEE, 2023.

[21] Deng, Shan, et al. "A comprehensive model for ferroelectric FET capturing the key behaviors: Scalability, variation, stochasticity, and accumulation." 2020 IEEE symposium on VLSI technology. IEEE, 2020.

[22] Parekh, Varun Darshana, et al. "A Study on the Impact of Temperature-Dependent Ferroelectric Switching Behavior in 3D Memory Architecture." 2025 38th International Conference on VLSI Design and 2024 23rd International Conference on Embedded Systems (VLSID). IEEE, 2025.

[23] Ferreira, João Dinis, et al. pLUTo: Enabling massively parallel computation in DRAM via lookup tables." 2022 55th IEEE/ACM International Symposium on Microarchitecture (MICRO). IEEE, 2022.

[24] J. Henry S. Warren, Hacker's Delight. Addison-Wesley, 2013.

[25] J. Han, C.-S. Park, D.-H. Ryu, and E.-S. Kim, "Optical image encryption based on xor operations," Optical Engineering, vol. 38, no. 1, pp. 47–54, 1999.

[26] A. Peleg and U. Weiser, "Mmx technology extension to the intel architecture," IEEE micro, vol. 16, no. 4, pp. 42–50, 1996.

[27] C.-Y. Chan and Y. E. Ioannidis, "Bitmap index design and evaluation," in Proceedings of the 1998 ACM SIGMOD international conference on Management of data, 1998, pp. 355–366.

[28] R. Kohut and B. Steinbach, "Boolean neural networks," Transactions on Systems, vol. 2, pp. 420–425, 2004.

[29] Seshadri, Vivek, et al. "RowClone: Fast and energy-efficient in-DRAM bulk data copy and initialization." Proceedings of the 46th Annual IEEE/ACM International Symposium on Microarchitecture. 2013.

979-8-3315-9813-6/25 $31.00 © 2025 IEEE

A Low-Power 4-bit Tracking-Type Analog-to-Digital Converter in SKY130 Process

Esteban Astudillo[1], Eduardo Holguín[1], Esteban Garzón[1,2], Luis Miguel Prócel[1]

[1] *Instituto de Micro y Nanoelectrónica, Universidad San Francisco de Quito*, Quito, Ecuador

[2] Department of Computer Engineering, Modeling, Electronics and Systems, Università della Calabria, 87036 Rende, Italy

Email: eastudillo@alumni.usfq.edu.ec, {eholguin,lprocel}@usfq.edu.ec, esteban.garzon@unical.it

Abstract—This paper presents the design and full-custom layout implementation of a 4-bit Tracking-Type Analog-to-Digital Converter (TT-ADC) using the SKY130 130 nm CMOS process. The proposed architecture mainly integrates a rail-to-rail analog comparator and a multiplexed resistor-string Digital-to-Analog Converter (DAC), combined with a synchronous controller and an output register. Unlike traditional tracking ADCs, this work introduces a fully integrated mixed-signal design optimized for both bandwidth and power efficiency, and evaluated under process-temperature-voltage variations accounting for layout parasitics. Simulations show that the proposed TT-ADC presents a bandwidth of 150 MHz while consuming only 505 μW of power. Compared to prior 4-bit implementations, the proposed design achieves over 2× improvement in bandwidth and an 87% reduction in power consumption. The area footprint is about 54.9 μm × 29.3 μm, making it highly suitable for energy-constrained, high-speed embedded applications.

Index Terms—ADC, TT-ADC, SKY130, Open-Source, Rail-to-Rail Comparator, Resistor-String DAC, Synchronous Up/Down Counter, Layout.

I. INTRODUCTION

Analog-to-Digital Converters (ADCs) are fundamental components in modern electronic systems, enabling the digitization of continuous-time analog signals for subsequent digital processing. ADCs are integral to a broad range of applications, including temperature sensing, biomedical signal acquisition, audio systems, industrial automation, and machine learning [1], [2], [3], [4], [5], [6]. Owing to diverse application-specific requirements, ADCs present a trade-off among resolution, power efficiency, and silicon area. For example, in the context of machine learning, higher resolution ADCs, i.e., increased bit precision, are often desired to enhance inference accuracy in compute-in-memory architectures [5]. In embedded and Internet of Things (IoT) platforms, selecting an appropriate ADC architecture is essential to meet stringent design constraints, such as low-power consumption, limited silicon footprint, and sufficient resolution for reliable signal acquisition [7].

Various ADC topologies have been proposed to meet distinct performance and integration goals. Flash, Delta-Sigma, and Successive Approximation Register (SAR) ADCs represent the most commonly adopted architectures across commercial and academic designs [8], [9], [10]. These ADCs differ in their resolution, sampling rate, power consumption, and hardware complexity [11]. Flash ADCs are favored for high-speed applications at the cost of larger area and power budgets, while Delta-Sigma and SAR ADCs offer improved resolution and energy efficiency for bandwidth-constrained systems.

Among the less-explored alternatives, Tracking-Type ADCs (TT-ADCs) have gained interest because of their minimal circuit complexity and real-time tracking behavior. In contrast to conventional ADC schemes, TT-ADCs operate by incrementally adjusting an up/down counter in response to the difference between an analog input and a DAC-generated reference signal [12], [13]. This operation results in a continuous digital representation of the input signal, making the architecture particularly attractive for systems requiring low-to-moderate resolution and medium bandwidth.

Despite the architectural simplicity and favorable characteristics of TT-ADC for embedded systems, existing implementations are mainly limited to schematic-level or HDL-based validation. For instance, the 4-bit converter presented in [14] demonstrates basic functionality using Verilog-based behavioral models for the digital components and eSim simulations of analog blocks using SKY130 PDK. However, this implementation omits layout-level realization and overlooks key considerations such as parasitic coupling, process variation, and robust synchronization across analog-digital boundaries. In [12], a 10-bit TT-ADC addresses some physical-level aspects. However, these designs target different application domains and do not emphasize compact, low-bit architectures suitable for minimal-area analog front-ends.

This work presents a fully integrated 4-bit TT-ADC architecture designed in SKY130 130 nm CMOS technology. The proposed design combines a rail-to-rail analog comparator, a multiplexed resistor-string DAC, and a synchronous digital controller with an output register to ensure stable and continuous digital readout. The proposed TT-ADC, comprehensively evaluated across process corners and accounting for layout parasitics, presents a bandwidth of 150 MHz while consuming 505 μW. The proposed 4-bit design achieves over 2× higher bandwidth and 87% lower power consumption compared to prior works.

II. PROPOSED ARCHITECTURE

Fig. 1 shows the top-level view of the proposed 4-bit TT-ADC. It comprises three main subsystems: the analog front-end (comparator and DAC), the digital control logic, and the output register. The ADC scheme operates through a continuous feedback mechanism in which the output of a comparator drives an "up/down" counter based on the difference between the analog input and a DAC-generated reference voltage. A dedicated controller detects signal transitions and triggers a register to capture the output synchronously.

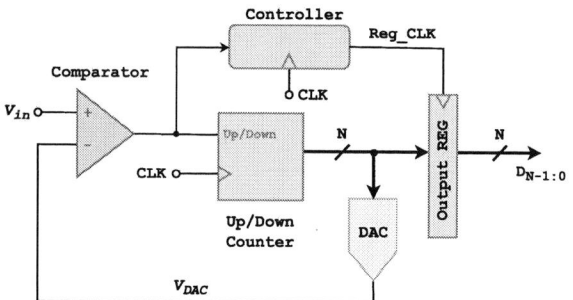

Fig. 1: Top-level block diagram of the proposed 4-bit TT-ADC.

Note that two main features distinguish the present work from prior Tracking-type-based schemes such as [14]. First, a transition-aware controller selectively clocks the output register, allowing for clean data sampling at signal slope changes. Second, the entire system is designed and validated at the layout level across different process corners.

A. Analog Front-End

The analog front-end includes a full-swing comparator and a resistor-string DAC. The comparator, shown in Fig. 2a, comprises three operational amplifiers (OpAmp), each optimized for a specific input voltage range. This configuration ensures full operation across the entire input range: the PMOS-input stage is fully active when the analog signal is below $V_{DD}/2$, while the NMOS-input stage takes over above $V_{DD}/2$ [15], [16]. To determine the active range, a third OpAmp compares the input against a fixed $V_{DD}/2$ reference. Its output drives a 2:1 multiplexer that selects between the two main OpAmp outputs, enabling rail-to-rail comparison across the entire input span.

The DAC is designed using a resistor-string architecture, preferred for its simplicity, low layout complexity, and monotonicity in low-bit designs. Fig. 2b shows the DAC scheme composed of sixteen equal-valued resistors that divide the supply voltage into 4-bit levels, selected through an analog multiplexer (MUX) controlled by the counter output [17].

B. Digital Control and Logic

The digital circuitry consists of a synchronous up/down counter, a transition-sensitive control unit, and an output register. The counter, shown in Fig. 3, presents a 4-bit feedback loop where the direction of counting is determined by the comparator output. A multiplexer selects either '0x1' or '0xF', which is added to the current value and latched at each clock edge [18].

To ensure glitch-free sampling of the counter output, the output register is clocked only upon transitions (both rising and falling edges) of the comparator signal. The controller thus triggers the register to latch the current counter value under these conditions. This edge-sensitive behavior—based on the current comparator output C_{now} and its previous state C_{prev}-generates the clock signal for the register according to:

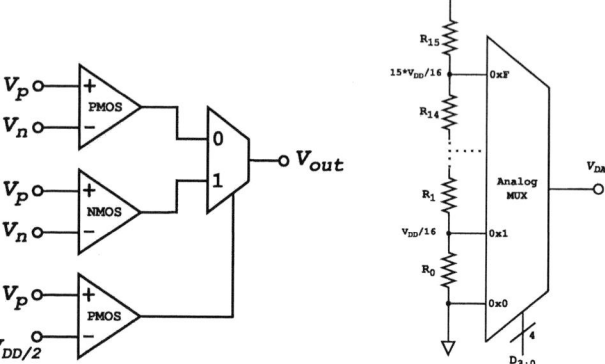

(a) Proposed Analog Full Swing Comparator.

(b) Resistor-string DAC with Analog MUX.

Fig. 2: Analog Circuit Design

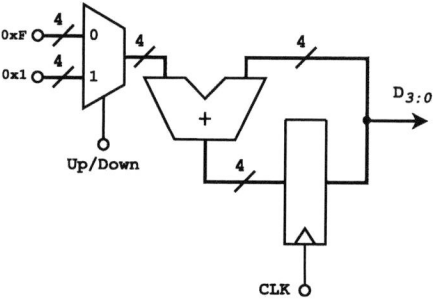

Fig. 3: Synchronous 4-bit up/down counter.

$$\text{Reg}_{\text{CLK}} = (C_{\text{now}} \wedge \overline{C_{\text{prev}}}) \vee (\overline{C_{\text{now}}} \wedge C_{\text{prev}}), \quad (1)$$

Notably, existing state-of-the-art architectures [8], [9] do not implement sequential logic to reliably latch digital levels.

We define the sampling rate of the converter as:

$$f_s = f_{\text{CLK}}/(2^N - 1) \quad (2)$$

Where N is the number of bits (i.e., the resolution), and f_{CLK} is the clock frequency-limited by the analog bandwidth of the circuit, i.e., the maximum allowable clock rate before the digital output presents an error.

Specifically, f_s arises from the tracking behavior of the TT-ADC scheme, in which the digital output follows the input signal level by incrementing or decrementing by one least significant bit per clock cycle. The worst-case scenario occurs when the input signal changes in a strictly monotonic fashion, either a full-scale ascending or descending ramp, thereby requiring the converter to traverse all $2^N - 1$ possible levels.

III. EVALUATION RESULTS

The proposed TT-ADC was designed using the SKY130 130 nm CMOS open-source Process Design Kit (PDK) [19], combining full-custom analog layout with digital logic. Specifically, the digital circuitry (synchronous counter, controller, and output register) was synthesized using the PDK standard cell libraries. The results rely on post-layout simulations, i.e., accounting for parasitics, and are evaluated across process, voltage, and temperature variations.

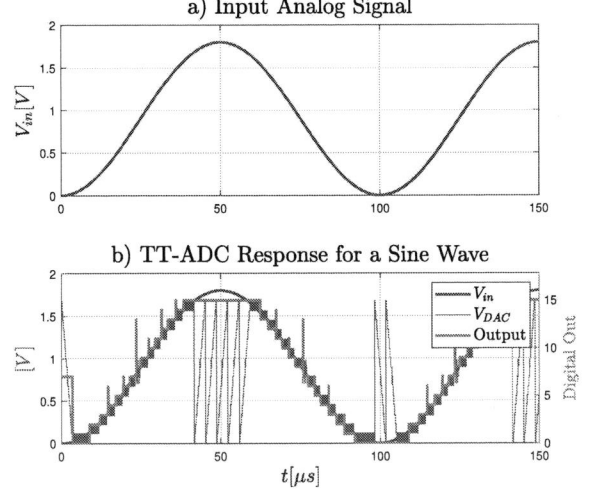

Fig. 4: Post-Layout Simulation Results: TT-ADC response under a sinusoidal input signal.

We validate the functional behavior of the proposed 4-bit TT-ADC under various input stimuli, considering the full voltage swing, typical-typical process variations, an average operating temperature of 25°C, and a moderate input frequency range: sinusoidal, rising ramp, and falling ramp. Figure 4(a) shows the simulation results using an analog sinusoidal input signal, while Figures 5(a) and 6(a) show the rising and falling ramp signals as input stimuli, respectively. Figure 4(b), Figure 5(b), and Figure 6(b) show the TT-ADC response to the analog input stimuli. As the analog input signal goes through the TT-ADC, the DAC output V_{DAC} (refer to magenta curves) may change in response to a small variation in V_{in}, leading to oscillations in V_{DAC}. The output register, clocked separately by the controller, captures and stabilizes the counter output (refer to cyan curves). Therefore, the TT-ADC consistently produces monotonic digital codes aligned with the input waveform, indicating correct tracking behavior.

A. Process, Voltage, and Temperature Variations

Table I summarizes the figures of merit (bandwidth and power consumption) for the typical-typical (TT), slow-slow (SS), and fast-fast (FF) process corners. These correspond to operating conditions of $V_{DD} = 1.8$ V and $T = 25$°C for TT, $V_{DD} = 1.65$ V and $T = 80$°C for SS, and $V_{DD} = 1.95$ V and $T = 0$°C for FF. The results indicate that the converter maintains functional operation across all corners, with bandwidth ranging from 33 MHz (SS) to 314 MHz (FF) while consuming below 1 mW in all cases. Although post-layout results show moderate degradation due to parasitics (14.4% power increase in FF corner and 50% bandwidth reduction in SS corner), the proposed TT-ADC remains well within target specifications, i.e., for data acquisition applications [11].

B. Comparison

The TT-ADC was benchmarked against flash and delta-sigma schemes [8], [9], considering the same bit resolution

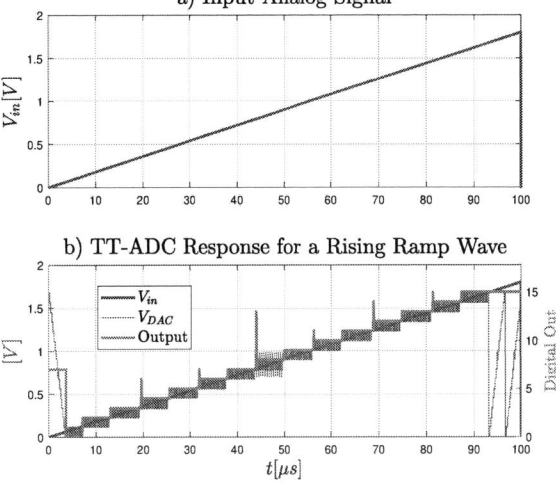

Fig. 5: Post-Layout Simulation Results: TT-ADC response under a rising ramp as input signal.

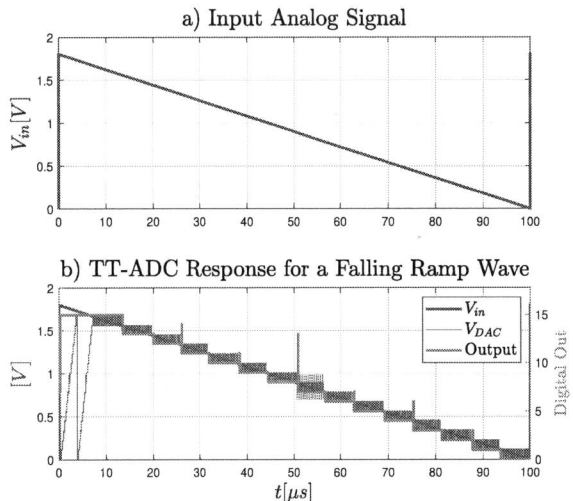

Fig. 6: Post-Layout Simulation Results: TT-ADC response under a falling ramp as input signal

TABLE I: Proposed TT-ADC For Different Process Corners

Corner				Figure of Merit	
nMOS	pMOS	V_{DD}	Temp.	Bandwidth	Power Consumption
T	T	1.8 V	25°C	150MHz	0.505mW
S	S	1.65 V	80°C	33MHz	0.284mW
F	F	1.95 V	0°C	314MHz	0.961mW

and technology node. Table II summarizes the main figures of merit and comparison. Compared to other schemes, the proposed converter achieves a better balance in terms of performance and efficiency. Flash ADCs offer high bandwidth but at the expense of power consumption (24.26 mW), while Delta-Sigma ADC presents reduced power but suffer from limited bandwidth. Specifically, the proposed TT-ADC scheme achieves 150 MHz bandwidth with only 0.505 mW power, offering a significant improvement in energy efficiency

979-8-3315-9813-6/25 $31.00 © 2025 IEEE

TABLE II: Figures of merit and comparison

Topology	Flash [8]	Delta-Sigma [9]	Tracking [14]	Tracking Proposed
Technology	180 nm	130 nm	130 nm	130 nm
Resolution	4-bit	4-bit	4-bit	4-bit
Bandwidth	500 MHz	104 MHz	69 MHz	150 MHz
Sampling Rate	500 MHz	104 MHz	4.6 MHz	10 MHz
Power Consumption	24.26 mW	3 mW	3.9 mW	0.505 mW
Power Supply	1.8 V	1.5 V	1.8 V	1.8 V
Area	0.05 mm^2	0.3 mm^2	–	1606.5 μm^2

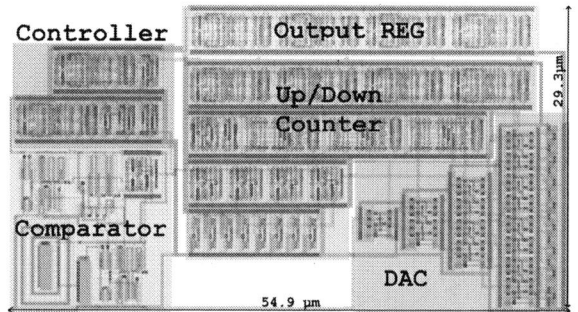

Fig. 7: Final layout of the proposed TT-ADC in SKY130 CMOS technology.

compared to other designs.

Figure 7 shows the layout of the proposed TT-ADC, highlighting the main blocks. The TT-ADC presents an area footprint of about 1600 μm^2, from which the major contributor is the "Up/Down Counter" circuitry with an area footprint of about 452 μm^2. Overall, TT-ADC presents about 99% less area footprint than the Flash and Delta-Sigma ADCs in Table II.

The proposed tracking-type ADC scales efficiently: the comparator and control unit do not increase in power or area with resolution, while the DAC grows exponentially with 2^N and the counter and output register scale linearly. Consequently, the sampling rate decreases roughly by a factor of $2^N - 1$, as shown in equation 2.

IV. CONCLUSION

This work presents a compact and energy-efficient 4-bit tracking ADC in SKY130 130 nm CMOS technology. The architecture was validated through post-layout simulations, demonstrating accurate tracking of sine and ramp inputs, and reliable behavior under process variations. The TT-ADC is also evaluated across TT, SS, and FF corners, with power consumptions below 1 mW and bandwidth ranging from 33 MHz to 314 MHz. When compared to existing ADC types, the proposed converter achieves an improved trade-off between bandwidth and power consumption, with 150 MHz bandwidth, 0.505 mW power consumption, and about 1600 μm^2 of area footprint. These results establish the proposed ADC as a suitable candidate for integration into low-power, high-speed applications such as IoT edge devices, sensor front-ends, and embedded mixed-signal systems.

ACKNOWLEDGMENTS

We extend our gratitude to Synopsys Chile for the invaluable support provided to USFQ in the use of their entire platform of design and simulation tools. Their support has been crucial for the development of our research and academic projects. E. Garzón was supported by the Italian Ministry of University and Research (MUR) under grant number SOE_20240000022.

REFERENCES

[1] S. Kumar, R. S. Gamad, and S. Dasgupta, "A resistive sensing and dual-slope adc based smart temperature sensor," *Analog Integrated Circuits and Signal Processing*, vol. 89, no. 1, pp. 163–172, 2016.

[2] Y. Zhang, W. Wang, X. Liu, Y. Wang, and Y. Zhang, "Design of an integrated temperature and humidity sensor based on high dynamic range utilization rate adc," *IEICE Electronics Express*, vol. 20, no. 1, p. 20230164, 2023.

[3] A. Aprile, M. Folz, D. Gardino, P. Malcovati, and E. Bonizzoni, "An area-efficient smart temperature sensor based on a fully current processing error-feedback noise-shaping SAR ADC in 180-nm CMOS," *IEEE Journal of Solid-State Circuits*, vol. 59, no. 3, pp. 716–727, 2023.

[4] A. Samiee, Y. Zhou, T. Zhou, and B. Jalali, "Deep Analog-to-Digital Converter for Wireless Communication," *arXiv preprint arXiv:2009.05553*, 2020.

[5] H. Jiang, S. Huang, W. Li, and S. Yu, "ENNA: An efficient neural network accelerator design based on ADC-free compute-in-memory subarrays," *IEEE Trans. on Circ. and Syst. I*, vol. 70, no. 1, pp. 353–363, 2022.

[6] Y. Huang, T. Ando, A. Sebastian, M.-F. Chang, J. J. Yang, and Q. Xia, "Memristor-based hardware accelerators for artificial intelligence," *Nature Reviews Electrical Engineering*, vol. 1, no. 5, pp. 286–299, 2024.

[7] L. M. Santana, D. L. de Oliveira, and L. d. A. Faria, "A 1V 5-bits Low Power Level Crossing ADC with OFF state in idle time for bio-medical applications in 0.18μ m CMOS," *arXiv preprint arXiv:2108.07564*, 2021. [Online]. Available: https://arxiv.org/abs/2108.07564

[8] I. S. A. Halim, S. L. M. Hassan *et al.*, "Comparative study of comparator and encoder in a 4-bit Flash ADC using 0.18 μm CMOS technology," in *2012 International Symposium on Computer Applications and Industrial Electronics (ISCAIE)*. IEEE, 2012, pp. 35–38.

[9] L. Dorrer, F. Kuttner, P. Greco, and S. Derksen, "A 3mW 74dB SNR 2MHz CT /spl Delta//spl Sigma/ ADC with a tracking-ADC-quantizer in 0.13 /spl mu/m CMOS," in *ISSCC. 2005 IEEE Int. Digest of Technical Papers. Solid-State Cir. Conf., 2005.*, 2005, pp. 492–612 Vol. 1.

[10] B. T. Reyes, L. Biolato, A. C. Galetto, L. Passetti, F. Solis, J. I. Giubilatto, L. A. Reyes, A. F. Bocco, and M. R. Hueda, "A 4GS/s 8-bit SAR ADC with an Energy-Efficient Time-Interleaved Architecture in 130nm CMOS," in *2020 Argentine Conf. on Elec.*, 2020, pp. 77–81.

[11] W. Kester, "Which ADC architecture is right for your application," in *EDA Tech Forum*, vol. 2, no. 4, 2005, pp. 22–25.

[12] S. Bramburger and D. Killat, "10-bit tracking adc with a multi-bit quantizer, variable step size and segmented current-steering dac," *Advances in Radio Science*, vol. 17, pp. 161–167, 2019.

[13] Y. Huang, "A tracking adc with transient-driven self-clocking for digital dc-dc converters," Ph.D. dissertation, Brandenburg Univ. of Tech., 2015.

[14] A. Kumar, K. Ghosh, S. Kar, and R. Paknikar, "Design of 4-bit servo tracking type ADC using Sky-Water SKY130 PDK and eSim," in *2023 International Conference on Artificial Intelligence and Applications (ICAIA) Alliance Technology Conference (ATCON-1)*, 2023, pp. 1–3.

[15] A. Verma, D. Sharma, R. K. Singh, and M. K. Yadav, "Design of Two-Stage CMOS Operational Amplifier," *Int. J. of Emerging Technology and Advanced Engineering*, vol. 3, no. 12, pp. 102–106, December 2013.

[16] A. Yadav, "A Review Paper on Design and Synthesis of Two-Stage CMOS Op-Amp," *International Journal of Advances in Engineering & Technology*, no. 1, pp. 677–688, January 2012.

[17] A. S. Kherde and P. R. Gumble, "An efficient design of r-2r digital to analog converter with better performance parameter in (90nm) 0.09- μ m cmos process," *International Journal of Innovative Technology and Exploring Engineering (IJITEE)*, vol. 3, no. 7, pp. 1–6, December 2013.

[18] S. Harris and D. Harris, *Digital design and computer architecture*. Morgan Kaufmann, 2015.

[19] Google and SkyWater Technology Foundry, "Skywater open source pdk," https://github.com/google/skywater-pdk, 2020.

Swift Synthesis of Approximate Hardware Accelerators Using Generative Adversarial Networks

Muhammad Awais[1], Hassan Ghasemzadeh Mohammadi[2], and Marco Platzner[1]

[1]Paderborn University, Paderborn, Germany
[2]Reneo group GmbH, Hamburg, Germany

Abstract—Deploying modern applications with significant resource demands is often challenging, but approximate designs offer a promising alternative by delivering high performance with minimal compromises in output quality. Traditionally, approximate hardware accelerators have been developed through search-based iterative frameworks, which suffer from long runtimes due to the exponential growth of the design space. A significant portion of the runtime is consumed by either invalid nodes or valid nodes that offer minimal improvements in performance metrics, such as runtime or power consumption. This severely limits the thorough exploration of the design space. In this paper, we introduce a novel approach for synthesizing approximate accelerators that leverages sparsity to reduce the complexity of the design space exploration problem. Our method employs a generative adversarial network (GAN) to rapidly generate a diverse set of high-quality design nodes, eliminating the need for costly node evaluations. This enables the swift creation of approximate accelerators generated for any given error threshold in a fraction of time as compared to a simulation-based framework. We conducted experiments on a suite of benchmarks from real-world domains, demonstrating that our methodology can generate approximate hardware designs with significant area and power savings, comparable to state-of-the-art search-based approaches. In a comparative evaluation against two leading methods, our approach achieved equal or better quality results for two out of four benchmarks while reaching up to 55% area savings, thus effectively demonstrating a new avenue for automated generation of approximate accelerators.

Index Terms—Approximate Computing, Accelerator Design, Machine Learning, Generative Adversarial Networks.

I. INTRODUCTION

In the aftermath of Dennard scaling, numerous efforts have targeted optimization of applications that process massive datasets with a focus on minimizing required computing resources. A prominent line of research has unfolded as approximate computing (AC) deviating from the conventional approach of accurate designs to obtain highly performant yet imprecise designs. AC has now gained distinguished traction across the computing stack, encompassing software to physical device levels. However since the deployment of hardware accelerator has recently seen an increase, many efforts have considered developing approximate variants of hardware accelerators [1, 2, 3, 4].

Hardware accelerators, as opposed to ad-hoc design methodologies for fundamental arithmetic components (adders and multipliers), are mainly acquired via an automated iterative process. This requires optimization of the original circuit for target metric by calibrating approximation knobs through iterative refinement. Many have adopted a search-based model to represent candidate designs [5, 6, 7, 8, 9, 10, 11, 12, 13, 14] during the design space exploration (DSE) process. However, the performance of such search-driven process is limited by the choice of search method and a quality evaluation technique. The DSE for approximate accelerators is mostly realized via greedy-based methods [15, 9, 6, 8, 16, 17], which, despite being capable of quickly generating approximate variants, tend to exhibit inferiority owing to disregarding initial good solutions. Alternatively, other works have adopted methods driven by statistical learning, such as MCTS or its variants [11, 5, 7, 11, 18]. Nonetheless they require reasonably large number of iterations before they start converging thus suffering poor scalability.

In search-based frameworks, design validation dominates the DSE runtime. Most works rely on circuit simulations using selectively sampled input vectors [6, 9, 10, 11], while others adopt formal verification, which incurs even greater runtime overheads [5, 19, 20, 21]. More recent approaches use analytical models trained on prior simulations to accelerate DSE [22, 7, 8]; however, these often yield conservative or inaccurate estimates, compromising exploration quality [4, 12]. Thus, quality evaluation remains the primary bottleneck in search-based design sampling.

This paper addresses the limitations of conventional search-based frameworks by discarding explicit search strategies altogether. Instead, we propose a generative adversarial network (GAN)-driven methodology that directly samples a sparse and relevant subset of the design space, focusing on nodes with high potential to improve the target metric. The framework leverages adversarial training, where a generator (G) network synthesizes candidate approximate designs, and a discriminator (D) network evaluates their validity, effectively distinguishing acceptable configurations from infeasible ones. This enables the rapid generation of high-quality approximate circuits while substantially reducing the computational burden of DSE. Specifically, our contributions are twofold. First, we introduce a GAN-based strategy for sparse design space generation of approximate circuits, enabling efficient production of diverse candidates within acceptable error bounds while omitting the majority of error-violating instances. Second, we demonstrate that this method facilitates fast DSE by replacing exhaustive and stochastic search techniques with a direct generation of high-quality solutions.

In the rest of the paper, we summarize the recent works in approximate accelerator synthesis domain in section II followed by a detailed description of our proposed method in Section III. Later in Section IV, we provide details of our experimental setup and discuss the results. Finally, Section V concludes the paper.

II. RELATED WORK

The generation of approximate hardware accelerators is typically formulated as an optimization problem that seeks feasible designs under defined error constraints. Witschen et al. [5] showed that the search algorithm's ability to rank intermediate solutions critically influences final design quality. Accordingly, we first classify prior work into two categories based on their design space exploration strategies, followed by a discussion of the quality validation techniques employed.

First, greedy-like approaches iteratively expand only the best candidate, progressing forward without backtracking. This aggressive pruning significantly reduces runtime [5, 9, 6, 8, 16, 17], but often discards promising designs with better target metrics due to the lack

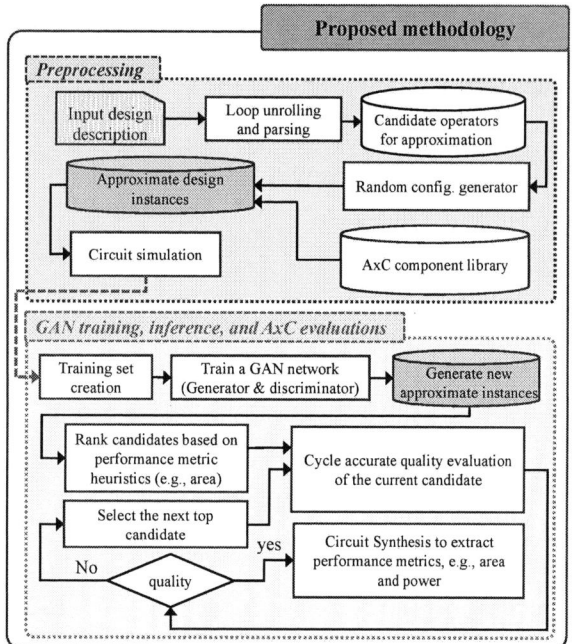

Fig. 1: Overall flow of the proposed GAN-based accelerator synthesis.

used, including its key components and the specifics of its training process.

A. Overall flow

Figure 1 illustrates key steps in our proposed methodology that adapts Generative Adversarial Networks (GANs) [29] to sparsify and accelerate the generation of approximate circuits. As the first step of our proposed methodology, we perform necessary preprocessing, where loops are unrolled and circuit components (e.g., adders and multipliers) are identified as candidate operators for approximation. A new approximate design is then generated by randomly substituting a selected number of these operators with their approximate counterparts from a given approximation library. After generating a sufficient number of approximate instances, quality estimation is carried out via circuit simulation. This step determines whether each generated instance qualifies as a valid or invalid circuit based on a user-specified error constraint (e.g., mean relative error).

The next phase (GAN training, inference, and AxC evaluation) begins with the formulation of a training dataset tailored to represent the design space of the circuit to be approximated. This dataset is critical as it consists of labeled samples used to guide the learning process. Initially, the Discriminator (D) is trained using these samples to distinguish between valid and invalid designs. Once trained, the Generator (G) enters the inference phase, where it progressively generates batches of approximate designs that adhere to the specified error constraint. These generated designs are then ranked using a heuristic estimation based on the target performance metric, which in this work is hardware area. The top-ranked circuit is selected as the most promising approximate circuit and is validated through simulation to ensure quality constraints are met. Key performance metrics such as area, power, and delay are then reported.

Unlike traditional exhaustive methods, which are impractical for large circuits due to the combinatorial explosion of configurations, our GAN-based strategy leverages learned distributions from representative datasets to efficiently generate high-quality approximate designs, significantly reducing computation time while upholding quality constraints.

B. Generative Adversarial Networks for Approximate Circuit Synthesis

Adapting Generative Adversarial Networks (GANs) [29] for the synthesis of approximate circuits necessitates the robust encoding of circuit quality constraints and the modification of the GAN training process to explicitly account for target performance metrics. We integrate the encoding of circuit quality directly into the adversarial training process, departing from conventional ML-oriented GAN applications. Each sample in the dataset not only provides a validity label but also includes a set of features that describe the state of the circuit components, as detailed in Figure 2. Specifically, each feature corresponds to a component within the circuit, indicating whether it has been approximated or remains precise. In cases where the component is approximated, the feature is assigned an enumerated identifier from a library of approximate components (e.g., [30]). Conversely, a precise component is denoted by a "0" in the feature list. This binary encoding of circuit component states ensures that the dataset captures the necessary design parameters for the GAN to effectively learn from and generate viable circuit configurations.

During the inference, the generator (Figure 2, step ④) produces discrete circuit configurations (while conforming to the same encoding scheme) that are compatible with standard synthesis tools. The generator's latent space is constrained to prioritize configurations that

of exploration beyond the current path. Notable examples in this category are the ABACUS [15] and, more recently, the JumpSearch [17] frameworks.

In the second category are the approaches that integrate learning to enhance design space exploration. Here, the intermediate designs are mostly generated via random approximate transformations. The designs are then ranked based on a reward formulation, and the search proceeds based on the obtained outcome [5, 23, 24, 7, 25]. The employed search methods include (but not limited to) variations of genetic algorithms [23], Monte Carlo Tree Search (MCTS) [24, 11], or even simulated annealing [5]. Additionally, the LDAX framework [7] uniquely combines a variant of Monte Carlo Tree Search (MCTS) for exploration with machine learning-based quality evaluation, resulting in extremely fast inference.

For accuracy evaluation, formal verification remains the most accurate yet computationally intensive approach, offering strong guarantees on maximum output error [20, 5, 19, 26, 21]. Simulation-based methods are widely adopted due to their scalability to large accelerators [9, 10, 11, 27, 6], but they suffer from significant runtime overhead. Recent works have also explored machine learning-based error models [7, 8, 28, 18], which, despite showing promise in practical domains, often fail to capture intricate structural-error correlations.

Irrespective of the employed search strategy, the dependence on accuracy evaluation during the search imposes a significant bottleneck on the design space exploration process, ultimately limiting the extent to which approximation can be effectively exploited. Hence, in this work, we propose a fundamentally different strategy that bypasses the search process entirely and directly generates high-quality design instances.

III. METHODOLOGY

This section presents our novel method for generating approximate accelerator design samples. We begin by outlining the overall methodology, followed by a detailed explanation of the GAN architecture

TABLE I: Benchmark accelerator circuits

Benchmark	Bit-width (I / O)	Quality§ metric	Area (μ^2)	Power (μW)
RGB2Gray	8/8	MRE (5)	2427	100
Addition tree	16/16	MRE (5)	454	1040
FIR filter	16/16	MRE (5)	7485	6068
Gaussian filter	8/8	PSNR (35dB)	7727	830

§ The value in the bracket represents the threshold for the quality constraint under the error metric.

minimize error propagation in target circuit topologies based on the simulated instances in the training set. The discriminator acts as a hardware quality checker, evaluating whether or not the generated configurations: (1) satisfy application-level error bounds (e.g., $< 5\%$ mean relative error, and (2) avoid incompatible component substitutions (e.g., cascaded approximate multipliers causing unbounded error)

As a result, this compliance ensures that the GAN produces synthesis-ready configurations adhering to practical design constraints, rather than merely generating statistically plausible but unsynthesizable solutions.

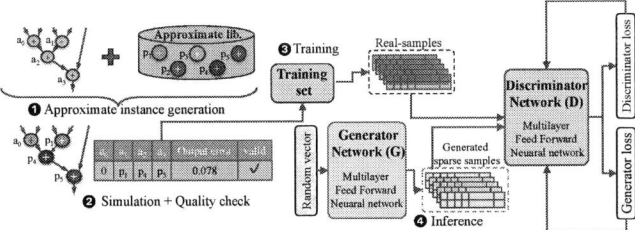

Fig. 2: GAN training flow to drive approximate accelerator DSE

C. Hardware-Constrained Adversarial Training

The adversarial training objective incorporates circuit quality constraints e.g., mean relative error. Let $C(\mathbf{x})$ denote a circuit configuration, $E(C)$ its error metric, and $\mathcal{L}_{\text{lib}}$ the set of valid component substitutions. The generator G and discriminator D optimize:

$$\min_{G} \max_{D} V(G, D) = \mathbb{E}_{C \sim p_{\text{valid}}(C)}[\log D(C)]$$
$$+ \mathbb{E}_{\mathbf{z} \sim p_{\mathbf{z}}(\mathbf{z})}\Big[\log\big(1 - D(G(\mathbf{z}))\big)\Big] \quad (1)$$
$$\text{s.t.} \quad E(G(\mathbf{z})) \leq \tau_{\text{error}}, \quad \forall \mathbf{z}$$
$$G(\mathbf{z})_i \in \mathcal{L}_{\text{lib}} \cup \{0\}, \quad \forall i$$

where τ_{error} is the maximum allowable error. The generator's loss $\mathcal{L}_G$ is augmented with quality violation penalties:

$$\mathcal{L}_G = -\mathbb{E}_{\mathbf{z}}[\log D(G(\mathbf{z}))] + \lambda_1 \max(0, E(G(\mathbf{z})) - \tau_{\text{error}})$$
$$+ \lambda_2 \sum_i \mathbb{I}(G(\mathbf{z})_i \notin \mathcal{L}_{\text{lib}} \cup \{0\}) \quad (2)$$

The regularization terms λ_1 and λ_2 enforce error tolerance and component validity, respectively, creating explicit trade-offs between approximation aggressiveness and electrical correctness. Training progresses until the generator produces configurations that consistently yield a meaningful number of quality-accepted, functionally valid circuits.

IV. EXPERIMENTS AND RESULTS

A. Experimental setup

We selected structurally and spatially diverse benchmarks from domains such as image and signal processing, as detailed in Table I.

TABLE II: Details of the GAN hyperparameters

Hyperparameter	Value
Latent dimensions	[100, 150]
Number of epochs	1000
Generator layer sizes	[[128, 256, 128], [64, 128, 64]]
Discriminator layer sizes	[[128, 256, 128], [64, 128, 64]]
Learning rates	[(0.0001, 0.0001), (0.0005, 0.0005)]

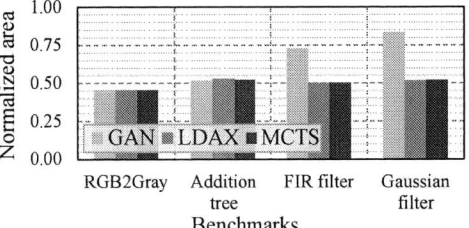

(a) Area results

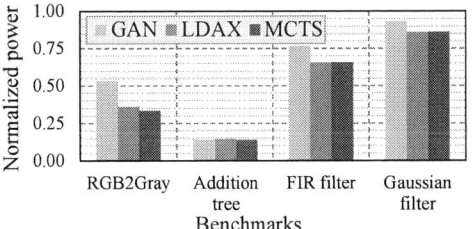

(b) Power consumption results

Fig. 3: Area and power consumption results for all benchmarks using a CMOS 22 nm technology library. Power estimates are based on switching activity generated from random input simulations and analyzed with *Synopsys Design Compiler*.

Each benchmark includes a testbench and dataset to enable meaningful design space exploration. Experiments were conducted on a virtual machine with 16 vCPUs (up to 3.5 GHz), 60 GB RAM, and an NVIDIA T4 GPU. The sample generation and evaluation flow was implemented in Python using PyTorch.

For each benchmark, 100,000 RTL-simulated samples were generated once to construct a reusable training dataset for multiple error thresholds. The generator (G) produced $m = 1,000$ samples across $n = 10$ batches per benchmark. The discriminator (D) was trained to distinguish valid from invalid designs. Table II lists GAN training hyperparameters.

Post-training designs generated by the GAN inference were sorted by estimated area using $\sum_{i=1}^{n} \text{Area}_i$, where n is the number of components and Area_i is the area of the i-th component. Valid configurations were selected and evaluated for area and power.

B. Results

In Figure 3, we present the hardware area and power consumption results for the benchmarks listed in Table I, comparing our GAN-based synthesis approach with two state-of-the-art search-based techniques: MCTS [24] and LDAX [7].

In terms of hardware area, the GAN-based approach performs comparably or even slightly better than both search-based methods for the *RGB2Gray* and *Addition tree* benchmarks, achieving up to 54.5% and 48.2% area savings, respectively. For the *FIR filter* and *Gaussian filter*, GAN achieves up to 27.0% and 16.7% area savings, respectively, while LDAX attains the highest area reduction in these cases with 49.8% and 48.3% savings.

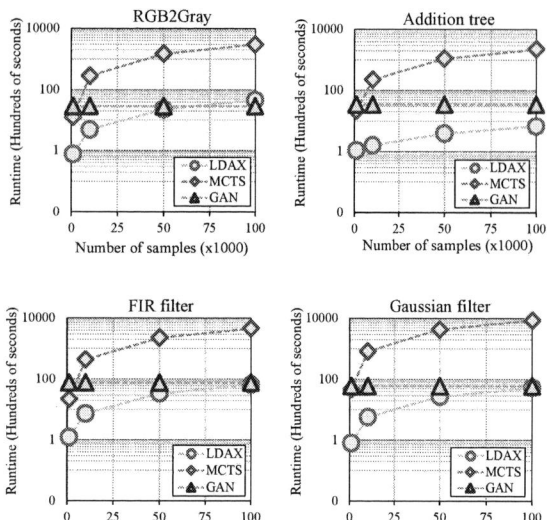

Fig. 4: Runtime comparison of LDAX, MCTS, and GAN across benchmarks for generating varying numbers of design samples.

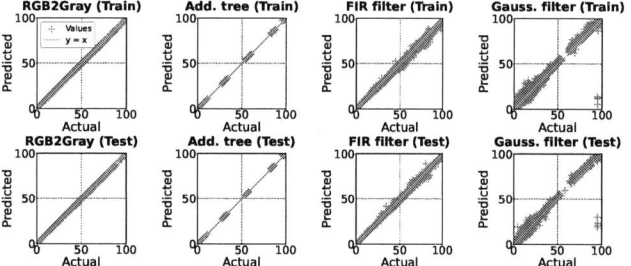

Fig. 5: Scatter plots showing the quality estimation model's performance on both train and test sets for all benchmark circuits

Regarding power consumption, GANs achieve similar reductions to LDAX and MCTS for the *Addition tree*. However, for the remaining benchmarks, GAN-based designs exhibit slightly higher power usage, reporting savings of 47.0%, 23.3%, and 6.7% for *RGB2Gray*, *FIR filter*, and *Gaussian filter*, respectively. Between the two search-based methods, LDAX and MCTS generally yield similar power results, with MCTS slightly outperforming LDAX in the case of *RGB2Gray*.

The runtime comparison is presented in Figure 4, illustrating the time required to generate varying numbers of sample configurations across different approaches. For search-based methods such as LDAX and MCTS, each configuration must undergo an explicit exploration and quality evaluation step—either through simulation or machine learning-based estimation—leading to a runtime that increases roughly linearly with the number of generated samples. In contrast, the GAN-based method reports total runtime as the sum of the training phase and the inference time for generating samples. To ensure fairness, the one-time cost of generating training data is excluded for both GAN and LDAX.

As shown in Figure 4, the GAN exhibits a relatively flat runtime as sample size increases, leading to improved overall performance. A crossover point near 100K samples marks where the GAN begins to outperform LDAX in runtime. This demonstrates that once trained, the GAN can generate thousands of high-quality approximate configurations in a fraction of the time required by LDAX or MCTS. It also underscores the superior scalability of the GAN-based approach, particularly for applications requiring fast and large-scale design space exploration.

The scatter plots in Figure 5 illustrate the performance of the discriminator network across both training and test datasets for all benchmark circuits. For the *RGB2Gray* and *Addition tree* benchmarks, the predictions align closely with the ground truth, as evidenced by the tight clustering of points along the diagonal. This indicates that the discriminator network accurately captures the relationship between the approximations and their associated errors. Notably, these two benchmarks also correspond to the highest area savings achieved by the GAN-based approach, suggesting a strong link between predictive accuracy and the quality of generated designs.

In the case of the *FIR filter*, the predictions are generally well-aligned but show slightly more dispersion compared to *RGB2Gray* and *Addition tree*, particularly in the test set. This moderate reduction in predictive accuracy might correspond with slightly lower savings in area and power. For the *Gaussian filter*, prediction quality is visibly lower, indicating that the discriminator struggles to generalize well on this dataset. This limitation is also reflected in the comparatively modest hardware savings for this benchmark.

C. Discussion on results

The results on practical benchmarks presented in the previous subsection highlight that GANs offer a fundamentally different operational advantage. Once trained, a GAN model can rapidly generate a large number of design candidates in a fraction of the time required by iterative search-based methods like LDAX and MCTS. This is particularly useful in scenarios where rapid prototyping is a priority, such as early-stage design space pruning or when seeding downstream fine-tuning algorithms with diverse, high-potential candidates. In such cases, GANs can serve as an efficient front-line tool, enabling fast identification of promising solutions before applying more resource-intensive optimization techniques.

Moreover, GANs show strong performance in benchmarks like *RGB2Gray* and the *Addition tree*, where they were able to match or even slightly outperform traditional methods in terms of area savings, achieving this in significantly less time. This demonstrates that GANs are not only efficient in runtime but also capable of generating high-quality approximations, particularly in design spaces where the underlying complexity is effectively captured by a highly accurate quality estimation model. Thus, GAN-based approximation provides a compelling alternative, especially when speed, scalability, and early-stage exploration are critical. Their ability to combine solid optimization performance with rapid design generation makes them a promising component in the approximate accelerator synthesis tool flow.

V. CONCLUSION

Automatically synthesizing approximate hardware accelerators has traditionally relied on iterative, search-based methods, but their scalability suffers due to time-intensive simulations during design exploration. In this paper, we propose a novel GAN-based methodology to bypass traditional search and rapidly generate diverse, high-quality approximate designs. This eliminates costly evaluations and enables fast, scalable synthesis for any error threshold, drastically shrinking time against simulation-based frameworks. On real-world benchmarks, we report up to 55% area and 85% power savings. Specifically, on *RGB2Gray* and *Addition tree*, GANs matched or exceeded traditional results in far less time, highlighting their value for early-stage approximation. This work offers several expansion paths: enriching the GAN training set with reinforcement learning for dynamic quality feedback, fine-tuning to integrate with modular state-of-the-art frameworks for stronger approximations, and leveraging efficient compute clusters to enable massive parallelism.

ACKNOWLEDGEMENTS

This work has been funded through the German Research Foundation (DFG) under grant number PL 471/9-1. The compute time and resources have been provided by the Paderborn Center for Parallel Computing (PC^2).

REFERENCES

[1] Hans Jakob Damsgaard et al. "Adaptive approximate computing in edge AI and IoT applications: A review". In: *Journal of Systems Architecture* (2024), p. 103114.

[2] Aleksandr Ometov and Jari Nurmi. "Towards approximate computing for achieving energy vs. accuracy trade-offs". In: *2022 Design, Automation & Test in Europe Conference & Exhibition (DATE)*. IEEE. 2022, pp. 632–635.

[3] Honglan Jiang et al. "Approximate arithmetic circuits: A survey, characterization, and recent applications". In: *Proceedings of the IEEE* 108.12 (2020), pp. 2108–2135.

[4] Ilaria Scarabottolo et al. "Approximate Logic Synthesis: A Survey". In: *Proceedings of the IEEE* 108.12 (2020), pp. 2195–2213.

[5] Linus Witschen et al. "CIRCA: Towards a Modular and Extensible Framework for Approximate Circuit Generation". In: *Microelectronics Reliability* 99 (Aug. 2019), pp. 277–290. DOI: 10.1016/j.microrel.2019.04.003.

[6] Yuqin Dou et al. "ENAP: An efficient number-aware pruning framework for design space exploration of approximate configurations". In: *IEEE Transactions on Circuits and Systems I: Regular Papers* (2023).

[7] Muhammad Awais, Hassan Ghasemzadeh Mohammadi, and Marco Platzner. "LDAX: A Learning-based Fast Design Space Exploration Framework for Approximate Circuit Synthesis". In: *Proceedings of the 31st ACM Great Lakes Symposium on VLSI (GLSVLSI)*. ACM. 2021, pp. 27–32. DOI: 10.1145/3453688.3461506.

[8] Vojtech Mrazek et al. "AutoAx: An Automatic Design Space Exploration and Circuit Building Methodology utilizing Libraries of Approximate Components". In: *Proceedings of the Design Automation Conference (DAC)*. IEEE. 2019, pp. 1–6.

[9] Ghasem Pasandi et al. "Deep-PowerX: A Deep Learning-based Framework for Low-power Approximate Logic Synthesis". In: *Proceedings of the ACM/IEEE International Symposium on Low Power Electronics and Design (ISLPED)*. ACM/IEEE. 2020, pp. 73–78.

[10] Jorge Castro-Godinez et al. "AxHLS: Design Space Exploration and High-level Synthesis of Approximate Accelerators using Approximate Functional Units and Analytical Models". In: *Proceedings of the International Conference On Computer Aided Design (ICCAD)*. IEEE. 2020, pp. 1–9.

[11] Muhammad Awais, Hassan Ghasemzadeh Mohammadi, and Marco Platzner. "A Hybrid Synthesis Methodology for Approximate Circuits". In: *Proceedings of the 30th ACM Great Lakes Symposium on VLSI (GLSVLSI)*. ACM. 2020, pp. 421–426. DOI: 10.1145/3386263.3406952.

[12] Linus Witschen et al. "Search space characterization for approximate logic synthesis". In: *Proceedings of the 59th ACM/IEEE Design Automation Conference (DAC)*. 2022, pp. 433–438.

[13] Marzieh Vaeztourshizi and Massoud Pedram. "An efficient error estimation technique for pruning approximate data-flow graphs in design space exploration". In: *2022 23rd International Symposium on Quality Electronic Design (ISQED)*. IEEE. 2022, pp. 102–107.

[14] Soheil Hashemi, Hokchhay Tann, and Sherief Reda. "BLASYS: Approximate Logic Synthesis using Boolean Matrix Factorization". In: *Proceedings of the Design Automation Conference (DAC)*. ACM. 2018, p. 55.

[15] Kumud Nepal et al. "ABACUS: A Technique for Automated Behavioral Synthesis of Approximate Computing Circuits". In: *Proceedings of the Design, Automation & Test in Europe Conference & Exhibition (DATE)*. IEEE. 2014, pp. 1–6.

[16] Yuqin Dou et al. "FPAX: A Fast Prior Knowledge-Based Framework for DSE in Approximate Configurations". In: *IEEE Transactions on Computer-Aided Design of Integrated Circuits and Systems* (2023).

[17] Linus Witschen et al. "Jump search: A Fast Technique for the Synthesis of Approximate Circuits". In: *Proceedings of the ACM Great Lakes Symposium on VLSI (GLSVLSI)*. ACM. 2019, pp. 153–158.

[18] Muhammad Awais, Hassan Ghasemzadeh Mohammadi, and Marco Platzner. "Design Space Exploration for Approximate Circuits via Checkpointing and DNN-Based Estimators". In: *IEEE Transactions on Very Large Scale Integration (VLSI) Systems* (2025).

[19] Linus Witschen et al. "MUSCAT: MUS-based circuit approximation technique". In: *2022 Design, Automation & Test in Europe Conference & Exhibition (DATE)*. IEEE. 2022, pp. 172–177.

[20] Morteza Rezaalipour et al. "A Parametrizable Template for Approximate Logic Synthesis". In: *2023 53rd Annual IEEE/IFIP International Conference on Dependable Systems and Networks Workshops (DSN-W)*. IEEE. 2023, pp. 175–178.

[21] Linus Witschen et al. "Search space characterization for approximate logic synthesis". In: *Proceedings of the 59th ACM/IEEE Design Automation Conference*. 2022, pp. 433–438.

[22] Ilaria Scarabottolo et al. "A formal framework for maximum error estimation in approximate logic synthesis". In: *IEEE Transactions on Computer-Aided Design of Integrated Circuits and Systems* 41.4 (2021), pp. 840–853.

[23] M. Barbareschi et al. "A genetic-algorithm-based approach to the design of DCT hardware accelerators". In: *ACM Journal on Emerging Technologies in Computing Systems* 18.3 (2022), pp. 1–25.

[24] Muhammad Awais, Hassan Ghasemzadeh Mohammadi, and Marco Platzner. "An MCTS-based Framework for Synthesis of Approximate Circuits". In: *Proceedings of the 26th IFIP/IEEE International Conference on Very Large Scale Integration (VLSI-SoC)*. IEEE. 2018, pp. 219–224. DOI: 10.1109/VLSI-SoC.2018.8645026.

[25] Muhammad Awais, Hassan Ghasemzadeh Mohammadi, and Marco Platzner. "DeepApprox: Rapid Deep Learning based Design Space Exploration of Approximate Circuits via Check-pointing". In: *2024 IEEE Computer Society Annual Symposium on VLSI (ISVLSI)*. IEEE. 2024, pp. 88–93.

[26] Chandan Kumar Jha et al. "cecApprox: Enabling Automated Combinational Equivalence Checking for Approximate Circuits". In: *IEEE Transactions on Circuits and Systems I: Regular Papers* 71.7 (2024), pp. 3282–3293.

[27] Marzieh Vaeztourshizi, Mehdi Kamal, and Massoud Pedram. "EGAN: A Framework for Exploring the Accuracy vs. Energy Efficiency Trade-off in Hardware Implementation of Error Resilient Applications". In: *Proceedings of the International Symposium on Quality Electronic Design (ISQED)*. IEEE. 2020, pp. 438–443.

[28] Salim Ullah, Siva Satyendra Sahoo, and Akash Kumar. "CLAppED: A design framework for implementing cross-layer approximation in FPGA-based embedded systems". In: *2021 58th ACM/IEEE Design Automation Conference (DAC)*. IEEE. 2021, pp. 475–480.

[29] Ian Goodfellow et al. "Generative adversarial networks". In: *Communications of the ACM* 63.11 (2020), pp. 139–144.

[30] Vojtech Mrazek et al. "EvoApproxSb: Library of Approximate Adders and Multipliers for Circuit Design and Benchmarking of Approximation Methods". In: *Proceedings of the Conference on Design, Automation & Test in Europe (DATE)*. IEEE, 2017, pp. 258–261.

Deus Ex LLMs: AI vs Humans in Post-Quantum Cryptographic Hardware Code Generation

Ethan Cornett[1,†], Rahul Magesh[1,†], Sharath Pendyala[1], Elif Bilge Kavun[2], Aydin Aysu[1]

[1]HECTOR Research Lab, Department of Electrical and Computer Engineering, North Carolina State University, USA

Email: edcornet@ncsu.edu, rmagesh@ncsu.edu, spendya@ncsu.edu, aaysu@ncsu.edu

[2]Barkhausen Institut, Germany

Email: elif.kavun@barkhauseninstitut.org

[†]These authors contributed equally to this work.

Abstract—Emerging Post-Quantum Cryptographic (PQC) schemes such as FALCON demand highly optimized hardware implementations to meet strict area and execution time constraints on embedded devices. Traditional hardware designs rely heavily on expert-crafted Register Transfer Level (RTL) or High Level Synthesis (HLS) code, which is time-consuming and error-prone. In this work, we explore the use of large language models (LLMs) for accelerating the development of cryptographic hardware, focusing on FALCON's performance-critical `SamplerZ` subroutine. We propose a design flow that iteratively leverages LLMs to generate, refine, and evaluate synthesizable C code using HLS tools. We analyze generated designs across a range of models (e.g., GPT-4, Claude, Gemini, Grok), compare them with prior hand-crafted RTL designs, and report implementation metrics including Area-Delay Product (ADP) and synthesis convergence. Alongside achieving implementations within 4% execution time and 30% area of expert-tuned code, our results demonstrate that LLMs can discover novel hardware optimizations. We finally identify key challenges in prompt engineering, numerical stability, and testbench overfitting, and provide actionable recommendations for future AI-assisted hardware design frameworks.

Index Terms—AI, LLMs, Post-Quantum Cryptography, HLS, FALCON

I. INTRODUCTION

The advent of LLMs is transforming the landscape of code development, including code for hardware design. LLMs have demonstrated remarkable capabilities in automating tasks such as code generation, simulation, and formal verification, thereby streamlining the hardware development process [1]–[9]. These models are instrumental in accelerating the design of hardware components, including arithmetic units and signal processing modules, by generating efficient and optimized code.

Despite recent advances, applying LLMs to complex hardware subroutines remains challenging. To investigate this, we focus on the `SamplerZ` subroutine from the FALCON post-quantum signature scheme [10], which involves discrete Gaussian Sampling with intricate control flow and complex arithmetic. Its complexity and lack of algorithmic optimizations make SamplerZ an excellent test case for evaluating LLMs' creative potential in hardware code generation and exploring AI-driven design in underexplored domains.

In this paper, we conduct a case study comparing human and LLM-assisted hardware design for `SamplerZ` and its supporting subroutines. Using tools such as OpenAI's GPT [11], Anthropic's Claude 3 [12], DeepSeek [13], Google's

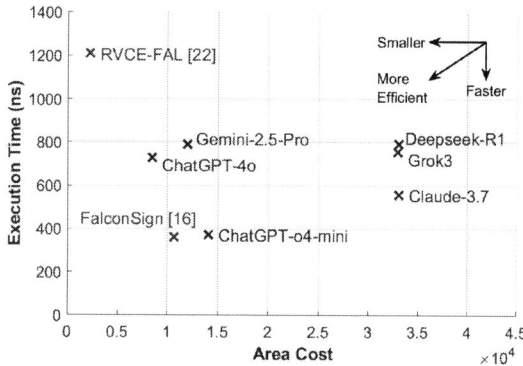

Fig. 1. Comparison of AI-generated vs human-written hardware designs of SamplerZ. LLMs can effectively help explore design space, provide insights into potential higher-level optimizations, and generate code performing similar to hand-coded, expert-driven RTL.

Gemini [14], and X's Grok [15], we restructure the program and incorporate hardware-specific pragmas. We develop a prompt-guided, feedback-driven flow that leverages synthesis results to iteratively refine LLM-generated outputs and adapt the software-oriented code for hardware synthesis. Through extensive experiments, we demonstrate that LLMs can produce functionally correct and resource-efficient hardware kernels for FALCON comparable with expert-tuned code.

Figure 1 breaks down performance and area-cost results across different LLMs and their comparison to hand-coded RTL published at premier hardware security conferences. These results reveal three key notions. First, LLMs can help explore design space and investigate area-performance trade-offs quickly. Second, LLMs can generate code closely matching expert-designed RTL; for example, our design achieves execution time within 5% and area around 30% of the state-of-the-art FALCON implementation [16]. Third, by investigating the improvements of LLM-generated code, human designers can gain insights about optimizations. In our scenario, we identified a feasible high-level optimization to improve the `ApproxExp` and `BerExp` subroutines. Our contributions are:

- We present the first known attempt to apply AI tools in generating hardware-compatible code for the `SamplerZ` subroutine, addressing a gap in current literature.
- By selecting a complex subroutine, we assess the capabilities and limitations of LLMs in handling non-traditional

hardware designs.

- We provide empirical data comparing the performance and efficiency of AI-assisted code generation against manual coding efforts, offering insights into the practical applications of LLMs in hardware development.
- We identify the challenges of applying LLMs to emerging cryptographic hardware design, and the opportunities they offer for faster and improved hardware development.

II. RELATED WORKS & BACKGROUND

Current hardware implementations of FALCON [16]–[24] identify the recursive Fast Fourier Sampling (FFSampling) subroutine as a compute bottleneck. Its core subroutine, SamplerZ, involves rejection sampling over discrete Gaussians, which presents significant implementation challenges for hardware accelerators. As SamplerZ is responsible for a large portion (over 65%) of the signature generation's execution cycles, optimizing its hardware implementation is crucial [16]. This section overviews the FALCON signature scheme and its discrete Gaussian sampler, SamplerZ.

A. FALCON Signature Scheme

Digital signature schemes are cryptographic mechanisms that ensure the authenticity, integrity, and non-repudiation of digital messages or documents. These schemes are widely used in secure communication protocols, such as SSL/TLS to prevent forgery and tampering.

FALCON is a lattice-based digital signature scheme designed for post-quantum cryptography, offering resistance against attacks from quantum computers. It comprises three main processes which leverage the NTRU lattice [25] and rely on the Fast Fourier Transform (FFT) [26]: key generation, signature generation, and signature verification. FALCON is highly efficient for embedded systems and is designed to meet the small signature sizes of constrained applications. FALCON offers two parameter sets: FALCON-512 and FALCON-1024, targeting NIST security Levels I and V, respectively. While other parameter differences exist, only the minimal standard deviation ($\sigma_{\min}$) impacts SamplerZ through the $\sigma_{\min}/\sigma'$ division, with $\sigma_{\min}$ a fixed constant for each FALCON variant.

B. Discrete Gaussian Sampler

The SamplerZ routine produces an integer from a discrete Gaussian distribution [27], [28] with center μ and deviation σ'. Algorithm 1 shows a function-inlined version of SamplerZ from [22] for simplicity. The function UniformBits refers to a source of uniform randomness, which is used throughout the algorithm for sampling and rejection decisions.

The algorithm first samples an integer z_0 from a fixed half-Gaussian distribution with standard deviation $\sigma_{\max} = 1.8205$ (lines 4–8), using a precomputed reverse cumulative distribution table (RCDT). A rejection step then evaluates whether to accept the sample. It computes an approximation of the acceptance probability $\text{ccs} \cdot e^{-x}$ (lines 11–18), where ccs is a scaling factor $\sigma_{\min}/\sigma'$, and x is derived from a centered squared distance penalty. The exponential term is evaluated in a time-constant form using a loop over precomputed constants

---

Algorithm 1 SamplerZ – Discrete Gaussian Sampling

Require: Floating-point values μ, σ' such that $\sigma' \in [\sigma_{\min}, \sigma_{\max}]$

Ensure: A sampled integer $z \in \mathbb{Z}$

1: $r \leftarrow \mu - \lfloor \mu \rfloor$
2: $\text{ccs} \leftarrow \sigma_{\min}/\sigma'$
3: **while** true **do**
4:     $u \leftarrow \text{UniformBits}(72)$
5:     $z_0 \leftarrow 0$
6:     **for** $i = 0$ to 17 **do**
7:         $z_0 \leftarrow z_0 + [\![u < \text{RCDT}[i]]\!]$
8:     **end for**   } BaseSampler
9:     $b \leftarrow \text{UniformBits}(8) \ \& \ 0x1$
10:    $z \leftarrow b + (2 \cdot b - 1) \cdot z_0$
11:    $x \leftarrow \frac{(z-r)^2}{2\sigma'^2} - \frac{z_0^2}{2\sigma_{\max}^2}$
12:    $s \leftarrow \lfloor x/\ln(2) \rfloor$
13:    $d \leftarrow x - s \cdot \ln(2)$
14:    $s \leftarrow \min(s, 63)$
15:    $y \leftarrow C[0]$
16:    **for** $i = 0$ to 12 **do**
17:       $y \leftarrow C[i] - (d \cdot y)$ } ApproxExp
18:    **end for**
19:    $y \leftarrow \lfloor \text{ccs} \cdot y \cdot 2^{63} \rfloor$
20:    $v \leftarrow (2 \cdot y - 1) \gg s$
21:    $p \leftarrow 64$
22:    **repeat**
23:       $p \leftarrow p - 8$
24:       $w \leftarrow \text{UniformBits}(8)$
         $-((v \gg p) \ \& \ 0xFF)$
25:    **until** $w \neq 0$ or $p \leq 0$
26:    **if** $w < 0$ **then**
27:       **return** $z + \lfloor \mu \rfloor$
28:    **end if**
29: **end while**

BerExp

C [29]. Finally, the sample is accepted using a byte-wise rejection loop (lines 19–25). The use of floating-point hardware is limited to the exponential approximation and comparison logic, while the half-Gaussian sampler is integer-based.

III. DESIGN METHODOLOGY AND WORKFLOW

The limited availability of RTL training data for LLMs, estimated to be over $40\times$ less than that of C/C++ [30], combined with the compelling advantages of LLM-driven HLS—including up to 38% area reduction and 64% fewer execution cycles compared to human-driven approaches [31]—motivates our choice of generating HLS code driven by LLMs. Figure 2 shows our design flow, consisting of following stages:

1) **AI Code Generation:** We extract SamplerZ and its components (BaseSampler, BerExp, ApproxExp) from the FALCON spec and NIST reference C code. Along with prototypes, parameter definitions, and Known Answer Tests (KATs) [32], these form the correctness baseline.

   Using this, we prompt LLMs to generate synthesizable, HLS-compatible C code, emphasizing algorithm

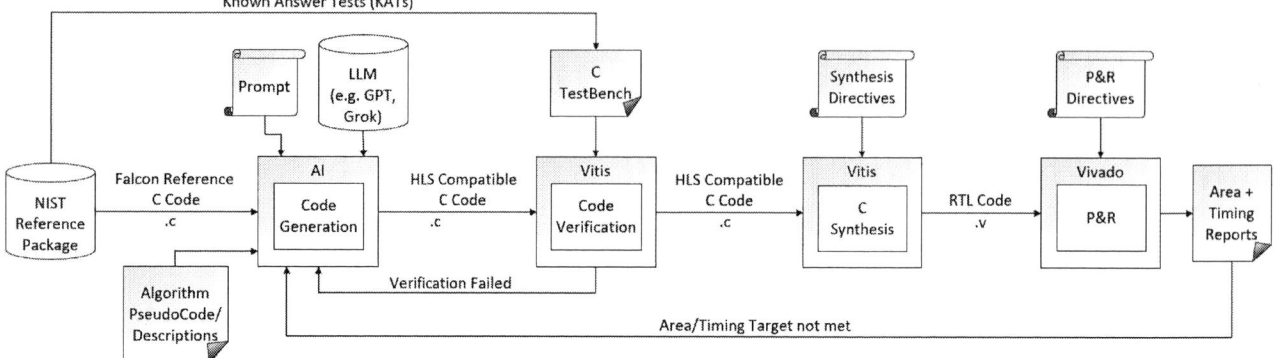

Fig. 2. AI-assisted HLS design flow from reference C-code to final hardware evaluation.

structure, control flow, and hardware synthesis features (pipelining, loop unrolling, fixed/double precision). Figure 3 shows this prompt. The resulting C files are the initial HLS design.

2) **Code Verification:** We use a hand-coded C testbench with KATs vectors to check the generated HLS code's functional correctness in AMD Vitis 2023.2 [33] (C-simulation), comparing outputs to expected KAT results. If tests fail, we refine the design by looping back to prompt/code generation until all KATs pass.

3) **C Synthesis to RTL:** After C verification, we use Vitis to synthesize the code into RTL (Verilog), producing synthesizable RTL and latency reports. This process uses the same synthesis directives for optimizations across all implementations we tested.

4) **Vivado Place and Route (P&R)** We import the HLS-generated RTL into Xilinx Vivado 2023.2 (through Vitis) for synthesis and implementation on our target FPGA (Zynq Ultrascale+ xczu7ev-ffvc1156-2-e) [34]. Vivado generates area utilization and timing reports. This process uses the same P&R directives for optimizations across all implementations we tested. If implementation fails to meet targets (timing, utilization), we refine the HLS code or directives and re-run Vitis from C Simulation to FPGA Implementation.

In this workflow, the LLM generates HLS code from the FALCON reference, Vitis manages simulation and C-to-RTL synthesis, and Vivado handles FPGA implementation. Feedback after simulation and PR guides refinements, ensuring a correct and optimized FPGA design.

IV. RESULTS

Table I presents the post-implementation FPGA results for the SamplerZ designs generated by each LLM, alongside the current state-of-the-art results reported by Ouyang et al. [16] and Yu et al. [22]. ChatGPT-o4-mini achieves execution time within 4% and Area Cost around 30% of Ouyang et al.'s work [16], and attains 60% higher clock frequency while requiring only five design iterations (20 hours) to produce functionally correct, synthesizable RTL code.

Despite being prompted to balance execution time and area–delay product, the LLMs prioritized execution time (Fig. 1). Consequently, unlike the area-optimized design by Yu

```
Goal: Generate HLS-compatible C code for four
functions ApproxExp, BerExp, BaseSampler,
SamplerZ optimized equally for latency and
area-delay product.
Return: For each function, provide .c and .h
files using exact prototypes:
• uint64_t ApproxExp(double x, double ccs);
• int BerExp(uint8_t random_byte, double x,
  double ccs);
• int BaseSampler(ap_uint<72> random_bits);
• int SamplerZ(double mu, double sigma_prime,
  fpr sigma_min, ap_uint<72> BaseSampler_rand,
  uint8_t b_rand, uint8_t BerExp_rand);
Requirements:
• Distribute the random 88 bits in SamplerZ:
  First 9 bytes for BaseSampler, next byte for
  variable b in SamplerZ, and the last byte for
  BerExp using little-endian format.
• Make sure to use the exact function prototypes
  as listed above.
• Make sure to use MAXIMUM precision for all
  calculations to get the correct functionality
• Make sure that each function can be called
  correctly by the function in the level above
  it in the hierarchy.
• Make sure that all functions are HLS compatible
  with Vitis.
Context:
• I will attach the following  files for Context
  SamplerZ.c (source code) and images of all of
  the algorithms broken into steps for easier
  optimization of each step.
• I will also include a file that defines the
  function headers that should be implemented.
```

Fig. 3. HLS generation prompt.

et al. [22], which consolidates operations onto a single shared floating-point unit (FPU) to minimize area, the LLM-assisted implementations did not employ FPU sharing, resulting in higher DSP utilization and increased area cost. Execution time remains paramount for Signature Generation due to the data dependencies between consecutive steps in the Signature Generation and FF-Sampling algorithms [10], [16], [17].

While other designs implement SamplerZ in signature generation [17], [19], [20], [23], they do not provide FPGA Utilization Metrics for SamplerZ. The work of Ouyang et al. [16] is the fastest and that of Yu et al. [22] is the smallest.

LLM-assisted design provided several notable benefits throughout our design flow:

• **Reduced Design Turnaround Time:** Rapid code gen-

TABLE I
SAMPLERZ: HLS SYNTHESIS RESULTS ACROSS LLM MODELS

Metric/Model	ChatGPT-4o	ChatGPT-o4-mini	Claude3.7	Deepseek-R1	Gemini-2.5-Pro	Grok3	[16][1]	[22]
LUTs	7299	7683	11438	14995	7524	14646	9368	4761
FFs	10096	8266	12538	17529	7959	17411	4527	692
DSPs	52	108	273	265	82	265	76	9
BRAMs	0	0	6	0	6	0	0	0
Area Cost[2]	**8412**	14013	33079	33076	11970	32974	10690	**2198**
Target Freq (MHz)	200	200	200	200	200	200	-	-
Max Freq (MHz)	231	**297**	243	222	185	233	**185**	86
Latency (Cycles)	168	111	136	176	146	176	67	104
Execution Time (ns)	727	**374**	560	793	789	775	**362**	1209
ADP ($\times 10^6$)	6.12	**5.24**	18.51	26.22	9.45	24.91	4.63	**2.66**
Design Iterations[3]	20	5	3	2	13	6	-	-

[1]: The utilization of the ChaCha20 PRNG block from [16] has been subtracted to enable a direct comparison. [2]: The Area Cost calculation is defined as Area Cost = (LUTs $\div$ 4) + (FFs $\div$ 8) + (DSPs $\times$ 102.4) + (BRAMs $\times$ 116.2), which is based on [35], [36], which provides the conversion of 1 LUT/FF/DSP/BRAM to an equivalent number of slices. [3]: Each iteration is equivalent to approximately 4 hours of design time.

eration and refinement enabled shorter iteration cycles, allowing designers to explore more ideas with minimal manual effort. For example, Deepseek-R1 reached functional correctness in 8 hours, whereas the average across all models was 33 hours.

- **Effective Handling of Low-Complexity Subroutines:** LLMs performed reliably on subroutines with minimal loop-carried data dependencies such as BaseSampler, consistently delivering functional, synthesizable RTL within short development times.

- **Discovery of Novel Optimizations:** Unexpectedly, ChatGPT-o4-mini replaced ApproxExp and parts of BerExp (lines 14–21 of Algorithm 1) with the standard exp() function. This change preserved functional correctness, simplified the dataflow, and significantly reduced area utilization by leveraging Xilinx's optimized floating-point IP. The resulting latency reduction stems from eliminating operations in the critical datapath (lines 14–16 and 21), which were required for integer-based exponentiation in the original design. Although the reference C implementation avoids floating-point exponentiation to ensure constant-time behavior, the fixed-latency nature of the Xilinx IP core allows a hardware-based replacement without violating timing guarantees.

- **Architectural Insight:** Algorithmic transformations suggested by LLMs, such as replacing complex integer arithmetic with simpler floating-point primitives, can inspire analogous optimizations in hand-written RTL, paving the way for hybrid co-optimization.

V. CHALLENGES IDENTIFIED AND LESSONS LEARNED

While LLMs accelerate design space exploration and reduce development time, we encountered several challenges when applying them to cryptographic hardware design:

- **Precision Mismatch:** LLM-generated code defaults to single-precision floating-point arithmetic, whereas the FALCON reference implementation requires double-precision (64-bit). This discrepancy compromises numerical accuracy and may undermine security unless function prototypes and headers are explicitly adjusted.

- **Arithmetic Representation Inconsistency:** Without explicit constraints, LLMs may alternate between floating-point and fixed-point arithmetic across functions or prompt iterations, leading to data-representation inconsistencies and unpredictable numerical behavior.

- **Incorrect Function Signatures:** LLMs frequently propose incomplete or incorrect function prototypes unless these are fully specified in the prompt, resulting in interface mismatches during synthesis or simulation.

- **Testbench Generation and Overfitting:** LLMs often produce testbenches that neither cover the full input range nor exercise non-trivial cases. Consequently, designs may overfit, resulting in loops being eliminated or fully unrolled when test vectors execute them only once. We mitigated this by using manually defined KATs.

- **Unsupported Bit-Widths:** LLMs sometimes suggest non-standard integer types (e.g., 72-bit integer), which are invalid in C. We resolved this by decomposing 72-bit randomness inputs into 24-bit chunks.

- **Skipped σ'^{-1} Computation:** In certain instances, LLMs used σ'^{-1} directly without computing the reciprocal of σ', introducing mathematical inconsistency.

- **Misinterpretation of Complex Expressions:** LLMs occasionally misinterpret or change expressions, such as the argument z in line 12 of SamplerZ, thereby affecting the correctness of the rejection-sampling algorithm.

VI. CONCLUSION

AI-assisted design can reduce development time for FALCON's performance-critical cryptographic subroutines to just 8 hours for functionally correct, synthesizable hardware. Within 20 hours, our implementations achieve execution time within 4% and area around 30% of state-of-the-art, human-optimized hardware. AI-driven exploration also revealed novel optimizations for the FALCON Gaussian Sampler. Ultimately, hybrid workflows that combine AI-generated suggestions with expert review and specialized RTL/FPGA knowledge can provide better results, as they address complex, domain-specific challenges and support effective, iterative refinement throughout the hardware design process.

REFERENCES

[1] Z. Zhou, L. Qiao, X. He, Y. Wang, Y. Dong, and S. Shi, "Empowering hardware development with generative ai: A review of large language models for hardware design automation," *Electronics*, vol. 14, no. 1, p. 120, 2023.

[2] S. Liu, W. Fang, Y. Lu, J. Wang, Q. Zhang, H. Zhang, and Z. Xie, "RTLCoder: Fully open-source and efficient LLM-assisted RTL code generation technique," *arXiv preprint arXiv:2312.08617*, 2023. [Online]. Available: https://arxiv.org/abs/2312.08617

[3] V. Gopinath and R. Sen, "An empirical study of LLMs for HLS-aided cryptographic design," 2024, https://arxiv.org/abs/2408.10428.

[4] K. Xu, R. Qiu, Z. Zhao, G. L. Zhang, U. Schlichtmann, and B. Li, "LLM-aided efficient hardware design automation," *arXiv preprint arXiv:2410.18582*, 2024. [Online]. Available: https://arxiv.org/abs/2410.18582

[5] F. Cui, Y. Xiao, K. Zhou, and Y. Liang, "An empirical comparison of LLM-based hardware design and high-level synthesis," *Proceedings of the 2025 ACM/SIGDA International Symposium on Field Programmable Gate Arrays*, 2025. [Online]. Available: https://dl.acm.org/doi/10.1145/3706628.3708861

[6] C. Xiong, C. Liu, H. Li, and X. Li, "HLSPilot: LLM-based high-level synthesis," in *Proceedings of the 43rd IEEE/ACM International Conference on Computer-Aided Design*, ser. ICCAD '24. New York, NY, USA: Association for Computing Machinery, 2025. [Online]. Available: https://doi.org/10.1145/3676536.3676781

[7] M. Abdollahi, S. F. Yeganli, M. A. Baharloo, and A. Baniasadi, "Hardware design and verification with large language models: A scoping review, challenges, and open issues," *Electronics*, vol. 14, no. 1, 2025. [Online]. Available: https://www.mdpi.com/2079-9292/14/1/120

[8] F. Cui, Y. Xiao, K. Zhou, and Y. Liang, "An empirical comparison of LLM-based hardware design and high-level synthesis," in *Proceedings of the 2025 ACM/SIGDA International Symposium on Field Programmable Gate Arrays*, ser. FPGA '25. New York, NY, USA: Association for Computing Machinery, 2025, p. 53. [Online]. Available: https://doi.org/10.1145/3706628.3708861

[9] Y. Li, Y. Wang, Z. Wang, Y. Wang, and Z. Wang, "Machine learning for FPGA electronic design automation," *IEEE Transactions on Computer-Aided Design of Integrated Circuits and Systems*, vol. 42, no. 10, pp. 3456–3469, 2023. [Online]. Available: https://ieeexplore.ieee.org/document/10776975

[10] P.-A. Fouque, J. Hoffstein, P. Kirchner, V. Lyubashevsky, T. Pornin, T. Prest, T. Ricosset, G. Seiler, W. Whyte, Z. Zhang *et al.*, "Falcon: Fast-Fourier lattice-based compact signatures over NTRU," *Submission to the NIST's post-quantum cryptography standardization process*, vol. 36, no. 5, pp. 1–75, 2018.

[11] OpenAI, "Chatgpt (may 2025 version) [large language model]," 2025. [Online]. Available: https://chat.openai.com/

[12] Anthropic, "Claude 3 (march 2025 version) [large language model]," 2025. [Online]. Available: https://claude.ai/

[13] D. AI, "Deepseek (may 2025 version) [large language model]," 2025. [Online]. Available: https://deepseek.com/

[14] Google, "Gemini (may 2025 version) [large language model]," 2025. [Online]. Available: https://gemini.google.com/

[15] xAI, "Grok (may 2025 version) [large language model]," 2025. [Online]. Available: https://grok.x.ai/

[16] Y. Ouyang, Y. Zhu, W. Zhu, B. Yang, Z. Zhang, H. Wang, Q. Tao, M. Zhu, S. Wei, and L. Liu, "FalconSign: An Efficient and High-Throughput Hardware Architecture for Falcon Signature Generation," *IACR Transactions on Cryptographic Hardware and Embedded Systems*, vol. 2025, no. 1, p. 203–226, Dec. 2024. [Online]. Available: https://tches.iacr.org/index.php/TCHES/article/view/11927

[17] E. Karabulut and A. Aysu, "A Hardware-Software Co-Design for the Discrete Gaussian Sampling of FALCON Digital Signature," *2024 IEEE International Symposium on Hardware Oriented Security and Trust (HOST)*, pp. 90–100, May 2024, conference Name: 2024 IEEE International Symposium on Hardware Oriented Security and Trust (HOST) ISBN: 9798350373943 Place: Tysons Corner, VA, USA Publisher: IEEE. [Online]. Available: https://ieeexplore.ieee.org/document/10545399/

[18] J. Qiu and A. Aysu, "SHIFT SNARE: Uncovering secret keys in FALCON via single-trace analysis," Cryptology ePrint Archive, Paper 2025/146, 2025. [Online]. Available: https://eprint.iacr.org/2025/146

[19] Y. Kim, J. Song, and S. C. Seo, "Accelerating Falcon on ARMv8," *IEEE Access*, vol. 10, pp. 44446–44460, 2022, conference Name: IEEE Access. [Online]. Available: https://ieeexplore.ieee.org/document/9762260

[20] Y. Lee, J. Youn, K. Nam, H. H. Jung, M. Cho, J. Na, J.-Y. Park, S. Jeon, B. G. Kang, H. Oh, and Y. Paek, "An Efficient Hardware/Software Co-Design for FALCON on Low-End Embedded Systems," *IEEE Access*, vol. 12, pp. 57947–57958, 2024, conference Name: IEEE Access. [Online]. Available: https://ieeexplore.ieee.org/document/10496572

[21] M. Schmid, D. Amiet, J. Wendler, P. Zbinden, and T. Wei, "Falcon Takes Off - A Hardware Implementation of the Falcon Signature Scheme," *IACR Cryptol. ePrint Arch.*, 2023. [Online]. Available: https://www.semanticscholar.org/paper/Falcon-Takes-Off-A-Hardware-Implementation-of-the-Schmid-Amiet/25690dd103f05b55ceeba78c00d646703b5e3dad

[22] X. Yu, Y. Sun, Y. Zhao, H. Kuang, and J. Han, "RVCE-FAL: A RISC-V Scalar-Vector Custom Extension for Faster FALCON Digital Signature," in *2024 Design, Automation & Test in Europe Conference & Exhibition (DATE)*, Mar. 2024, pp. 1–6, iSSN: 1558-1101. [Online]. Available: https://ieeexplore.ieee.org/document/10546713

[23] D. T. Nguyen and K. Gaj, "Fast Falcon Signature Generation and Verification using ARMv8 NEON Instructions," in *Progress in Cryptology - AFRICACRYPT 2023: 14th International Conference on Cryptology in Africa, Sousse, Tunisia, July 19–21, 2023, Proceedings*. Berlin, Heidelberg: Springer-Verlag, 2023, p. 417–441. [Online]. Available: https://doi.org/10.1007/978-3-031-37679-5_18

[24] D. Soni, K. Basu, M. Nabeel, N. Aaraj, M. Manzano, and R. Karri, *Hardware Architectures for Post-Quantum Digital Signature Schemes*. Springer International Publishing, 2020. [Online]. Available: https://doi.org/10.1007/978-3-030-57682-0

[25] J. Hoffstein, J. Pipher, and J. H. Silverman, "NTRU: A ring-based public key cryptosystem," in *Algorithmic Number Theory*, J. P. Buhler, Ed. Berlin, Heidelberg: Springer, 1998, pp. 267–288.

[26] L. Ducas and T. Prest, "Fast Fourier Orthogonalization," in *Proceedings of the 2016 ACM International Symposium on Symbolic and Algebraic Computation*, ser. ISSAC '16. New York, NY, USA: Association for Computing Machinery, Jul. 2016, pp. 191–198. [Online]. Available: https://dl.acm.org/doi/10.1145/2930889.2930923

[27] J. Howe, A. Khalid, C. Rafferty, F. Regazzoni, and M. O'Neill, "On practical discrete gaussian samplers for lattice-based cryptography," *IEEE Transactions on Computers*, vol. 67, no. 3, pp. 322–334, Mar. 2018.

[28] L. Kong, S. Li, and R. Liu, "High-performance constant-time discrete gaussian sampling," *IEEE Transactions on Computers*, vol. 70, no. 7, pp. 1019–1033, Jul. 2021.

[29] R. K. Zhao, R. Steinfeld, and D. Liu, "FACCT: FAst, Compact, and Constant-Time discrete gaussian sampler over integers," *IEEE Transactions on Computers*, vol. 69, no. 1, pp. 126–137, Jan. 2020. [Online]. Available: https://doi.org/10.1109/TC.2019.2940949

[30] J. Gai, H. Chen, Z. Wang, H. Zhou, W. Zhao, N. Lane, and H. Fan, *Exploring Code Language Models for Automated HLS-based Hardware Generation: Benchmark, Infrastructure and Analysis*. New York, NY, USA: Association for Computing Machinery, 2025, p. 988–994. [Online]. Available: https://doi.org/10.1145/3658617.3697616

[31] Y. Liao, T. Adegbija, and R. Lysecky, "Are LLMs any good for high-level synthesis?" in *Proceedings of the 43rd IEEE/ACM International Conference on Computer-Aided Design*, ser. ICCAD '24. New York, NY, USA: Association for Computing Machinery, 2025. [Online]. Available: https://doi.org/10.1145/3676536.3699507

[32] T. Prest, "falcon.py," https://github.com/tprest/falcon.py.git, December 2022, accessed: 2025-05-30.

[33] AMD, *Vitis HighLevel Synthesis User Guide*, Advanced Micro Devices, Inc., Mar. 2025, uG1399 (v2024.2). [Online]. Available: https://docs.amd.com/r/en-US/ug1399-vitis-hls

[34] AMD, *UltraScale Architecture Overview*, 2023, document Number: DS890. [Online]. Available: https://docs.amd.com/v/u/en-US/ds890-ultrascale-overview

[35] W. Liu, S. Fan, A. Khalid, C. Rafferty, and M. O'Neill, "Optimized Schoolbook Polynomial Multiplication for Compact Lattice-Based Cryptography on FPGA," *IEEE Transactions on Very Large Scale Integration (VLSI) Systems*, vol. 27, no. 10, pp. 2459–2463, 2019.

[36] "7 Series FPGAs Data Sheet: Overview (DS180)," 2020. [Online]. Available: https://docs.amd.com/v/u/en-US/ds180_7Series_Overview

Open-Source Approach to IC Development: Validation Against Measurements of Selected Devices from the IHP-Open-PDK

1st Krzysztof Herman
Department of Technology
The Leibniz Institute for High Performance Microelectronics
Frankfurt Oder, Germany
herman@ihp-microelectronics.com

2nd Dietmar Warning
Department of Technology
The Leibniz Institute for High Performance Microelectronics
Frankfurt Oder, Germany
warning@ihp-microelectronics.com

Abstract—This paper presents a comparative study between the measurement data provided in the IHP-Open-PDK and the simulation results obtained using open-source tools. Our primary objective is to evaluate how accurately open source simulators, specifically Ngspice and Xyce, can replicate real-world performance when used in conjunction with open-source device models developed by IHP. To ensure a meaningful comparison, we recreated the measurement conditions within our simulation environment, carefully aligning the test scenarios and circuit setup. The study focuses on key discrepancies observed between our open-source simulation results and reference simulations performed with proprietary EDA tools, as documented in the IHP-Open-PDK repository. We analyze these differences in detail and discuss possible causes, including modeling limitations, simulator-specific behaviors, and configuration variations. Our findings highlight areas where open source tools currently diverge from industry-standard results and suggest targeted improvements for future tool and model development. By systematically revealing these differences, our work contributes valuable feedback to the open-source hardware design community. The insights gained from this study will help developers and users of open-source PDKs and simulation tools to improve model accuracy, improve tool compatibility, and ultimately promote broader adoption of open-source methodologies in integrated circuit design workflows.

Index Terms—open-source silicon, OpenPDK,

I. INTRODUCTION

The IHP-Open-PDK, described in the papers [1] and [2], has been announced at the beginning of the 2023 and continues in a preview mode, what means being in a development phase. The first stable release of the PDK should be available at the end of the 2025. The fundamental motivation of IHP is not only to provide reliable solutions for education and training, but also to advance open-source silicon solutions toward production-level readiness. The design kit itself targets mainly RF designs since the SG13G2 process delivers high frequency silicon germanium bipolar devices operating up to 350 GHz. Beside the HBT's (Heterojunction Bipolar Transistor) the PDK showcase also low and high voltage CMOS devices, a varicap, ESD structures, diodes, tap devices, a few poly silicon resistors and a MiM capacitor. Since the design kit contains also digital cells; standard cells, input/output

cells and SRAM macro blocks, it can be used also for the development of digital and mixed signal circuits. Inside the PDK one can find: device symbols for open-source schematic capture tools, model libraries for simulators, technology files and parametric cells for layout editors, DRC and LVS rule decks, and also reference libraries for digital design flows. The versatile capabilities of the PDK point to a wide range of potential applications, with particular emphasis on RF and analog/mixed-signal circuit design. One of the most important and indispensable part of the design kit are device models, which should be accurate and compatible with the open-source EDA simulators like Ngspice or Xyce. This paper aims to demonstrate that open-source solutions are capable of producing reliable results, consistent with measurements from fabricated silicon. Over the last five years, multiple OpenMPW shuttles have been carried out, employing all three available and manufacturable PDKs—SKY130, GF180, and IHP-SG13G2. Although designers received the fabricated silicon, very little information has been reported regarding actual measurement data obtained from open-source circuits or system. An unquestionable leader here is the TinyTapeout project described here [3], which offers low cost access to a small piece of area in order to develop a tiny circuit. There are also some research papers published recently [4], [5], which report details about the post silicon behaviour of the chips designed using open-source EDA and PDK. Regardless of the information channel—be it scientific journals, conference proceedings, press releases, or social media—the ultimate proof of the open-source EDA and PDK concept lies in a working design. This article aims to provide a detailed comparison of measurement results with simulation data derived from open-source methodologies and models.

II. METHODOLOGY

From a timeline perspective, the SKY130 process was released in 2020 as the first open-source PDK suitable for manufacturing. Two years later, a dedicated GitHub repository was established to supplement the PDK with measurement data for the available devices. In 2023, the Leibniz Institute for

979-8-3315-9813-6/25 $31.00 © 2025 IEEE

High Performance Microelectronics (IHP) released the IHP-Open-PDK, thereby making the details of the SG13G2 process publicly available. from the outset, IHP included measurement data for active devices such as HBTs (Heterojunction Bipolar Transistor) and MOSFETs, accompanied by measurement reports and comparisons with proprietary simulators. In booth PDKs the measurements data were published using MDM (multi dimensional measurements) data format. These ASCII files are usually divided into informative header section and data section. The analysis of the MDM file provides detailed information on how to reproduce the same data in simulation. In the public domain, several Python modules are also available to facilitate importing and processing the data using this popular language. The approach used in this paper is shown on the Fig. 1. Based on a test structure and associated MDM

respectively, which are available in the open-source domain through the following github repository [8]. After analysis of the measurements data, especially sections available in the headers we have reproduced respective test benches in order to run simulation, which mimics the measurement setup. The first test bench shown on the Fig. 2 presents a setup for measuring noise spectral density on a given resistor R for different bias conditions forced by bias I_P current. The inductor $L = 1H$ blocks AC noise component and together, with the shunt capacitor $C = 1F$ force the condition that the total noise signal produced on the resistor can be measured as a current I_1. Since Ngspice does not support current noise as a direct output of the noise analysis and Xyce supports it partially we have introduced a CCSV source $H1$ in order to transform the current I_1 into a voltage V_{out} since an unity gain of the CCSV source. The second comparison relates to

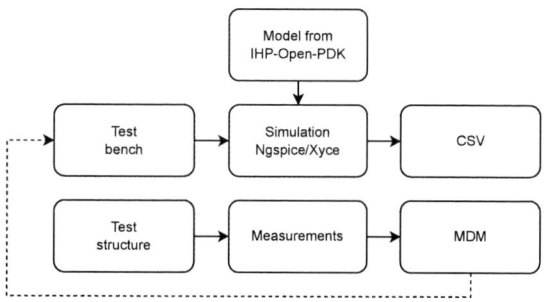

Fig. 1. Current noise spectral density evaluated using Ngspice and Xyce simulators.

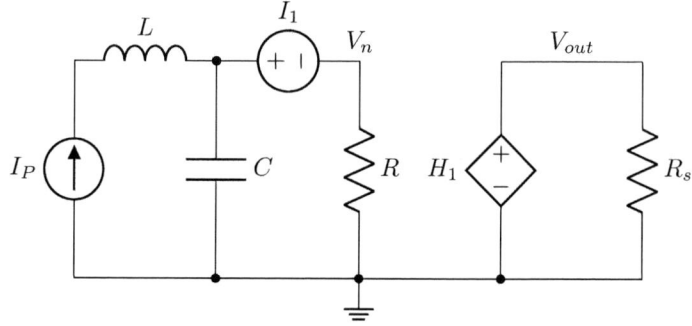

Fig. 2. Current noise density evaluation test bench used to reproduce the measurements setup.

file we were reproducing a test bench using open-source EDA tools in order to verify the quality of the results reproduced during the simulation. The EDA tool chain consists of Xschem schematic capture, which streams simulator dependent netlist. We have used two open-source simulators: (1) Ngspice and (2) Xyce available on the sourceforge and github platforms respectively. Both simulators claim to be compatible with the SPICE language; however, there are several differences that have been addressed during the preparation of the PDK libraries. One method of achieving compatibility is the use of behavioral models for devices instead of relying on built-in models. Unfortunately both compilers do not have a native support for Verilog-A defined devices, however there are existing open-source tools, which bridge these gaps reported in this article [6]. In the case of Ngspice simulator it can support Verilog-A compiled binaries using a non standard OSDI (Open Source Device Interface) supported by Ngspice. There also exists a tool, namely OpenVAF, which is an open-source Verilog-A compiler which generates OSDI binaries supported by open source simulators. Similarly Xyce simulator supports Verilog-A defined devices using plugins (shared objects) compiled from the source code using an open-source ADMS compiler. In this article we have been analyzing two devices: an unsalicided poly silicon resistor and second generation, ultra fast SiGe HBT transistor described in this paper [7] . We have applied Verilog-A models, namely R3_CMC and VBIC

the simulation and the measurements of a maximum transit frequency F_T of a HBT device, namely npn13G2, which is a second generation SiGe device elaborated by IHP and institute featuring f_T up to 350 GHz. The MDM file, in the data section, contains the scattering S-parameter vectors as a function of frequencies up to 65 GHz, which is a physical limitation of the measurement setup. The S-parameters were measured in a common emitter configuration for a different bias conditions. In order to extrapolate the f_T frequency, which is out of the limits of the measurement setup we assume that the amplifier's current gain h_{21} at 30 GHz is already situated on the falling slope of the $h_{21}(f)$ function decreasing $-3dB/dec$. The aforementioned assumption enables us to apply gain bandwidth product equation to calculate f_T as $h_{21}@30GHz \cdot 30GHz$. Since the measurements in the mm-Waves range is affected by the parasitic elements introduced by the test structures the measurements data set contains also the S-parameter values for de-embedding the measurements results from effects introduced by the parasitics. More about de-embedding techniques can be found in this article [9]. Having the de-embedded set of S-parameters we have transformed it into H-parameters according the methods described in this paper [10] in order to extract the value of $h_{21}@30GHz$ for different bias conditions. The simulation test

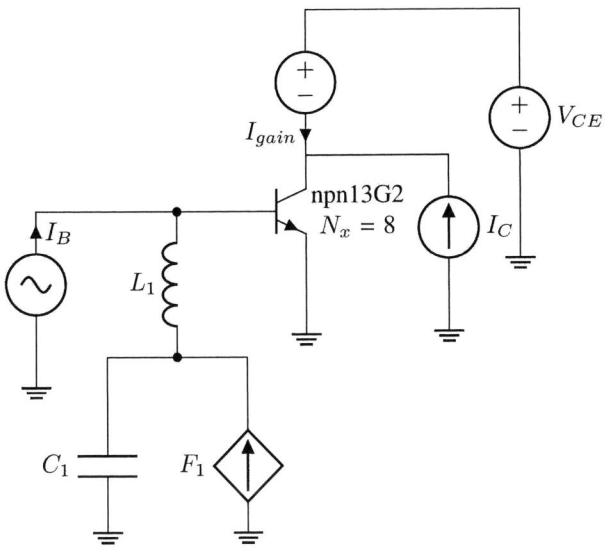

Fig. 3. A simulation test bench for evaluation of the transit frequency of the HBT devices.

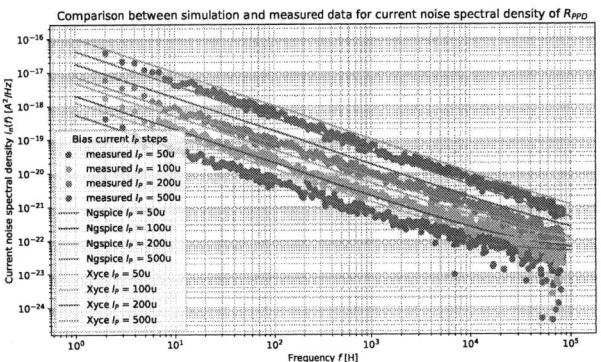

Fig. 4. Current noise spectral density evaluated using Ngspice and Xyce simulators.

The results of transit frequency estimation can be found on the Fig. 5, where the current value was varied from from $10~\mu A$ to $60~mA$. The respective V_{CE} voltage was changed from 0.4 V to 1.6 V with 0.4 V step.

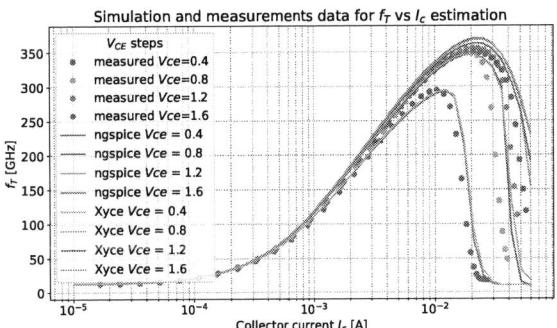

Fig. 5. Transit frequency estimation for bipolar devices using open-source model and Ngspice and Xyce simulators.

bench is shown on the Fig. 3. It applies base AC current I_B and measures output current I_{gain} since during AC analysis I_C source remains open, while V_{CE} is seen as a short. In the case of both simulator we have used the measurement feature in order to estimate the value of the current gain. The following listings present the respective code sections for Ngspice and Xyce simulators.

```
* Ngspice
ac dec 100 1G 1000g
meas ac gain_at find vm(vgain#branch) at=30GHz
let ft = {$ac}.gain_at*30
```

```
* Xyce
.ac dec 100 1G 1000g
.measure ac gain_at find IM(vgain) at=30GHz
```

Both simulators have different way to report the output data. In the case of Ngspice we have used the internal control interface of Ngspice, which permits to declare variables. Once accumulated the variables (vectors) can be written to an CSV file. Since the Xyce simulator does not have this feature it reports the measured value into the output log, which we post process creating a separate CSV file. After preparing the data sets we have used python language in order to process the measured and simulated data.

III. RESULTS

The thermal and flicker noise comparison for the poly silicon resistor R_{PPD} having respective width and length equal to $w = 0.5\mu m$ and $l = 0.5\mu m$, what implies a value of 396Ω is shown on the Fig. 4. The simulation range of frequencies encompasses the range from 1 Hz to 100 kHz and the bias current varies from $50\mu A$ to $500\mu A$. We have plotted measurements results as a points and the results of the simulations using continues lines of different colors.

IV. DISCUSSION

Analyzing the results shown on the Fig. 4 we can observe a good match between the measurements data and the results obtained using Ngspice simulator. Although both simulators use the same model card, where the values of the parameters responsible for flicker noise modelling are equal to $kfn = 4.60110^{11}$, $afn = 1.886$, $bfn = 0.9963$, the simulation results obtained using Xyce tool overestimate the value of current noise spectral density. The flicker noise on the resistors is modeled using the Compact modeling Council CMC approach described in the paper [11]. Also it is worth to mention that Xyce simulator does not have built-in resistor model, which would support flicker noise estimation so the approach presented here is the only way to calculate flicker noise contribution. In the region of the frequencies, where the total noise spectral density is dominated by the thermal noise, both simulators converge to the same value.

In the case of transit frequency evaluation of the HBT device we have obtained good alignment with the measurements in

979-8-3315-9813-6/25 $31.00 © 2025 IEEE

case of both simulators, what is shown on the Fig. 5. Although the simulation results are slightly off in the high-current range—where the maximum performance of the HBT device begins to degrade—they are still useful for estimating the bias conditions to target maximum performance. The results of the simulation can be compared to a similar evaluation, where a proprietary tool was used to simulate the $f_T = f(I_C)$ dependency using the following link. The Ngspice and Xyce simulators have native support for the VBIC model, however the structure of the library (model card) can be different in both cases. Since we applied Verilog-A VBIC model we can easily switch between the simulators using the same or almost identical model card.

Since the simulators deliver the same results, one might ask: Which one should I use? The answer depends on the chosen metrics. In the open-source domain, one of the key aspects of a tool or project is whether it is actively maintained. Many excellent open-source projects are no longer supported, which can lead to bottlenecks in the long term.

Among the available tools and PDKs—particularly the IHP Open PDK—the optimal choice from this perspective is to use Xschem and Ngspice in combination with OpenVAF, as these projects are under active development and offer clear roadmaps. In the case of Xyce, the project is active; however, the development pace is slower compared to Ngspice. Additionally, ADMS, which is a Verilog-A compiler for Xyce, is not supported at the time of writing this article. Nevertheless, these tools can complement each other—offering, for example, a rich control interface in the case of Ngspice, and robust, concise parametric analysis capabilities in the case of Xyce, where a .STEP command is available. From the PDK development perspective maintaining support for two set of tools is an additional workload nevertheless since the flows based on Ngspice and Xyce can be complementary it can be beneficial for the end user, who will dispose a broader spectrum of possibilities during the integrated circuit development.

V. Conclusions

This study presents the comparison between the simulation results and the measurements for two devices from the IHP-Open-IHP PDK. We have shown, that the open source EDA tools together with open source models from the PDK are capable to deliver reliable results. The quality of results plays an indisputable role in the industrial adoption of open-source solutions within the silicon technology domain. The successful proliferation of open-source silicon into commercial domains depends on fostering greater trust across the entire ecosystem. This article also showcases a method for homogenizing PDK libraries, where the Verilog-A models contribute to the process by flattening the differences between the two simulators. The study further exemplifies an approach to analyzing datasets from the GitHub repository, thereby encouraging interested researchers to reproduce the results or leverage the available data for additional insights.

Acknowledgment

The authors would like to thank all the contributors to open source silicon solutions and also acknowledge the funding agency BMFTR for providing generous support to this study under the project FMD-QNC grant No. (16ME0831).

References

[1] K. Herman, N. Herfurth, T. Henkes, S. Andreev, R. Scholz, M. Müller, M. Krattenmacher, H. Pretl, and W. Grabinski, "On the Versatility of the IHP BiCMOS Open Source and Manufacturable PDK: A step towards the future where anybody can design and build a chip," *IEEE Solid-State Circuits Magazine*, vol. 16, no. 2, pp. 30–38, 2024.

[2] K. Herman, R. Scholz, and S. Andreev, "Reflections on the first european open source pdk by ihp-experiences after one year and future activities," in *2024 31st International Conference on Mixed Design of Integrated Circuits and System (MIXDES)*. IEEE, 2024, pp. 19–22.

[3] M. Venn, "Tiny tapeout: A shared silicon tape out platform accessible to everyone," *IEEE Solid-State Circuits Magazine*, vol. 16, no. 2, pp. 20–29, 2024.

[4] J. Marin, C. A. Rojas, A. H. Wilson-Veas, N. Salvador, J. Gak, N. Calarco, M. Miguez, and A. R. Oliva, "Open-source multilevel converter power ic design and test," *IEEE Design & Test*, vol. 41, no. 6, pp. 19–27, 2024.

[5] R. H. Yang and Y. Xia, "An open-source 12-bit 10-ks/s incremental adc in 130-nm cmos," *IEEE Design & Test*, vol. 41, no. 6, pp. 28–35, 2024.

[6] Á. Bűrmen, T. Tuma, I. Fajfar, J. Puhan, Ž. Rojec, M. Kunaver, and S. Tomažič, "Free software support for compact modelling with verilog-a," *Informacije MIDEM*, vol. 54, no. 4, 2024.

[7] B. Heinemann, R. Barth, D. Bolze, J. Drews, G. Fischer, A. Fox, O. Fursenko, T. Grabolla, U. Haak, D. Knoll *et al.*, "Sige hbt technology with f t/f max of 300ghz/500ghz and 2.0 ps cml gate delay," in *2010 International Electron Devices Meeting*. IEEE, 2010, pp. 30–5.

[8] Dietmar Warning, "Verilog-A Models for Circuit Simulation," https://github.com/dwarning/VA-Models, 2025, accessed: 2025-05-28.

[9] X. S. Loo, K. S. Yeo, and K. Chew, "On-wafer microwave de-embedding techniques," in *Microwave Systems and Applications*. IntechOpen, 2017, pp. 101–120.

[10] D. A. Frickey, "Conversions between s, z, y, h, abcd, and t parameters which are valid for complex source and load impedances," *IEEE Transactions on microwave theory and techniques*, vol. 42, no. 2, pp. 205–211, 1994.

[11] W. C. Pflanzl and E. Seebacher, "1/f noise temperature behaviour of poly resistors," in *Proceedings of the 19th International Conference Mixed Design of Integrated Circuits and Systems-MIXDES 2012*. IEEE, 2012, pp. 297–299.

Toward Multi-Person Breath Rate Estimation via mmWave Radar

Cristian Turetta*, Christian Farina*, Chiara Bozzini‡, Morteza Varasteh†, and Graziano Pravadelli‡

*Dep. of Computer Science, University of Verona, Italy, name.surname@studenti.univr.it
†School of Computer Science and Electronic Engineering, University of Essex, UK, m.varasteh@essex.ac.uk
‡Dep. of Engineering for Innovation Medicine, University of Verona, Italy, name.surname@univr.it

Abstract—The measurement of breath rate (BR) is essential for comprehensive human health monitoring across a wide range of scenarios. Several studies in the literature have explored the estimation of BR using millimeter wave (mmWave) technology. However, these approaches typically focus on a single subject at a time. To enable multi-person estimation, researchers have often relied on data fusion with camera systems or employed specialized hardware configurations. On the contrary, this paper proposes a methodology that employs only one Frequency Modulated Continuous Wave (FMCW) radar to estimate the BR of multiple subjects stationary in the environment. The proposed methodology includes a pre-processing pipeline to refine the radar-captured signals, followed by frequency-domain analysis to distinguish between subjects. Finally, phase variations in the reflected signals caused by chest movements are analyzed to estimate the BR. Advantages and limitations of the approach are discussed on the basis of an experimental campaign.

Index Terms—Breath rate estimation, vital-sign estimation, mmWave radar, multi-person monitoring

I. INTRODUCTION

Monitoring breath rate (BR) is a key factor in assessing an individual's physiological state across various scenarios, including clinical diagnostics, sports performance, and home-based healthcare. Traditionally, BR has been measured using contact-based techniques such as spirometry, chest-worn sensors, facial masks, or nasal cannulas. While these methods are generally precise, they can interfere with natural movement and behavior due to their intrusive design and constraints like battery limitations and dependence on external devices [1], [2].

Advancements in monitoring technologies have increasingly focused on reducing individual's discomfort and enabling unobtrusive tracking of physiological signals. This shift has driven significant progress in the development of contactless sensing approaches, particularly through the use of Radio Frequency (RF) technologies that fill a significant gap in health monitoring practices, providing the capability to measure physiological parameters non-invasively [3]–[5].

In the context of RF-based methods, several studies have explored millimeter wave (mmWave) for vital signs detection

This study was carried out within the MICS (Made in Italy – Circular and Sustainable) Extended Partnership and received funding from Next-Generation EU (Italian PNRR – M4 C2, Invest 1.3 – D.D. 1551.11-10-2022, PE00000004). CUP MICS D43C22003120001 - Cascade funding project CollaborICE, and within the PRIN 2022 project "WE.SMOOTH.PD: a WEarable-based System to MOnitor motor functions and levodopa levels for THerapy optimization in Parkinson's Disease" funded by European Union - Next Generation EU, Mission 4 Component 1 CUP 2022EJM345.

TABLE I: Characteristics of SOTA approaches for BR estimation through mmWave radar technologies together with the MAE and MAPE.

Ref.	Year	Hardware	MAPE	MAE	# Sbj	Movements
[7]	2016	Dedicated HW	N/A	0.43	2	No
[8]	2022	TI AWR1642	1.33%	N/A	1	No
[9]	2024	TI AWR1642	4.44%	N/A	1	No
[10]	2024	TI AWR1243	26.72%	2.48	1	Yes, freely
[11]	2023	TI AWR1843 + Camera	N/A	0.5	3	Yes, in place

MAE: Mean Absolute Error – MAPE: Mean Absolute Percentage Error

in different settings. These approaches make use of Frequency Modulated Continuous Wave (FMCW) radars, such as the AWR1xxx series from Texas Instruments, which typically operate in high-frequency bands (e.g., 77–81 GHz), to detect subtle human body movements, such as those produced by respiration and heartbeat on the chest [6]. The periodic chest displacements caused by these physiological activities induce slight variations in the reflected radar signals, which can be analyzed to estimate BR, and also heart rate (HR).

Consequently, in the recent past, some approaches have been proposed to perform vital sign estimation through mmWave, the current state-of-the-art (SOTA) is summarized in Table I. In a pioneering work of 2016, Yang *et al.* [7] propose a system that uses 60 GHz mmWave signals for vital sign monitoring. Their setup is composed by two antennas, a transmitter and a receiver, placed at a distance d inside the environment. Thanks to this setup, authors were able to obtain a wide field of view and detailed spatial information, which made them achieve a Mean Absolute Error (MAE) of 0.43. Authors in [8] performed vital sign monitoring of a person sitting on a chair inside an office using FMCW operating at 77 GHz. Furthermore, they introduced a wavelet-based denoise approach to the mmWave data. Sadeghi *et al.* [9] proposed a FMCW radar dataset, with ground truth annotation provided by a Polar H10 chest band. Besides, the authors also provided benchmark results, obtaining a Mean Absolute Percentage Error (MAPE) of 7.33% and 4.44% respectively for HR and BR estimation. In [10], the authors present a methodology to estimate BR of a subject freely moving inside and environment, achieving a MAPE of 26.72%, which correspond, in average, to an error of 2.48 breaths per minute. Wang *et al.* [11] proposed a data fusion-based approach. In particular, they use a Kinect V2 camera to enhance the mmWave estimation, allowing multiple subjects to monitor the environment and achieve promising results.

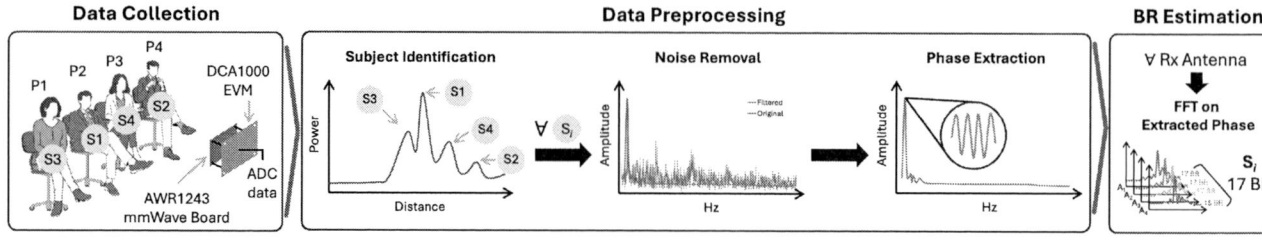

Fig. 1: Methodology Overview

Overall, the previous methodologies, while promising, present some limitations: either the BR estimation is limited to a single subject [8]–[10], or they use a dedicated hardware setup to monitor multiple subjects [7], [11]. Furthermore, using other data sources, such as video streaming, like in [11], can limit the usability and privacy-aware characteristics of the mmWave technology.

To address the previous limitations, this paper introduces a novel mmWave-based technique for estimating the BR of multiple subjects simultaneously present within an environment, without relying on dedicated hardware or data fusion from multiple sources. The proposed methodology operates under the assumption that subjects remain stationary. While this may appear restrictive, it is highly relevant in practical scenarios such as monitoring individuals in senior residences or hospital rooms, particularly during sleep. In particular, the main contributions of our work are:

- A signal processing pipeline capable of identifying and isolating stationary individuals simultaneously present within the radar's field of view. The pipeline incorporates spatial filtering techniques to differentiate also between closely positioned subjects.
- A BR estimation technique based on Fast Fourier Transformation (FFT) spectrum analysis that accurately extracts respiratory frequencies even in the presence of potential cross-subject interference. Our algorithm incorporates adaptive noise suppression and peak validation mechanisms.
- An analysis of the proposed method through experimental validation, comparing our radar-derived estimations against ground truth data collected using Hexoskin Pro Kits [12], a clinically validated wearable respiratory monitoring system.

The remainder of the paper is organized as follows: Section II presents the proposed methodology; Section III discusses the obtained results; and Section IV provides the concluding remarks.

II. METHODOLOGY

With reference to Figure 1, this section describes the methodology designed to perform BR estimation of multiple persons inside an environment. The methodology is composed of three main parts: i) data collection, ii) preprocessing pipeline, and iii) BR estimation.

A. Data Collection Setup

Our data collection setup employs the Texas Instruments (TI) AWR1243BOOST mmWave board for capturing radar signals. This board interfaces with the DCA1000EVM data capture adapter, enabling real-time streaming of ADC data to a host computer system. We managed the entire data acquisition process through TI's mmWave Studio software package, which allows for the configuration of capture parameters. We configured the system with two transmitting antennas and four receiving antennas to optimize respiratory movement detection, exploiting an enhanced Signal-to-Noise Ratio (SNR) and wider angular resolution. The measurement rate was set at 25 Hz, meaning it collects 25 frames per second, to adequately sample breath-related motion. Each frame is composed of 99 chirps that capture 374 analog-to-digital (ADC) samples. In practice, each frame collected from the mmWave board can be represented by a 3D matrix where the first dimension corresponds to the number of chirps, the second represents the number of ADC samples, and the last denotes the number of antennas. A chirp refers to a frequency-modulated signal whose frequency increases over time during a fixed time interval, known as the chirp duration. The mmWave radar continuously transmits a sequence of chirps and listens to the reflected signals from objects within its sensing environment. The transmitted signal can be modeled as a linearly increasing frequency sweep, covering a bandwidth of 4 GHz, which enables fine range resolution. When the transmitted chirp hits a moving or stationary object, the signal is reflected back with a delay proportional to the subject/object's distance. Additionally, if the object moves (e.g., due to breathing-induced chest motion), the reflected signal experiences a phase shift and a small frequency offset due to the Doppler effect. By analyzing time delay and frequency shift across successive chirps, it is possible to estimate both the range and the motion characteristics of the target. Therefore, each collected chirp contains spatial and temporal information crucial for detecting respiratory movements.

B. Preprocessing Pipeline

To estimate the BR of multiple subjects from the FMCW, the collected data flow through the following preprocessing pipeline.

1) Data Preparation: From the ADC samples collected by the AWR1243BOOST, we organized the radar data into a 4D tensor with the following dimensions: (frames × chirps × ADC samples × antennas). Then, we applied the proposed methodology using a time windowing approach, with window sizes of, for example, 30, 60, or 120 seconds. Each window contains a fixed number of frames, determined by the window duration. For example, a 30-second window at a sampling rate of 25 Hz contains 750 frames (30 s × 25 Hz). To reduce data complexity while retaining meaningful temporal information, we applied a temporal aggregation technique. Specifically, for each frame, we grouped the ADC samples (third dimension) into clusters—where each cluster represents the full set of samples from a single chirp—and averaged the rows within each cluster. In this context, a "row" refers to a single time sample from the ADC. Therefore, each cluster includes all ADC samples from a frame, and the number of rows per cluster equals the number of ADC samples. This averaging process reduces the data dimensionality but preserves the temporal structure of the radar signal, which is essential for analyzing motion patterns such as breathing. The aggregation, as well as the steps reported in the following of this section, were applied separately to the data from each receiving antenna.

2) Subject Identification: As reported above, FMCW radars transmit chirps, i.e., continuous signals whose frequency increases linearly over time. When these signals reflect off an element (either persons or objects) and return to the radar, there is a time delay between the transmitted and received signals. This delay results in a frequency difference, called the *beat frequency*, which is directly proportional to the element's distance. The Range FFT is applied to the Intermediate Frequency (IF) signal obtained after mixing the received signal with the transmitted signal. This process transforms the time-domain signal into the frequency domain. The spectral elements that demonstrate significantly higher amplitude values, which are graphically visualized as peaks relative to the surrounding frequency bins, correspond to distinct physical element present in the environment. The corresponding peak will always appear at the same index if an element is static. Using this logic, we can remove peaks corresponding to static elements by subtracting consecutive frames. As a result, we have a tensor containing peaks that show only the moving elements, i.e., the chests of the subjects stationary present in the environment.

3) Noise Removal: We perform an average on the ADC samples of the chirps belonging to the same frame to remove hardware and environmental noise, such that we can better recognize the moving entities according to their respective peaks. Then, the peaks that exceed a predetermined threshold are retained, and their indices are monitored to track the movement of the chest.

Since the chest movement is composed of inspiration and expiration phases, continuously shifting for one to the other, the distance between the chest and the radar behaves like a continuous function, thus each breath of a subject results in a cluster of closely spaced peaks. Among those, we select only the index associated with the highest peak as a representative of the cluster.

4) Phase Extraction: Using the information gathered in Section II-B3, we extract the phase information from the raw data tensor for each detected individual. At this point, we perform phase unwrapping and phase differentiation for each individual. Phase unwrapping ensures a continuous phase representation by eliminating artificial discontinuities, while phase differentiation extracts the instantaneous frequency variations, making it possible to isolate relevant physiological signals with greater accuracy [13], [14]. We then implement a band-pass filtering operation with cutoff frequencies specifically calibrated to encompass the standard human respiratory frequency spectrum, from 0.16 to 0.50 Hz, corresponding to 10-30 breaths per minute.

C. Breath Rate Estimation

The final stage in our processing pipeline is the estimation of BR for each detected subject. We perform frequency domain analysis on the extracted phase signal using the FFT. This transformation converts the time-domain respiratory oscillations into the frequency domain, where respiratory patterns appear as distinct spectral peaks. The respiratory rate is identified as the dominant frequency component in the expected physiological range for human breathing (e.g., 0.25 Hz corresponds to 15 breaths per minute). To enhance robustness against noise and movement artifacts, we leverage our multi-antenna configuration. The BR estimation process is performed independently for each of the four antennas in our radar array, resulting in four separate respiratory rate estimations (BR_1, BR_2, BR_3, BR_4) for each detected subject. We then employ a consensus mechanism to determine the final breathing rate BR_f as follows:

$$BR_f = \text{mode}(BR_1, BR_2, BR_3, BR_4) \tag{1}$$

This approach selects the most frequently occurring breathing rate estimate across all antennas. In cases of a tie (when two different rates occur with equal frequency), we select the median value to minimize the impact of outlier measurements.

III. EXPERIMENTAL RESULTS

To evaluate the proposed approach we first collected data as outlined in Section II-A, through the TI AWR1243BOOST radar operating at 77-81 Hz. For all experiments, we used the following settings: two transmitting antennas, four receiving antennas, 374 samples per chirp, 99 chirp loops, a measurement rate of 25 Hz, an ADC sample rate of 3×10^6 and a chirp slope of 3.013×10^{13}.

During the recording, multiple subjects were present in the environment, in particular, sitting in different positions facing the mmWave board. The data collection involved 4 volunteers (3 males, 1 female), breathing normally. Considering the fact that the subjects' bodies should not overlap, we chose to position them diagonally, as shown in Figure 1. All the subjects were free from respiratory, cardiac, or any other diseases that may alter the normal breathing. Each recording we performed lasts two minutes.

We collected ground truth data for breath rate using Hexoskin Pro Kit [12], which is a wearable system designed for continuous monitoring of physiological parameters. This shirt integrates Respiratory Inductance Plethysmography (RIP) bands positioned around the chest and abdomen, which detect thoracic and abdominal movements associated with breathing cycles.

TABLE II: MAPE and MAE in breath rate estimation, at varying of the number of subjects and the window size.

# Sbj	# Exp	30 sec window		60 sec window		120 sec window	
		MAPE	*MAE*	*MAPE*	*MAE*	*MAPE*	*MAE*
1	8	4.6%	0.83	2.4%	0.44	2.9%	0.55
2	7	5.7%	0.96	6.4%	1.07	2.9%	0.53
3	8	3.9%	0.70	3.1%	0.56	4.5%	0.81
4	4	4.8%	0.75	6.5%	1.03	2.1%	0.34
Average		4.7%	0.79	4.6%	0.77	3.3%	0.59

A. Analysis of the Results

To evaluate the proposed methodology, we estimated the BR varying the number of subjects sitting in the room and using different time window sizes to find the best configuration. Table II reports the achieved results in terms of estimation accuracy. The first column indicates the number of subjects simultaneously sitting in the room, while the second column shows the number of experiments conducted under the corresponding setup. The remaining columns present the MAPE and MAE values for three different window sizes: 30, 60, and 120 seconds, respectively. The MAE and MAPE are defined as follows:

$$\text{MAPE} = 100 \frac{1}{n} \sum_{t=1}^{n} \left| \frac{A_t - F_t}{A_t} \right| \qquad \text{MAE} = \frac{1}{n} \sum_{t=1}^{n} |A_t - F_t|$$

where A_t is the ground truth value, F_t is the estimated value, and n is the number of observations.

The results demonstrate that window size has a significant impact on the estimation of the BR. Short windows (e.g., 30 seconds) make it more challenging to capture complete respiratory cycles for frequency analysis, resulting in higher MAE and MAPE values. In our experiments, since subjects maintained a constant respiratory rhythm during data collection, the optimal window size was found to be 120 seconds, yielding an average MAPE of 3.3% and an average MAE of 0.59. However, one could argue that longer windows may smooth out short-term variations by averaging different BR values, potentially reducing estimation accuracy when the subject experiences sudden changes in breathing rate. While this aspect warrants further investigation, such abrupt changes are relatively infrequent within the population. On the other side, the results obtained using 30 and 60-second time windows are getting worse, respectively scoring a MAPE of 4.6% and 4.7% and a MAE of 0.77 and 0.79. With very short time windows (e.g., 15 seconds, not reported in the table), the algorithm often failed to detect significant peaks, resulting in highly inaccurate BR estimations. We faced some variability in optimal window sizes for different subject counts, particularly

evident in the 3-subject scenario. However, there is not a specific trend on accuracy degradation on the BR estimation error when several people are present simultaneously in the monitored environment.

Compared to state of the art methods exploiting TI AWRxxx devices (see Table I), the approach proposed in this paper achieves a comparable MAE and MAPE while simultaneously monitoring more subjects (up to 4). Optimal performance was obtained, in particular, when monitoring 4 subjects concurrently, achieving a MAE of 0.34 and MAPE of 2.1%.

B. Limitations and Future Work

The proposed approach presents some limitations related to the position of individuals involved and their movements. Specifically, it was not possible to accurately estimate the breathing rate of subjects who were not facing the mmWave board. In such cases, chest movements produced minimal phase variation, rendering respiratory detection ineffective. For similar reasons, the board must be positioned at chest height to maximize sensitivity to respiratory motion. Moreover, since the algorithm differentiates subjects based on their distance from the board, individuals must be placed at distinct and sufficiently spaced positions to prevent signal interference caused by adjacent chest movements. This issue was particularly evident during measurements involving three subjects, where the central individual consistently showed the highest estimation error due to interference from neighboring movements. Another limitation is the requirement for subjects to remain still, as any body movement introduces significant noise and degrades the quality of the collected data.

While these limitations may affect the general applicability of mmWave technology for BR estimation in dynamic or uncontrolled environments, there are specific scenarios where these constraints are naturally satisfied. For instance, in hospital rooms, sleep laboratories, or elderly-care facilities, individuals often remain relatively stationary for extended periods and can be positioned in a controlled manner with respect to the mmWave board. These environments also allow for proper sensor alignment at chest level, minimize motion artifacts, and reduce the likelihood of interference from nearby individuals. As such, they represent ideal use cases where the proposed approach can be effectively applied.

Future work will focus on improving robustness in less controlled environments, enhancing subject identification in crowded settings, and incorporating motion compensation techniques to extend applicability to a broader range of real-world scenarios.

IV. CONCLUSIONS

Exploiting FMCW radar, this paper presents an approach to estimate the breath rate of up to four people simultaneously sitting in an environment. We tested the proposed methodology across different settings, achieving in average, for the most challenging case of 4 simultaneous subjects, a MAPE of 2.1%, with a deviation from ground truth of only 0.34 breaths, by considering an observation time window of 120 seconds.

REFERENCES

[1] T. Hussain, S. Ullah, R. Fernández-García, and I. Gil, "Wearable sensors for respiration monitoring: A review," *Sensors*, vol. 23, no. 17, p. 7518, 2023.

[2] H. C. Bidsorkhi, N. Faramarzi, B. Ali, L. R. Ballam, A. G. D'Aloia, A. Tamburrano, and M. S. Sarto, "Wearable graphene-based smart face mask for real-time human respiration monitoring," *Materials & Design*, vol. 230, p. 111970, 2023.

[3] A. M. M. L. C. C. A. C. X. W. O. S. D. B. J. M. B. L. Le Ngu Nguyen, Praneeth Susarla, "Monitoring long-term cardiac activity with contactless radio frequency signals," *Information Fusion*, vol. 110, no. 1, pp. 1–18, 2024.

[4] P. Susarla, A. Mukherjee, M. L. Cañellas, C. Á. Casado, X. Wu, O. Silvén, D. B. Jayagopi, M. B. López *et al.*, "Non-contact multimodal indoor human monitoring systems: A survey," *Information Fusion*, vol. 110, p. 102457, 2024.

[5] A. Singh, S. U. Rehman, S. Yongchareon, and P. H. J. Chong, "Multi-resident non-contact vital sign monitoring using radar: A review," *IEEE Sensors Journal*, vol. 21, no. 4, pp. 4061–4084, 2020.

[6] J. Gong, X. Zhang, K. Lin, J. Ren, Y. Zhang, and W. Qiu, "Rf vital sign sensing under free body movement," *Proceedings of the ACM on Interactive, Mobile, Wearable and Ubiquitous Technologies*, vol. 5, no. 3, pp. 1–22, 2021.

[7] Z. Yang, P. H. Pathak, Y. Zeng, X. Liran, and P. Mohapatra, "Monitoring vital signs using millimeter wave," in *Proceedings of the 17th ACM international symposium on mobile ad hoc networking and computing*, 2016, pp. 211–220.

[8] M. Xiang, W. Ren, W. Li, Z. Xue, and X. Jiang, "High-precision vital signs monitoring method using a fmcw millimeter-wave sensor," *MDPI Sensors*, vol. 22, no. 19, p. 7543, 2022.

[9] E. Sadeghi, K. Skurule, A. Chiumento, and P. Havinga, "Comprehensive mm-wave fmcw radar dataset for vital sign monitoring: Embracing extreme physiological scenarios," *arXiv preprint arXiv:2405.12659*, 2024.

[10] C. Turetta, M. Varasteh, S. Kolozali, and G. Pravadelli, "Leveraging mmwave for contactless breath rate estimation of moving subjects," in *2024 IEEE International Conference on Digital Health (ICDH)*, 2024, pp. 33–39.

[11] Y. Wang, Z. Wang, J. A. Zhang, H. Zhang, and M. Xu, "Vital sign monitoring in dynamic environment via mmwave radar and camera fusion," *IEEE Transactions on Mobile Computing*, vol. 23, no. 5, pp. 4163–4180, 2023.

[12] Hexoskin, "Hexoskin Pro Kit," https://hexoskin.com/products/hexoskin-pro-kit-mens, 2025, accessed: 2025-04-29.

[13] Y. Chen, J. Yuan, and J. Tang, "A high precision vital signs detection method based on millimeter wave radar," *Scientific Reports*, vol. 14, no. 1, p. 25535, 2024.

[14] B. Zhang, B. Jiang, R. Zheng, X. Zhang, J. Li, and Q. Xu, "Pi-vimo: Physiology-inspired robust vital sign monitoring using mmwave radars," *ACM Trans. Internet Things*, vol. 4, no. 2, May 2023. [Online]. Available: https://doi.org/10.1145/3589347

Application and Detection of Hardware Trojans Applied to Valid Data States of NCL Combinational Circuits

João Pedro Pereira Magalhães
IESTI
Federal University of Itajubá
Itajubá – MG, Brazil
joaopedropereiramagalhaes@unifei.edu.br

Tales Cleber Pimenta
IESTI
Federal University of Itajubá
Itajubá – MG, Brazil
tales@unifei.edu.br

Diogo Leonardo Ferreira Silva
Institute of Technological Sciences
Federal University of Itajubá, Itabira campus
Itabira – MG, Brazil
diogoleonardof@unifei.edu.br

Abstract—The globalization of digital circuit manufacturing has reduced costs and enabled large-scale production. However, reduced visibility into the supply chain can lead to supply issues and the insertion of malicious modifications into the original design. Hardware Trojans are alterations introduced into circuits to degrade performance, change functionality, or leak information. This paper proposes the application of the detection method for Null Convention Logic (NCL) combinational circuits, targeting the valid data states ('01' and '10'). Furthermore, it reviews existing methods for detecting Hardware Trojans in NCL architectures. The propose method investigated is based on probabilistic transition calculations. Within this framework, the output probabilities of each logic element—specifically, the likelihood of transitioning to a '10' or '01' state—are computed based on the corresponding input probabilities. Consequently, should unauthorized modifications be present, the algorithm identifies them by detecting statistical divergences between the probabilistic values calculated for the golden (original) circuit and the circuit under test.

Index Terms—hardware trojan, null convention logic, probabilistic transition.

I. INTRODUCTION

Since the emergence of digital circuits in the mid-20th century, their application across various fields—medical, automotive, computational, and military, among others—has driven demands for greater energy efficiency, robustness, performance, and reliability. These requirements are closely tied to circuit scalability, as global synchronization in complex devices, such as Very Large Scale Integration (VLSI) circuits, becomes increasingly costly and challenging [1].

Asynchronous circuits, such as NULL Convention Logic (NCL), bypass global synchronization by using a local handshake-based communication protocol, overcoming limitations of traditional digital circuits [2]. Due to these advantages, the International Technology Roadmap for Semiconductors (ITRS) predicted in 2013 that by 2027, 54% of the semiconductor industry would adopt such circuits [3]. Similarly, in 2018, the International Roadmap for Devices and Systems (IRDS) highlighted asynchronous circuits as a key solution for reducing power consumption [4].

The increasing use of NCL circuits in industrial applications raises security concerns, as malicious actors in the manufacturing chain could modify them to extract information or induce failures at critical moments. These modifications, known as Hardware Trojans (HT), alter original circuits by inserting additional logic elements, leading to unexpected outputs, partial or total failures, or even confidential data leakage [5], [6].

The development of methods for identifying Hardware Trojans is a well-established research area. Commercial approaches, such as those based on Automatic Test Pattern Generation (ATPG), are widely employed. Nevertheless, the structure of NCL circuits, which operate on Dual-Rail logic (requiring two signals to represent a single bit), presents challenges. The doubled number of signals relative to Boolean circuits increase the analytical complexity for these conventional algorithms. Accordingly, this paper aims to review existing HT identification methods applicable to NCL circuits and to propose another detector tailored for combinational NCL circuits by analyzing their valid data states ('01' and '10').

The proposed methodology for HT identification relies on a probabilistic transition analysis of the circuit's logic elements. The method calculates the probability of a logic gate's output being '1' or '0' by considering the inherited probabilities from preceding elements. Ultimately, a report is generated detailing the probabilistic divergences identified between the netlists of the golden (original) circuit and the circuit under test.

This paper is organized as follows: Section II introduces the characteristics of Hardware Trojans; Section III reviews the state-of-the-art in HT detection for NCL Circuits; Section IV details the proposed detection program; Section V discusses the obtained results; and finally, Section VI presents the conclusions and suggestions for future work.

II. TROJAN DESIGN

Hardware Trojans consist of two components: the trigger circuit and the payload. The trigger circuit is composed of input, output, or internal signals of the target circuit, activated under rare operating conditions. This ensures the Trojan is triggered only at specific moments during circuit operation [6]–[8].

The payload modifies the original circuit according to the attacker's intent, potentially causing data leakage, output manipulation, or partial/total circuit malfunction [9].

An example is a Hardware Trojan targeting an AND gate. The Trojan is designed to activate when inputs A and B are equal. Its trigger mechanism is an XNOR gate, which activates an XOR payload.

When the trigger is active (i.e., A = B), the payload inverts the AND gate's original output. For example, if A = B = '1', the AND output (1) is flipped to 0. Conversely, when the trigger is inactive (A $\neq$ B), the payload passes the AND output unchanged, concealing the Trojan's presence.

III. DETECTION METHODS FOR HARDWARE TROJANS IN NCL CIRCUITS

Guimarães et al. [10] propose a side-channel analysis technique for Hardware Trojan (HT) detection in Quasi-Delay-Insensitive (QDI) circuits. The method measures the global supply current (I_{DD}) to isolate the transient current peaks (I_{DDT}) and the propagation delay (Δt) of each pipeline stage. An HT alters these parameters, enabling the differentiation between a golden and an infected circuit, even when accounting for Process Variations (PV). The main limitation is that this analysis is restricted to post-silicon and is not applicable at the Register-Transfer-Level (RTL) design stage.

Expanding on this approach, Guazzelli et al. [11] utilize the complete current signature waveform as a feature set to train a machine learning algorithm, specifically a One-Class Support Vector Machine (OC-SVM). Trained exclusively on samples from trojan-free circuits, the model learns the boundary of nominal behavior and classifies any device whose current signature deviates as infected. This increases detection sensitivity and allows for the identification of smaller HTs without requiring additional on-chip circuitry.

In the context of hybrid MSMA (Macro Synchronous Micro Asynchronous) circuits, Lodhi et al. [12] propose HT detection by creating a unique timing signature from the asynchronous block's handshake cycle. The total latency of this cycle is compared against a golden reference to identify anomalies. Subsequently, Lodhi et al. [13] suggested a formal verification method using the nuXmv model checker to analyze functional and timing properties and identify vulnerabilities through the generated counterexamples.

Finally, Ponugoti et al. [14] present an RTL-level formal verification method that focuses on a specific vulnerability in NCL circuits: the exploitation of illegal inputs. QDI circuits using dual-rail encoding have an undefined state ('11') that is typically overlooked. Their work defines trojan models activated exclusively by these illegal inputs. To detect them, formal properties are verified using the Z3 SMT solver, checking whether the circuit's behavior with legal inputs diverges from its behavior with illegal inputs, thus proving the method's efficacy against this class of HTs.

IV. TROJAN DETECTION VIA PROBABILISTIC TRANSITION

The Trojans implemented in this section adhere to the approach for combinational circuits described in Section II. Furthermore, it is assumed that the malicious agent possesses full knowledge of the circuit being modified.

A. Design of NCL Circuits and Trojans

The implementation of the NCL circuits in this section (both Trojans and target circuits) is derived from the Boolean expressions of each circuit's truth table. However, due to the inherent characteristics of NCL circuits—such as dual-rail representation, completeness, and observability—the logic gates must be described in accordance with these properties.

- **Dual-rail representation:** Each logical bit is represented by two signals, where '01' corresponds to a logical '0', '10' corresponds to a logical '1', and '00' represents the null state (NULL), indicating that data is not yet available.
- **Completeness:** The transition from NULL to DATA at the outputs can only occur after all inputs have also completed this transition. Likewise, the transition from DATA back to NULL only happens when all inputs have transitioned from DATA to NULL.
- **Observability:** Any logical transition within an NCL circuit, whether from DATA to NULL or NULL to DATA, must be reflected at its outputs.

To ensure compliance with these properties, the NCL circuit synthesis method described in [1] was utilized. In this method, NCL logic gates are synthesized from fundamental M-of-N gates. Specifically, Huffman logic [15] was adopted for the synthesis of these gates.

```
1 inpt 1 0
2 inpt 1 0
3 and 1 2
      1 2
4 out 0 1
      3
```

Fig. 1. Netlist model

B. The Detection Algorithm

For the detection of Trojans in NCL circuits, a probabilistic transition calculation method is proposed, based on the same process used for Boolean circuits [16], [17]. This algorithm was chosen because NCL circuits, being described in dual-rail, have twice the number of inputs compared to their Boolean counterparts. Consequently, algorithms based on the analysis of a partial or complete set of test vectors become infeasible as the circuit scales. Moreover, the proposal to consider only the valid data states ('01' and '10') allows for the direct use of the equations presented in [16] without requiring adaptations, considering the NCL circuit characteristics presented in Section IV-A.

Figure 1 illustrates the netlist description format, while Algorithm 1 details the detection procedure. Each element is defined by a positional identifier, followed by its type (input, output, or logic element), the number of output connections (fan-out), and finally, the number of inputs (fan-in). For

979-8-3315-9813-6/25 $31.00 © 2025 IEEE

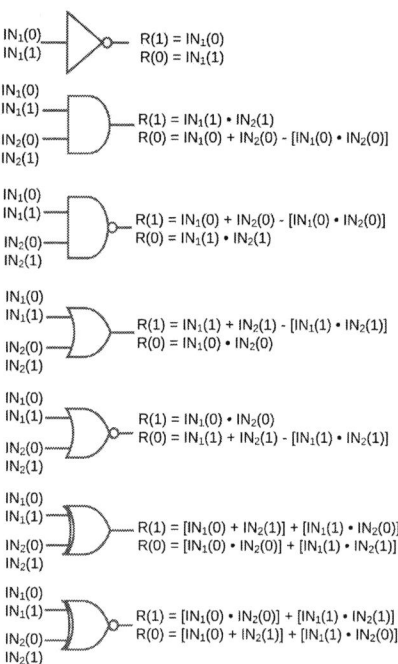

$$R(1) = IN_1(0)$$
$$R(0) = IN_1(1)$$

$$R(1) = IN_1(1) \cdot IN_2(1)$$
$$R(0) = IN_1(0) + IN_2(0) - [IN_1(0) \cdot IN_2(0)]$$

$$R(1) = IN_1(0) + IN_2(0) - [IN_1(0) \cdot IN_2(0)]$$
$$R(0) = IN_1(1) \cdot IN_2(1)$$

$$R(1) = IN_1(1) + IN_2(1) - [IN_1(1) \cdot IN_2(1)]$$
$$R(0) = IN_1(0) \cdot IN_2(0)$$

$$R(1) = IN_1(0) \cdot IN_2(0)$$
$$R(0) = IN_1(1) + IN_2(1) - [IN_1(1) \cdot IN_2(1)]$$

$$R(1) = [IN_1(0) + IN_2(1)] + [IN_1(1) \cdot IN_2(0)]$$
$$R(0) = [IN_1(0) \cdot IN_2(0)] + [IN_1(1) \cdot IN_2(1)]$$

$$R(1) = [IN_1(0) \cdot IN_2(0)] + [IN_1(1) \cdot IN_2(1)]$$
$$R(0) = [IN_1(0) + IN_2(1)] + [IN_1(1) \cdot IN_2(0)]$$

Fig. 2. Probability calculation for each logic gate

Algorithm 1 Pseudocode for Hardware Trojan Detection Based on Probabilistic Transitions

Require: Golden Reference Netlist N_{gold}, Netlist Under Test N_{test}
Ensure: Trojan insertion point P_{trojan}, or `null` if no divergence is found.
1: **function** CALCULATEPROBABILITIES(N)
2:    Let T be a map from element identifier to its transition probability.
3:    **for** each primary input i in N **do**
4:       $P_{i(out=0)} \leftarrow 0.25$
5:       $P_{i(out=1)} \leftarrow 0.25$ ▷ Based on dual-rail logic's four equally likely states
6:    **end for**
7:    **for** each element e in N in topological order (inputs to outputs) **do**
8:       Let I_e be the set of inputs to element e.
9:       Calculate $P_{e(out=0)}$ and $P_{e(out=1)}$ using probabilities from elements in I_e.   ▷ Using equations from Fig. 2
10:       $T[e] \leftarrow P_{e(out=0)} \times P_{e(out=1)}$ ▷ Calculate transition probability
11:    **end for**
12:    **return** T
13: **end function**

14: $T_{gold} \leftarrow$ CalculateProbabilities(N_{gold})   ▷ Process the golden reference netlist
15: $T_{test} \leftarrow$ CalculateProbabilities(N_{test})   ▷ Process the netlist under test
16: $P_{trojan} \leftarrow$ `null`
17: **for** each primary output o in N_{gold} (and corresponding o' in N_{test}) **do**
18:    Let $Paths_{gold}$ be all paths from inputs to o.
19:    Let $Paths_{test}$ be all paths from inputs to o'.
20:    **for** each corresponding pair of paths ($Path_{gold} \in Paths_{gold}$, $Path_{test} \in Paths_{test}$) **do**
21:       **for** each corresponding element pair ($e \in Path_{gold}$, $e' \in Path_{test}$) **do**
22:          **if** $T_{gold}[e] \neq T_{test}[e']$ **then**
23:             $P_{trojan} \leftarrow$ identifier of e' ▷ Divergence found
24:             ▷ The Trojan is the last differing element before reconvergence.
25:             **goto** EndComparison
26:          **end if**
27:       **end for**
28:    **end for**
29: **end for**

30: **return** P_{trojan}

---

example, for element 3 (an AND gate), the fan-out is 1 and the fan-in is 2. The subsequent line indicates, by identifier, which elements connect to its inputs—in this case, elements at positions 1 and 2 (the circuit inputs). Finally, element 4, an output, receives the output of element 3 as its connection.

The probability calculation for logic level '0' or '1' defines the likelihood of the analyzed element's output being equal to '0' or '1', according to the probabilities of its inputs. The probabilistic transition calculation, which is the product of the probability of the output being '0' and the probability of the output being '1', highlights the switching activity of a logic element. With this value, it is possible to infer the circuit's rare nodes and paths—that is, the elements and connections that switch least frequently. This parameter is important, as rare node and path conditions in combinational circuits are commonly chosen for HT insertion, since their activation depends on a specific and infrequent input condition [6].

V. RESULTS OBTAINED

The implementation of the Trojan detection algorithm for the circuits described in Section IV was carried out in C++. The circuits themselves were described using the method presented in Section IV and synthesized using Quartus Prime Lite (version 23.01), in conjunction with the Questa Starter simulator (version 2023.3).

Figure 3 presents the implementation of the circuit under analysis: a 4-bit NCL Gray code encoder. Adopting the Trojan insertion methodology at rare nodes and analyzing the golden circuit column of Table III, the elements with the lowest transition probability are elements 11 through 13, which correspond to the OR logic gates in Figure 3. By altering the logic gate at position 13 to an XOR gate, the probabilistic transition value for this element will differ from that of the golden circuit, given the divergence in their respective logic equations.

During the algorithm's comparison stage, a comparison between output 16 of the golden circuit and the modified circuit reveals a divergence. This discrepancy indicates that an unauthorized modification exists somewhere along the logic path leading to that output. By tracing the entire logic path of the circuit, element 13 is identified as the point where the divergence begins. Therefore, the algorithm is not only capable of detecting the presence of a Trojan but also pinpoints its precise location at the RTL stage by traversing the logic paths. Furthermore, tracing the paths from the outputs back to the inputs ensures full coverage of the analyzed circuit.

Table II presents a comparison of time and test coverage between the developed algorithm and formal verification via testbench for the Gray code encoder circuit and two other NCL circuits: a 4-bit Barrel Shifter [18] and a 5-bit ALU (Arithmetic Logic Unit). Since the algorithm aims to verify valid data cases ('01' and '10'), the developed testbenches analyze the circuits only for these same cases, with the null data case being added between valid data points to maintain the circuits' completeness property. The execution time for each method was obtained from the average of 100 runs. This number was chosen due to the low coefficient of variation observed in all scenarios. Analyzing the results in Table II, it can be inferred that both verification methods were assertive in

TABLE I
COMPARATIVE TABLE OF HT DETECTION TECHNIQUES

Technique	Reference	Destructive?	Detection Phase	Extra Circuitry?
Probabilistic Transition	This work	No	Pre-silicon	No
Side-Channel	[10]	No	Post-silicon	No
Side-Channel + OC-SVM	[11]	No	Post-silicon	No
Timing Signature / Formal Verification (MSMA)	[12], [13]	No	Post and Pre-silicon	Yes
Formal Verification (Illegal Inputs)	[14]	No	Pre-silicon	No

TABLE II
COMPARISON BETWEEN THE PROPOSED ALGORITHM VS SIMULATION VIA TESTBENCH

	Testbench			Transition Probability		
	Gray Enconder	*Barrel Shifter*	*ALU*	*Gray Enconder*	*Barrel Shifter*	*ALU*
Average Execution Time (ms)	133.264	610.758	5,354.498	71.894	74.743	86.628
Standard Deviation (ms)	9.359	54.536	147.755	7.772	7.884	8.961
Coefficient of Variation (%)	7.02	8.93	2.76	10.81	10.55	10.34
Coverage of valid cases (%)	100	100	100	100	100	100

finding the implanted trojans. However, the proposed method using probabilistic transitions requires less time for a complete analysis in all comparisons, proving to be a more viable option as the complexity of NCL circuits increases.

Finally, Table I presents a comparison between the method discussed in this paper and other Trojan detection techniques found in the literature. It is evident that, among the existing methods, detection for valid data-state scenarios is often concentrated in the post-silicon phase. The proposed method thus provides a complementary alternative for testing combinational NCL circuits at the design/RTL stage, ensuring that the described circuit is delivered without malicious modifications and maintaining its reliability throughout the entire fabrication process.

TABLE III
COMPARISON OF CIRCUIT TRANSITION PROBABILITIES

Element	Transition Probability	
	Original	*Modified*
1 - 4	0.0625	0.0625
5 - 10	0.0273438	0.0273438
11 - 12	0.0231781	0.0231781
13	0.0231781	**0.0106812**
14	0.0231781	0.0231781
15	0.0231781	0.0231781
16	0.0231781	**0.0106812**
17	0.0625	0.0625

VI. CONCLUSIONS AND FUTURE WORK

This paper introduced a method for identifying Hardware Trojans in NCL combinational circuits via probabilistic transition calculation. The increasing adoption of NCL circuits across various industry sectors heightens the risk of attacks by

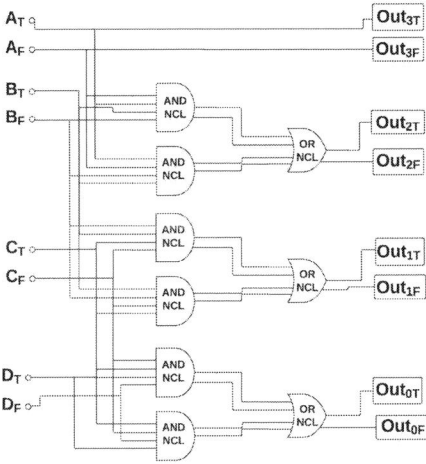

Fig. 3. Schematic Gray Enconder NCL

malicious actors, who may exploit vulnerabilities to exfiltrate sensitive information or compromise system functionality.

The probabilistic transition algorithm has demonstrated its value as a complementary approach to existing NCL digital circuit analysis methods. When used in conjunction with other detection techniques, it enhances the overall reliability of the circuit across diverse operational conditions and throughout the manufacturing process.

Future work will focus on extending the application of this method to sequential NCL circuits.

ACKNOWLEDGMENT

The authors would like to thank the Federal University of Itajubá (UNIFEI) for its structural and technical support and the funding agencies FAPEMIG, CNPQ, and CAPES.

REFERENCES

[1] D. L. D. Oliveira, O. Verducci, V. L. V. Torres, R. Moreno, and L. D. A. Faria, "Synthesis of qdi combinational circuits using null convention logic based on basic gates," *Advances in Science, Technology and Engineering Systems*, vol. 3, pp. 308–317, 2018.

[2] K. Fant and S. Brandt, "Null convention logic/sup tm/: a complete and consistent logic for asynchronous digital circuit synthesis," in *Proceedings of International Conference on Application Specific Systems, Architectures and Processors: ASAP '96*. IEEE Computer Soc. Press, pp. 261–273. [Online]. Available: http://ieeexplore.ieee.org/document/542821/

[3] "The international technology roadmap for semiconductors: 2013 international technology roadmap for semiconductors 2013 edition design," Semiconductor Industry Association, Tech. Rep., 2013.

[4] "The international roadmap for devices international roadmap for devices and systems 2018 update medical devices market driver," IEEE, Tech. Rep., 2019.

[5] R. Karri, J. Rajendran, K. Rosenfeld, and M. Tehranipoor, "Trustworthy hardware: Identifying and classifying hardware trojans," *Computer*, vol. 43, pp. 39–46, 10 2010. [Online]. Available: http://ieeexplore.ieee.org/document/5604161/

[6] R. S. Chakraborty, S. Narasimhan, and S. Bhunia, "Hardware trojan: Threats and emerging solutions," in *2009 IEEE International High Level Design Validation and Test Workshop*. IEEE, 11 2009, pp. 166–171. [Online]. Available: http://ieeexplore.ieee.org/document/5340158/

[7] Z. Zhou, U. Guin, and V. D. Agrawal, "Modeling and test generation for combinational hardware trojans," in *2018 IEEE 36th VLSI Test Symposium (VTS)*. IEEE, 4 2018, pp. 1–6. [Online]. Available: https://ieeexplore.ieee.org/document/8368626/

[8] V. Govindan and R. S. Chakraborty, *Logic testing for hardware trojan detection*. Springer International Publishing, 11 2017, pp. 149–182.

[9] M. Tehranipoor and F. Koushanfar, "A survey of hardware trojan taxonomy and detection," *IEEE Design & Test*, pp. 1–1, 12 2009.

[10] L. A. Guimaraes, T. F. de Paiva Leite, R. P. Bastos, and L. Fesquet, "Non-intrusive testing technique for detection of trojans in asynchronous circuits," in *2018 Design, Automation & Test in Europe Conference & Exhibition (DATE)*. IEEE, 3 2018, pp. 1516–1519. [Online]. Available: http://ieeexplore.ieee.org/document/8342255/

[11] R. A. Guazzelli, M. G. Trindade, L. A. Guimarães, T. F. de Paiva Leite, L. Fesquet, and R. P. Bastos, "Trojan detection test for clockless circuits," *Journal of Electronic Testing: Theory and Applications (JETTA)*, vol. 36, pp. 23–31, 2 2020.

[12] F. K. Lodhi, S. R. Hasan, O. Hasan, and F. Awwad, "Hardware trojan detection in soft error tolerant macro synchronous micro asynchronous (msma) pipeline," in *2014 IEEE 57th International Midwest Symposium on Circuits and Systems (MWSCAS)*. IEEE, 8 2014, pp. 659–662. [Online]. Available: https://ieeexplore.ieee.org/document/6908501

[13] ——, "Formal analysis of macro synchronous micro asychronous pipeline for hardware trojan detection," in *2015 Nordic Circuits and Systems Conference (NORCAS): NORCHIP & International Symposium on System-on-Chip (SoC)*. IEEE, 10 2015, pp. 1–4. [Online]. Available: http://ieeexplore.ieee.org/document/7364384/

[14] K. K. Ponugoti, S. K. Srinivasan, S. C. Smith, and N. Mathure, "Illegal trojan design and detection in asynchronous null convention logic and sleep convention logic circuits," *IET Computers and Digital Techniques*, vol. 16, pp. 172–182, 9 2022.

[15] C. J. Myers, *Asynchronous circuit design*. John Wiley & Sons, 2001.

[16] J. Popat and U. Mehta, "Transition probabilistic approach for detection and diagnosis of hardware trojan in combinational circuits," in *2016 IEEE Annual India Conference (INDICON)*. IEEE, 12 2016, pp. 1–6. [Online]. Available: http://ieeexplore.ieee.org/document/7838895/

[17] H. Salmani, M. Tehranipoor, and J. Plusquellic, "New design strategy for improving hardware trojan detection and reducing trojan activation time," in *2009 IEEE International Workshop on Hardware-Oriented Security and Trust*. IEEE, 2009, pp. 66–73.

[18] M. R. Pillmeier, M. J. Schulte, and E. G. W. III, "Design alternatives for barrel shifters," in *Advanced Signal Processing Algorithms, Architectures, and Implementations XII*, F. T. Luk, Ed., vol. 4791, International Society for Optics and Photonics. SPIE, 2002, pp. 436 – 447. [Online]. Available: https://doi.org/10.1117/12.452034

979-8-3315-9813-6/25 $31.00 © 2025 IEEE

Accelerating Machine Learning using RISC-V Vector Extension in a Manycore Platform

Willian Analdo Nunes, Antônio Vinicius Corrêa dos Santos, César Marcon, and Fernando Gehm Moraes
School of Technology, Pontifical Catholic University of Rio Grande do Sul – PUCRS – Porto Alegre, Brazil
willian.nunes@edu.pucrs.br, antonio.s001@edu.pucrs.br, cesar.marcon@pucrs.br, fernando.moraes@pucrs.br

Abstract—**This work addresses the acceleration of convolutional neural network (CNN) inference in manycore architectures using coarse and fine-grain parallelism. Prior approaches focus on dedicated accelerators or modified NoCs, limiting flexibility. This work proposes integrating a RISC-V processor extended with the vector extensions (RVV) as general-purpose processing elements in a NoC-based manycore. The implementation applies depthwise convolution mapped across PEs and uses auto-vectorization provided by the compiler. Experiments on a 4×4 manycore running the first AlexNet layer achieved up to 5.70x speedup and reduced execution cycles by 82.45% compared to a scalar single-core baseline.**

Index Terms—**Manycore, RISC-V Vector Extension, Hardware Acceleration, CNN Inference Acceleration.**

I. INTRODUCTION

Hardware accelerators, specialized components optimized for specific tasks, have regained prominence due to the increasing computational demands of complex tasks, such as video processing and machine learning (ML) [1]. Hardware acceleration can be achieved either by coarse-grain (MIMD) or fine-grain (SIMD) parallelism.

Coarse-grain acceleration is typically implemented using manycore architectures, where multiple processing cores are integrated into a single chip to enable concurrent task execution. In ML applications, such architectures accelerate complex tasks during inference, leading to faster and scalable solutions.

Fine-grain acceleration can be achieved through vector processing. RISC-V [2], an open-source instruction set architecture (ISA), includes the Vector Extension (RVV) [3], which enhances the base ISA by incorporating vector processing capabilities. The RVV extension introduces 32 vector registers whose widths are implementation-defined via a parameter called **VLEN**. RVV supports runtime-configurable element widths, allowing applications to operate on data elements of various sizes, ranging from 8 bits up to ELEN bits, where ELEN denotes the maximum element width supported by the hardware. Additional runtime configurations, such as vector length (VL) and mask registers, control vector instruction execution, enabling complex and flexible operations.

The *objective* of this work is to investigate the speedup to execute the inference phase of convolutional neural networks (CNNs) using both coarse- and fine-grain parallelism. Coarse-grain acceleration is achieved through a method known as depthwise convolution, combined with a mapping strategy that optimizes traffic within the Network-on-Chip (NoC). Fine-grain acceleration is accomplished via vectorized convolution processing, using the auto-vectorization capabilities provided by the GCC compiler. The implementation of these hardware acceleration techniques, and the resulting speedup evaluation, is the *original contribution* of this work.

II. RELATED WORK

The literature presents multiple works that employ NoCs for domain-specific acceleration (DSA). The reviewed works rely on NoCs to interconnect PEs specifically tailored for their target applications, often optimized for operations such as multiply-accumulate (MAC). Consequently, most NoC-based works focus on NoC architectures designed explicitly for DSA.

Ultra-NoC [4] employs a 13×13 mesh interconnect with 169 tiles to implement a unified routing scheme supporting unicast, multicast, and broadcast for diverse Deep Neural Network (DNN) data patterns. Each tile integrates a router, network interface, and a Processing Element (PE) comprising a multiplier, adder, and control logic. Ultra-NoC reduces memory access by 26–49% (AlexNet) and 30–56% (ResNet-50).

Eyeriss v2 [5] employs a hierarchical mesh NoC optimized for compact DNNs like MobileNet. Each PE handles 8-bit fixed-point data and includes local scratchpads, with global buffers (GLBs) forming a two-level memory hierarchy. Separate NoCs are dedicated to transferring inputs, weights, and partial sums. The architecture comprises 16 PE and 16 GLB clusters in an 8×2 array, each containing 12 PEs (3×4).

Meta's [6] MTIA v1 employs an 8x8 mesh NoC that can be subdivided to run independent jobs. Each PE is equipped with two RISC-V processor cores (one with vector extension) and accelerators for domain-specific applications.

Intra-NoC acceleration modifies NoC components (e.g., routers) for specific Neural Network (NN) tasks. NOVA [7] introduces a NoC-based vector unit that approximates non-linear operations and broadcasts values via enhanced routers with lookup-driven data routing. Tiwari et al. [8] integrate In-Network Accumulation (INA) into routers, adding partial sum accumulation blocks that cut runtime by 1.22x and boost power efficiency by 2.16x. Gao et al. [9] propose a dual-layer NoC accelerator for lightweight CNNs, comprising a NoC array, PE clusters, a loop controller, and an ISA-based instruction decoder. Each PE performs a MAC operation, with the controller managing memory and data flow.

The effective mapping of DNNs onto NoC-based accelerators is key to minimizing communication overhead. Zhao et al. [10] group convolutions and map them to PEs using a genetic algorithm on a PE communication graph (PCG). By assigning pooling and convolution to the same PE, and allocating one PE per channel in early layers, their method achieves up to 50% power reduction.

979-8-3315-9813-6/25 $31.00 © 2025 IEEE

Liang et al. [11] apply a modified Whale Optimization Algorithm for multi-objective CNN mapping, balancing energy, latency, and load. Despite its limitations for discrete spaces, it yields significant power and latency improvements.

Ye et al. [12] propose a task allocation scheme for resource-constrained NoCs, grouping CNN tasks under PE capacity constraints. A dynamic dense reverse mapping algorithm uses the PCG to reduce power by 31% and latency by 38%.

GAP-8 [13] uses a RISC-V controller allied with an 8-Core cluster with an additional dedicated accelerator for convolutions, the Hardware Convolution Engine. Although ARA2 [14] uses a core with an RVV accelerator in a multicore system, it lacks NoC interconnects, highlighting a gap in integrating RVV-based PEs within NoC architectures. Manycore designs often compensate for limited PE capability with fine-grain NoC router modifications, due to the use of MAC-centric PEs requiring NoC-level enhancements for performance.

Table I categorizes prior work. Unlike these, our approach uses general-purpose PEs with internal acceleration, enabling broader task flexibility. Our approach employs a NoC interconnecting RVV-enabled RISC-V cores, exploiting SIMD parallelism within vector cores and MIMD across the manycore fabric. The NoC facilitates efficient, concurrent inter-core communication without performance loss. We close a *gap* in CNN acceleration by integrating RVV-based processors into a NoC-based manycore, enabling support for diverse workloads with internally accelerated, general-purpose PEs.

TABLE I
RELATED WORK ON NN (DNN, CNN, ML) ACCELERATION.

	Architecture		**Target Application**
Ultra-NoC [4] (2024)	NoC Modification	Hybrid data transmission to leverage different data reuse methods	DNN
Eyeriss v2 [5] (2019)	Custom NoC Architecture	Separate NoC for different data types and clusters of PEs and GLBs	DNN on Mobile Devices
NOVA [7] (2024)	Intra-NoC Acceleration	Special router to support non-linear operations	CNN Attention Layers
Tiwari et al. [8] (2022)	Intra-NoC Acceleration	Routers support partial sum accumulation (In-Network Accumulation (INA))	DNN
Gao et al. [9] (2023)	Custom NoC architecture	Dual-layer NoC-based accelerator with dedicated custom ISA	CNN for Edge Computing
Zhao et al. [10] (2020)	Mapping method	PEs communication graph optimization algorithm (Genetic Algorithm)	CNN
Liang et al. [11] (2024)	Mapping method	PEs communication graph optimization algorithm (Whale Optimization Algorithm)	CNN
Ye et al. [12] (2023)	Mapping method	PEs communication graph optimization algorithm considering resource limitation	CNNs on resource-limited NoC
GAP-8 [13] (2018)	Manycore	RISC-V Controller + RISC-V 8-Core Cluster W/ Dedicated Accelerator	CNN on the edge
ARA2 [14] (2024)	Multicore	Multicore Cluster of RISC-V core W/ RVV Co-Processor	ML
This Work	General-Purpose PE with vector acceleration	Usage of a general purpose PE with a tightly coupled accelerator	ML (CNN)

III. CNN ACCELERATION METHOD

This section presents (*i*) the strategy for distributing a CNN through multiple PEs and (*ii*) the CNN mapping method.

The selected CNN mapping strategy exploits input channel independence to maximize parallelism via **depthwise convolutions** [15]. This technique assigns different input channels to separate PEs, enabling intra-layer parallelism. As each input channel and filter pair produces one output channel, output channel computations are also independent. Thus, a 1st PE layer performs depthwise convolutions, passing each output channel's feature map to a 2nd PE layer, aggregating the results. While the 2nd layer processes one set of outputs, the 1st layer can advance to the next, enhancing pipeline efficiency.

Equation (1) describes the standard convolution.

$$\mathbf{O}[c_o][x][y] = \mathbf{B}[c_o] + \sum_{k=0}^{C_i-1} \sum_{i=0}^{Width-1} \sum_{j=0}^{Height-1} (\mathbf{I}[k][Sx+i][Sy+j] * \mathbf{W}[c_o][i][j]) \quad (1)$$

where: c_o is the current output channel; x and y are the horizontal and the vertical positions; C_i is the total number of input and filter channels; Width and Height correspond to the filter size; S is the stride; O is the output; I is the input; W is the filter tensors; and B is the bias vector.

Equation 2 describes the convolution operation performed by the 1st PE layer, using the depthwise convolution method. It differs from the standard convolution by not having the outer summation, which sums the input channels.

$$\mathbf{O_p}[c_o][x][y] = \sum_{i=0}^{Width-1} \sum_{j=0}^{Height-1} (\mathbf{I}[K][Sx+i][Sy+j] * \mathbf{W}[c_o][i][j]) \quad (2)$$

where: O_p: partial output; K: input channel (fixed parameter in PEs that performs depthwise convolution).

To compensate for the removal of the outer summation, the 1st PE layer must include c_i processing elements (PEs), corresponding to the number of output channels. Each PE is assigned a fixed K parameter, which designates the specific input channel it is responsible for processing, and performs the computation defined in Equation (2).

The operation output O_p is partial and needs to be summed with the partial outputs from other input channels to generate the output feature map. The operation in Equation (2) is executed for every output channel in the 1st PE layer.

Equation (3) defines the convolution in the 2nd PE layer, finishing the operation started in Equation (2). It aggregates all partial outputs from the 1st PE layer per output channel, adds the bias **B**, and generates the final output feature map. This layer may also apply ReLU and Max-Pooling to balance the computational load. The resulting feature maps are then passed to a 3rd PE layer, initiating the next convolution via depthwise convolution.

$$\mathbf{O}[c_o] = \mathbf{B}[c_o] + \sum_{k=0}^{C_i-1} O_p[c_o][C_i] \quad (3)$$

The number of PEs varies with layer depth and manycore size. For example, the 1st convolutional layer of AlexNet has three input channels, so the 1st PE layer consists of three PEs, each handling one channel. Each PE performs 96 iterations of depthwise convolution, matching the output depth of the layer. After each iteration, results are sent to the 2nd PE layer.

The 2nd PE layer comprises one PE and is responsible for receiving the data sent by the 1st PE layer after the processing of each output channel. It then sums the partial feature maps, adds the bias, and performs the ReLU and Max-Pool operations. The output feature map is then sent to the 3rd PE layer, initiating the next convolutional layer.

979-8-3315-9813-6/25 $31.00 © 2025 IEEE

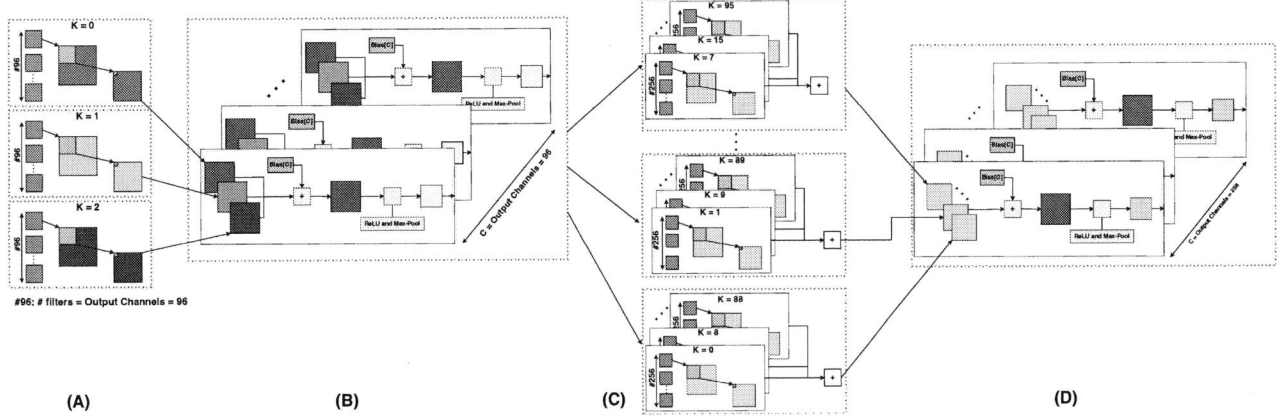

Fig. 1. 1st and 2nd layers of the depthwise convolution. Red dashed lines indicate the computation boundaries for each process. **(A)** P1 processes, each handling one input channel. **(B)** P2 process, which receives partially convolved channels from all P1 processes, accumulates them to produce the 1st convolution output, and applies ReLU and Max-Pool to generate the output feature map (in yellow). **(C)** P3 processes, each handling multiple input channels. **(D)** P4 process, which receives partially convolved data from P3, accumulates it, and applies ReLU and Max-Pool to produce the 2nd output feature map (in green).

In the 3rd PE layer, the input depth is 96 and the output depth is 256 (in the AlexNet example). Due to the high input depth, assigning a PE per input channel is impractical, as it would require an excessive number of PEs. Instead, each PE processes multiple input channels, with the number per PE determined by $\frac{96}{PEs_3}$, where PEs_3 is the number of allocated PEs. Each PE accumulates partial outputs from its assigned channels before passing them to the 4th layer.

The 4th PE layer is similar to the 2nd one. It is responsible for gathering the partial outputs, summing them, performing the ReLU and Pooling operations if required, and then sending the result to the 5th PE layer, which has a structure similar to the 3rd layer.

Four C-programs were developed using this partitioning approach, each named with a leading "P" followed by the corresponding PE layer number. For clarity, only four PE layers were implemented, covering the first two convolutional layers of AlexNet.

- *P1* - Entry-point program implementing the 1st PE layer. It processes a single input channel and forwards the result to the second program. Three instances are required, each running with a different K parameter (input channel).

- *P2* – Implements the 2nd PE layer. It receives partial outputs from P1 processes, sums them to form the output feature map, applies ReLU and Max-Pool, and forwards each output channel to the next PE layer.

- *P3* – Implements the 3rd PE layer. It processes multiple input channels (based on PE count and convolution depth), receives data from P2, performs depthwise convolution, and sends results to the next layer.

- *P4* – Implements the 4th PE layer. It sums partial outputs from P3 processes to produce the output feature map, applies ReLU and Max-Pool, and passes each output channel to the following PE layer.

Figure 1(A-B) presents *P1* (3 instances) and *P2* (with a single instance) programs handling the 1st convolutional layer of AlexNet. Figure 1(C-D) presents *P3* (with 8 instances) and *P4* programs handling the 2nd convolutional layer of AlexNet.

Figure 2 illustrates the **mapping** of tasks (programs) onto the manycore, based on the number of PEs assigned to each PE layer: 3 for the 1st, 1 for the 2nd, 8 for the 3rd, and 1 for the 4th. For clarity, the figure is divided into three parts. The 13 tasks are mapped onto the manycore using 12 out of 16 available PEs. Only the PE located at position 1x2 executes more than two tasks. This mapping was chosen to minimize the hop count between tasks.

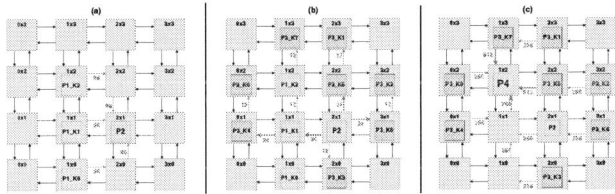

Fig. 2. Mapping in a 4x4 manycore. Red arrows represent communications, and the number associated represents the number of channels/packets sent. (a) three P1 instances communicating with the P2 process. (b) P2 process sending the channels to be processed by the eight P3 processes. (c) eight P3 processes sending data to the P4 process.

Figure 2(a) presents the three P1 instances and the P2 process. The P2 process is centralized to receive data from the maximum number of ports, avoiding packets using the same router. With the P1 and P2 mapping positions, the P2 process receives data from the South (S), West (W), and North (N) ports, while the East (E) is unused. Figure 2(b) presents the P2 process sending the first convolution layer output channels to the P3 processes (8x P3 instances). Each P3 process receives 12 channels to process. They are mapped to PEs surrounding the centralized PE to divide the packets between all the ports. The routing is not optimal given that the routing algorithm used by the NoC is the XY, which first routes the channel in the X-axis. Finally, Figure 2 (c) presents the P3 processes sending data to the P4 process. The P4 process is centralized and receives the packets in the following order: W, N, E, S, S, E, S, N. Although some ports are used sequentially, the NoC is not a bottleneck and can handle the data transmission without reducing the application's performance.

IV. RESULTS

This work adopts the Memphis-V manycore [16]–[19] publicly available at https://github.com/gaph-pucrs/Memphis-V, providing coarse-grain acceleration. Memphis-V is modeled at the RTL level using SystemVerilog. A subset of the RISC-V Vector Extension (RVV) [20], aimed at accelerating Convolutional Neural Networks (CNNs), has been integrated into the processor. This extension features a configurable vector length (VLEN). This processor is responsible for the fine-grain acceleration through the RVV extension. The CNN model is compiled using GCC 14.1 [21], with support to **automatic vectorization**. The compiler's auto-vectorization showed limitations by only vectorizing portions of the code, but still provided good results without the need to modify the source code. A solution is adopt manual vectorization.

This Section presents the AlexNet [22] 2-D CNN results. Experiments adopted a 4x4 manycore, which mapping can be seen at Figure 2. Simulations were performed using a scalar PE and 3 accelerated versions. The 3 accelerated versions include the RVV extension with VLENs of 64, 128, and 256 bits.

Table II presents the results of the P1 tasks on the manycore. Each process only computes one of the three RGB input channels and generates the 96 (partial) output channels of the 1st convolutional layer for that given input channel.

TABLE II
RESULTS ON RUNNING RGB CHANNELS COMPUTATION (P1) ON MEMPHIS-V WITH VECTOR EXTENSION.

VLEN	Computation			Communication	
	Avg cycles per channel	Speed-up	Total cycles (x96 channels)	Avg cycles per channel	Total cycles (x96)
-	4.74e+6	-	4.55e+8	1.19e+5	1.14e+7
64	4.75e+6	1.00x	4.56e+8	1.19e+5	1.14e+7
128	2.65e+6	1.79x	2.54e+8	1.34e+5	1.28e+7
256	2.37e+6	2.00x	2.27e+8	1.39e+5	1.34e+7

The average number of cycles required to compute each output channel is presented in the 2nd column of Table II. The scalar and 64-bit VLEN configurations exhibit comparable cycle counts. Implementations with larger VLENs achieve computational speedups, reducing the number of cycles required per output channel by up to **2x**. Communication time is lower by approximately one order of magnitude, as the computational load of the application significantly outweighs the volume of data transmitted. Overall, the 1st convolutional layer of AlexNet involves 105,415,200 multiplication operations. The P2 process performs fewer computations during execution, approximately one order of magnitude less than P1, and spends most of its time waiting for data from the P1 processes.

Table III compares the single and manycore versions of the 1st layer of the 2-D CNN benchmark, exploring the scalar and vectorized versions for multiple VLENs. In the single-core version, the speedup of using the vector extension reached 1.65x. Applying the manycore mapping technique, the speedup reached ≈**3x** using scalar PEs.

The best trade-off was the manycore version with a 128-bit VLEN, offering a speedup of **5.15x**, meaning a reduction of around 80% of the cycles compared to a scalar single-core and 1.72x of the scalar manycore.

TABLE III
PERFORMANCE ON THE ALEXNET FIRST LAYER, COMPARING SINGLE AND MANYCORE ARCHITECTURES.

Single-/Manycore	VLEN	Cycles	Speed-up	Cycle reduc. (%)
Single	-	1,418,604,031	-	-
Single	64	1,418,604,052	1.00x	0.00
Single	128	1,012,741,796	1.40x	28.61
Single	256	858,535,157	**1.65x**	39.48
Many	-	474,913,995	**2.99x**	66.52
Many	64	475,713,468	2.98x	66.47
Many	128	275,530,467	**5.15x**	80.58
Many	256	248,906,823	5.70x	82.45

Results for the 2nd AlexNet layer are omitted due to memory constraints on the manycore. Partial simulations validated the partitioning method's correctness, but speedup could not be measured. A potential solution is to store application parameters in external memory, minimizing local storage use by dynamically loading weights and biases as needed.

Direct comparison with related work is difficult due to different parameters and goals. Works that involve NoC architecture have very different PEs, as our work focuses on having a general purpose, while theirs focuses on having dedicated accelerators. Manycore works are more similar due to using RISC-V cores and the RVV extension, but they lack NoC interconnection.

The Authors of GAP-8 [13] report speedups ranging from 5.3x to 8.4x when using eight cores in the cluster. These results are comparable to those presented in this work in acceleration regarding the 1st layer of the CNN using a distributed approach, where with only four PEs, we achieved a speedup of up to 5.7x. For full CNNs, the performance improvement of GAP-8 ranged from 9.3x to 11.7x using 8-PEs. ARA2 [14] concluded that a multicore design with smaller Ara2 instances behaves better than a single-core larger Ara2. This work achieved similar results, as having multiple scalar cores achieved a better speedup (almost double) than the biggest single core with VLEN of 256 bits.

V. CONCLUSIONS AND FUTURE WORK

This work demonstrated that integrating RVV-enabled RISC-V cores into a NoC-based manycore enables CNN inference acceleration through combined SIMD and MIMD parallelism. The proposed method achieved a 5.70x speedup and an 82.45% reduction in execution cycles compared to a scalar single-core baseline. These results indicate that combining vectorized processing and manycore mapping strategies enhances the efficiency of CNN inference. Distributing a CNN layer to multiple PEs also helps keep memory usage low by reducing the number of parameters each PE must hold.

Future work includes: (i) support for additional CNN layers (and full network); (ii) integrate external memory to address local storage limitations; (iii) evaluate scalability on larger manycore configurations; (iv) explore the suitability of manycores and the vector extension for other machine learning structures beyond CNN.

ACKNOWLEDGMENTS

This work was financed in part by Coordenação de Aperfeiçoamento de Pessoal de Nível Superior (CAPES), Finance Code 001; Conselho Nacional de Desenvolvimento Científico e Tecnológico (CNPq), grants 308182/2023-5 and 305621/2024-6; Fundação de Amparo à Pesquisa do Estado do Rio Grande do Sul (FAPERGS), 23/2551-0002200-1.

REFERENCES

[1] B. Peccerillo, M. Mannino, A. Mondelli, and S. Bartolini, "A survey on hardware accelerators: Taxonomy, trends, challenges, and perspectives," *Journal of Systems Architecture*, vol. 129, pp. 1–51, 2022, https://doi.org/10.1016/j.sysarc.2022.102561.

[2] RISC-V Foundation, "The RISC-V Instruction Set Manual, Volume I: User-Level ISA, Document Version 20240411," 2024, https://drive.google.com/file/d/1uviu1nH-tScFfgrovvFCrj7Omv8tFtkp/view?usp=drive_link, January 2025.

[3] ——, "RISC-V "V" Vector Extension," 2021, https://github.com/riscv/riscv-v-spec/releases/tag/v1.0.

[4] K.-C. J. Chen, H.-H. Peng, and P.-C. Shen, "Ultra-NoC: Unified Low-Transmission Routing Assisted NoC for High-flexible DNN Accelerator," in *IEEE International System-on-Chip Conference (SOCC)*, 2024, pp. 1–5, https://doi.org/10.1109/SOCC62300.2024.10737754.

[5] Y.-H. Chen, T.-J. Yang, J. Emer, and V. Sze, "Eyeriss v2: A Flexible Accelerator for Emerging Deep Neural Networks on Mobile Devices," *IEEE Journal on Emerging and Selected Topics in Circuits and Systems*, vol. 9, no. 2, pp. 292–308, 2019, https://doi.org/10.1109/JETCAS.2019.2910232.

[6] Meta AI, "MTIA v1: Meta's first-generation AI inference accelerator," 2023, https://ai.meta.com/blog/meta-training-inference-accelerator-AI-MTIA/.

[7] M. Upadhyay, R. Juneja, W.-F. Wong, and L.-S. Peh, "NOVA: NoC-based Vector Unit for Mapping Attention Layers on a CNN Accelerator," in *ACM/IEEE Design, Automation Test in Europe Conference (DATE)*, 2024, pp. 1–6, https://doi.org/10.23919/DATE58400.2024.10546727.

[8] B. Tiwari, M. Yang, X. Wang, and Y. Jiang, "In-Network Accumulation: Extending the Role of NoC for DNN Acceleration," in *IEEE International System-on-Chip Conference (SOCC)*, 2022, pp. 1–6, https://doi.org/10.1109/SOCC56010.2022.9908106.

[9] J. Gao, Q. Shao, F. Deng, Q. Wang, N. Jing, and J. Jiang, "An NoC-based CNN Accelerator for Edge Computing," in *International Conference on ASIC (ASICON)*, 2023, pp. 1–4, https://doi.org/10.1109/ASICON58565.2023.10396346.

[10] Y. Zhao, F. Ge, C. Cui, F. Zhou, and N. Wu, "A Mapping Method for Convolutional Neural Networks on Network-on-Chip," in *International Conference on Communication Technology (ICCT)*, 2020, pp. 916–920, https://doi.org/10.1109/ICCT50939.2020.9295883.

[11] Z. Liang, F. Hu, B. Xu, and C. Wei, "Multi objective non dominated sorting whale optimization genetic algorithm for convolutional neural network-based on-chip networks," in *International Seminar on Artificial Intelligence, Networking and Information Technology (AINIT)*, 2024, pp. 653–656, https://doi.org/10.1109/AINIT61980.2024.10581699.

[12] J. Ye, F. Ge, and F. Zhou, "A Method of Mapping Convolutional Neural Networks on Resource-limited NoC Platform," in *International Conference on ASIC (ASICON)*, 2023, pp. 1–4, https://doi.org/10.1109/ASICON58565.2023.10396382.

[13] E. Flamand, D. Rossi, F. Conti, I. Loi, A. Pullini, F. Rotenberg, and L. Benini, "GAP-8: A RISC-V SoC for AI at the Edge of the IoT," in *IEEE International Conference on Application-Specific Systems, Architectures and Processors (ASAP)*, 2018, pp. 1–4, https://doi.org/10.1109/ASAP.2018.8445101.

[14] M. Perotti, M. Cavalcante, R. Andri, L. Cavigelli, and L. Benini, "Ara2: Exploring Single- and Multi-Core Vector Processing With an Efficient RVV 1.0 Compliant Open-Source Processor," *IEEE Transactions on Computers*, vol. 73, no. 7, pp. 1822–1836, 2024, https://doi.org/10.1109/TC.2024.3388896.

[15] A. G. Howard, M. Zhu, B. Chen, D. Kalenichenko, W. Wang, T. Weyand, M. Andreetto, and H. Adam, "MobileNets: efficient convolutional neural networks for mobile vision applications (2017)," *CoRR*, vol. abs/1704.04861, pp. 1–9, 2017, http://arxiv.org/abs/1704.04861.

[16] M. Ruaro, L. L. Caimi, V. Fochi, and F. G. Moraes, "Memphis: a framework for heterogeneous many-core SoCs generation and validation," *Springer Design Automation for Embedded Systems*, vol. 23, no. 3-4, pp. 103–122, 2019, https://doi.org/10.1007/s10617-019-09223-4.

[17] M. Ruaro, F. B. Lazzarotto, C. A. Marcon, and F. G. Moraes, "DMNI: A Specialized Network Interface for NoC-based MPSoCs," in *IEEE International Symposium on Circuits and Systems (ISCAS)*, 2016, pp. 1202–1205, https://doi.org/10.1109/ISCAS.2016.7527462.

[18] E. Wachter, L. L. Caimi, V. Fochi, D. Munhoz, and F. G. Moraes, "BrNoC: A broadcast NoC for control messages in many-core systems," *Microelectronics Journal*, vol. 68, pp. 69–77, 2017, https://doi.org/10.1016/j.mejo.2017.08.010.

[19] W. A. Nunes, A. E. Dal Zotto, C. d. S. Borges, and F. G. Moraes, "RS5: An Integrated Hardware and Software Ecosystem for RISC-V Embedded Systems," in *IEEE Latin American Symposium on Circuits and Systems (LASCAS)*, 2024, pp. 1–5, https://doi.org/10.1109/LASCAS60203.2024.10506171.

[20] W. A. Nunes and F. G. Moraes, "Accelerating Machine Learning with RISC-V Vector Extension and Auto-Vectorization Techniques," in *IEEE International Symposium on Circuits and Systems (ISCAS)*, 2025, pp. 1–5, https://doi.org/10.1109/ISCAS56072.2025.11043225.

[21] GCC 14, "GCC 14 Release Series Changes, New Features, and Fixes," 2024, https://gcc.gnu.org/gcc-14/changes.html.

[22] A. Krizhevsky, I. Sutskever, and G. E. Hinton, "Imagenet classification with deep convolutional neural networks," *Communications of the ACM*, vol. 60, no. 6, pp. 84–90, 2017, https://doi.org/10.1145/3065386.

TLGLock: A New Approach in Logic Locking Using Key-Driven Charge Recycling in Threshold Logic Gates

Abdullah Sahruri   Martin Margala

School of Computing and Informatics, University of Louisiana at Lafayette, Lafayette, LA, USA

{abdullah.sahruri1, martin.margala}@louisiana.edu

Abstract—**Logic locking remains one of the most promising defenses against hardware piracy, yet current approaches often face challenges in scalability and design overhead. In this paper, we present TLGLock, a new design paradigm that leverages the structural expressiveness of Threshold Logic Gates (TLGs) and the energy efficiency of charge recycling to enforce key-dependent functionality at the gate level. By embedding the key into the gate's weighted logic and utilizing dynamic charge sharing, TLGLock provides a stateless and compact alternative to conventional locking techniques. We implement a complete synthesis-to-locking flow and evaluate it using ISCAS, ITC, and MCNC benchmarks. Results show that TLGLock achieves up to 30% area, 50% delay, and 20% power savings compared to latch-based locking schemes. In comparison with XOR and SFLL-HD methods, TLGLock offers up to 3× higher SAT attack resistance with significantly lower overhead. Furthermore, randomized key-weight experiments demonstrate that TLGLock can reach up to 100% output corruption under incorrect keys, enabling tunable security at minimal cost. These results position TLGLock as a scalable and resilient solution for secure hardware design.**

Index Terms—**EDA, Threshold Logic Gates, Logic Locking, Key Embedding, AIG, Hardware Security, SAT-Based Attacks.**

I. INTRODUCTION

Logic locking is an essential technique for protecting integrated circuits (ICs) against reverse engineering and unauthorized use. The globalization of the IC supply chain, driven by cost reduction and rapid time-to-market, has amplified risks such as intellectual property (IP) piracy, overproduction, and unauthorized activation, especially in the presence of third-party vendors [1]. Traditional logic locking schemes like XOR/XNOR-based insertion offer baseline security but are vulnerable to SAT-based attacks that can efficiently recover secret keys [2].

To address these vulnerabilities, resilient schemes like SAR-Lock and Anti-SAT were proposed [2], [3]. However, these methods significantly increase area, delay, and power due to additional logic and key-validation circuitry. This trade-off between security and circuit performance poses a challenge, particularly for area- and power-constrained systems. Furthermore, techniques such as point function obfuscation, scan chain locking, and routing-based methods face challenges in output corruptibility, scalability, and design complexity [4].

Threshold Logic Gates (TLGs) offer a promising alternative for secure and efficient digital logic synthesis. TLGs, as seen in Fig. 1, replace multi-level CMOS logic with a single gate

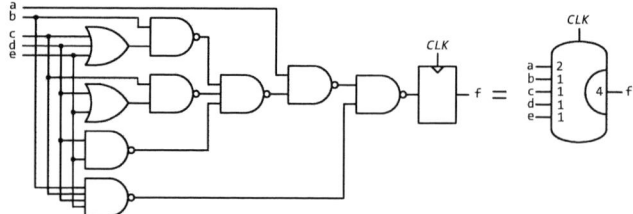

Fig. 1. A single Threshold Logic Gate (TLG) replacing a complex multi-level CMOS logic network.

by computing a weighted sum of binary inputs and comparing it to a threshold:

$$\text{Output} = \begin{cases} 1, & \text{if } \sum_{i=1}^{n} W_i \cdot X_i \geq T \\ 0, & \text{otherwise,} \end{cases} \tag{1}$$

where $X_j \in \{0,1\}$ are binary inputs, W_j are integer weights, and T is the threshold. This compact representation improves logic density and can be exploited for both performance and obfuscation.

More than 50 TLG designs have been reported [5] and used in ALUs [6], processing-in-memory [7], [8], and low-power systems [9]–[11]. CMOS realizations include traditional, conductance-based, and capacitive structures [12]–[14]. Among these, *LCTL* [15] and *CRTL* [16] offer complementary trade-offs: *LCTL* supports high fan-in with low static power but higher area, while *CRTL* achieves superior energy efficiency and speed via charge recycling.

These traits make TLGs well-suited for logic locking: embedding key inputs as weighted terms obfuscates functionality without extra gates. However, threshold-based locking remains underexplored in synthesis flow, key embedding, and power/delay/area (P/D/A) trade-offs. We introduce a threshold-locked methodology using both *LCTL* and *CRTL* (Fig. 2), implemented in Cadence 45 nm GPDK045. Keys are mapped as input variables with tunable weights, yielding a native, compact locking primitive. We develop a complete synthesis flow and evaluate on ISCAS'85, ISCAS'89, and ITC'99.

CRTL-based locking reduces area by up to **30%**, delay by **50%**, and power by **20%** versus *LCTL*; SAT solvers also time out (1 h) on large designs such as c7552, i10, and b17, indicating strong attack resilience. Overall, the approach

979-8-3315-9813-6/25 $31.00 © 2025 IEEE

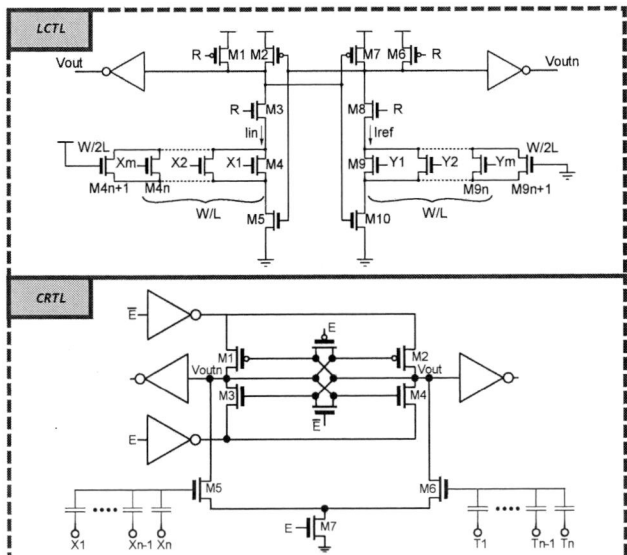

Fig. 2. TLG architectures: Latch-type Low Power Threshold Logic (*LCTL*) and Charge Recycling Threshold Logic (*CRTL*).

outperforms conventional locking while remaining compact and efficient. methods in both efficiency and security.

The rest of the paper is organized as follows: Section II discusses limitations in traditional logic locking. Section III introduces our TLG-based locking flow. Section IV presents experimental analysis on area, power, delay, and SAT resilience. Section V concludes the work.

II. OVERHEAD IN LOGIC LOCKING TECHNIQUES

Classic XOR/XNOR locking offers low cost but is highly vulnerable to SAT attacks with oracle access [17], [18]. SAT-resilient variants (e.g., SARLock, Anti-SAT, DisORC) increase effort via point-function structures yet inflate area and power [2], [19]; hybrids that mix XOR/XNOR with point functions further raise overhead [20]. Cyclic schemes frustrate SAT by introducing feedback, but incur substantial area and implementation/DRC challenges [21], [22]. LUT- and FSM-based obfuscation embed keys implicitly but require extra states or memory-like blocks, and scan access can expose internal signals. In short, existing approaches trade either security for cost or vice versa, motivating a compact alternative. We therefore embed the key directly into threshold-logic synthesis, aiming for strong protection with efficient hardware.

III. PROPOSED FLOW

The proposed flow in Fig. 3 embeds logic locking within threshold logic gates synthesized from And-Inverter Graphs (AIGs), integrating key inputs into the threshold structure to ensure secure circuit locking. The process begins at ❶ by analyzing the Boolean structure from BLIF or BENCH and ❷ computing threshold cuts capable of representing the Boolean functions using TLGs [24]. In ❸, after synthesis, TLGs are produced, each characterized by specific weights and a threshold based on input requirements. The flow supports, in ❹, merging two TLGs through linear combination, leveraging

analytic techniques from prior work [23] to optimize the circuit by combining gates and simplifying the overall design.

In ❺, key inputs are embedded into the design with their assigned weights set proportionally to the sum of the input weights, preserving functionality under the correct key while enforcing logic locking by exploiting the gate's threshold characteristics. Stage ❻ focuses on extracting the weights and thresholds from the synthesized and merged TLGs, preparing these parameters for assignment. In ❼, the extracted weights and thresholds are accurately mapped to their respective inputs, ensuring correct behavior of the TLG-based design.

Finally, in ❽, the flow concludes with synthesis of the locked TLG circuit and physical design verification. Performance in terms of delay, area, and power is extracted and assessed, and the process iterates until predefined specifications are met, ensuring that the TLG circuit design aligns with design goals. To further illustrate the steps, Algorithm 1

Algorithm 1 Threshold Logic Synthesis and Key Preparation

Require: Original circuit C_{orig}, #Keys N_k, Insertion % P
Ensure: Synthesized circuit C_{TLG}, Key vector K

1: $C_{\text{TLG}} \leftarrow \text{SYNTHESIZETLG}(C_{\text{orig}})$
2: Select $\mathcal{G}_{\text{lock}} \subseteq C_{\text{TLG}}$ with $|\mathcal{G}_{\text{lock}}| = \lceil |C_{\text{TLG}}| \cdot \frac{P}{100} \rceil$
3: $K \leftarrow \{k_1, \ldots, k_{N_k}\}$ ▷ Random key vector
4: **return** C_{TLG}, K

Algorithm 2 Key Integration and Logic Locking

Require: TLG circuit C_{TLG}, Key vector K, Thresholds T_j
Ensure: Locked circuit C_{locked}

1: **for all** $g_j \in \mathcal{G}_{\text{lock}}$ **do**
2:    Integrate key bits $\{k_{j1}, \ldots, k_{jm}\}$ with weights $\{v_{j1}, \ldots, v_{jm}\}$
3:    Update sum: $S_j = \sum w_i x_i + \sum v_{jl} k_{jl}$
4:    Set $g_j.\text{output} \leftarrow \mathbb{1}[S_j \geq T_j]$
5: **return** $C_{\text{locked}} \leftarrow \text{FORMAT}(C_{\text{TLG}})$

---

and Algorithm 2 formalize the TLG synthesis and locking processes. Algorithm 1 describes the initial synthesis of the original circuit netlist C_{orig} into a TLG-based circuit C_{TLG}, calculates gate properties (weights w_i, fan-in f_{in}, fan-out f_{out}), determines the percentage P of gates to modify for key integration, and generates a key vector K. Algorithm 2 embeds the key inputs within C_{TLG}: for each selected gate g_j, key inputs $\{k_{j1}, k_{j2}, \ldots\}$ are integrated with corresponding weights, the combined weighted sum is evaluated against threshold T_j, and the gate output is set based on whether the total meets or exceeds T_j. This produces the locked circuit C_{locked}, which is then formatted and validated as the final locked netlist.

The integration process described in Algorithm 2 is mathematically formalized by extending the standard weighted sum equation of TLGs to incorporate key inputs. The traditional operation of a TLG is defined as:

$$\text{Output} = \begin{cases} 1, & \text{if } \sum_{i=1}^{n} w_i \cdot x_i \geq T \\ 0, & \text{otherwise,} \end{cases} \quad (2)$$

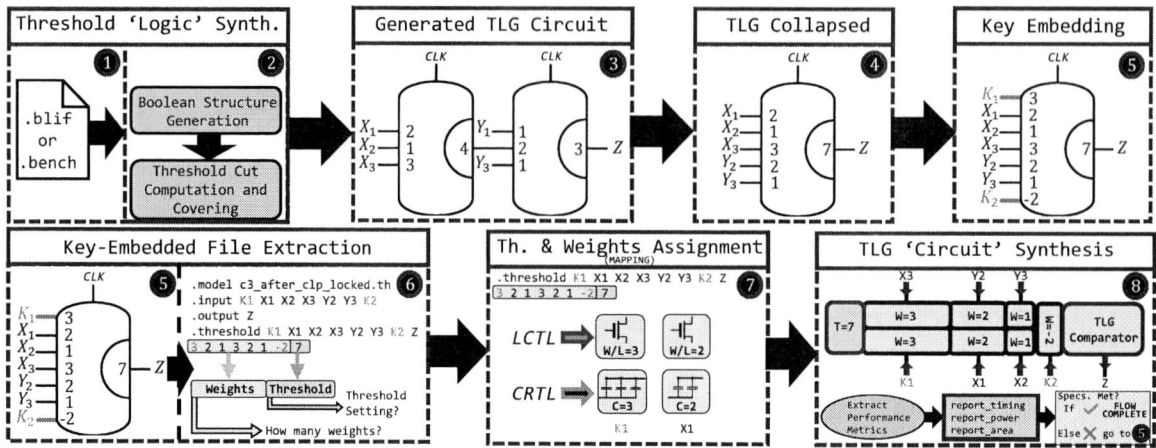

Fig. 3. Proposed flow: ❶ Threshold logic synthesis from a BLIF or BENCH file involving ❷ Boolean structure analysis and threshold cut computation. ❸ TLGs with respective weights and thresholds post-synthesis. ❹ Merging two TLGs via linear combination [23]. ❺ Embedding key inputs with weights proportional to the input weight sum. ❻ Extraction of weights and threshold values to be used in ❼ for assigning them into their corresponding inputs. ❽ TLG circuit synthesis and PnR (Place and Route) of the TLG circuits for performance extraction.

where x_i are the input signals, w_i are their associated weights, and T is the threshold value.

To integrate logic locking, we introduce key inputs k_j with corresponding weights v_j. The modified equation becomes:

$$\text{Output} = \begin{cases} 1, & \text{if } \sum_{i=1}^{n} w_i \cdot x_i + \sum_{j=1}^{m} v_j \cdot k_j \geq T \\ 0, & \text{otherwise.} \end{cases} \quad (3)$$

This modification ensures that the circuit produces the correct output only when the appropriate key vector $\mathbf{K} = [k_1, k_2, \ldots, k_m]$ is applied. An incorrect key vector $\mathbf{K}'$ introduces a deviation in the weighted sum:

$$\Delta = \sum_{j=1}^{m} v_j \cdot (k'_j - k_j), \quad (4)$$

which disrupts the threshold condition T, obfuscating the circuit's behavior and enhancing security. These equations form the basis for integrating logic locking in threshold logic circuits. When applied to specific architectures like *LCTL* and *CRTL*, this approach ensures efficient and secure operation under varying design constraints

By leveraging the properties of the *LCTL* and *CRTL* architectures, we ensure seamless integration of key-driven security. Both architectures use a fixed-weight approach, where each input contributes uniformly to the sum. This uniformity simplifies the integration of key inputs while ensuring that any deviation from the correct key disrupts the threshold conditions. This not only obfuscates the circuit's functionality but also prevents unintended outputs, preserving the logical integrity of the design.

Consider a TLG with three input nodes x_1, x_2, x_3, weights $w_1 = 1, w_2 = 1, w_3 = 1$, and threshold $T = 3$:

$$\text{Output} = 1 \quad \text{if} \quad 1 \cdot x_1 + 1 \cdot x_2 + 1 \cdot x_3 \geq 3. \quad (5)$$

With added key inputs k_1, k_2 and weights $v_1 = -2, v_2 = 3$, the equation becomes:

$$\text{Output} = 1 \quad \text{if} \quad 1 \cdot x_1 + 1 \cdot x_2 + 1 \cdot x_3 - 2 \cdot k_1 + 3 \cdot k_2 \geq 3. \quad (6)$$

The correct key vector $\mathbf{K} = [1, 1]$ neutralizes the effect of the key inputs, maintaining circuit functionality. Any incorrect key vector $\mathbf{K}'$ results in a non-zero Δ, altering the circuit's output and ensuring security against unauthorized use.

IV. RESULTS AND DISCUSSION

A. SAT Attack Analysis

We evaluate the scheme with `MiniSAT+` on ISCAS'85/'89, ITC'99, and MCNC; Table I reports solver outcomes and P/D/A for LCTL vs. CRTL. Large designs (`c1355`, `c7552`, `i10`, `b17`) consistently hit the 1-hour timeout, indicating strong SAT resistance; on `i10`, CRTL also lowers area and delay versus LCTL, highlighting scalability. Small designs (e.g., `c17`) are solved quickly, yet CRTL still leads across metrics; the trend holds on mid/large cases such as `des` and `b15`, where CRTL achieves notably lower area and delay.

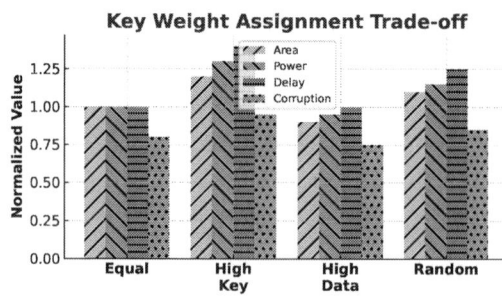

Fig. 4. Impact of key input weight assignment on power, delay, and corruption rate in a TLG. Balanced weights offer higher corruption but incur power/delay trade-offs.

Circuits with higher locking percentages, like `i8` at 80%, exhibited increased SAT solver complexity. Similarly, in `b17`,

TABLE I
SAT SOLVER PERFORMANCE METRICS FOR VARIOUS BENCHMARKS WITH *LCTL* AND *CRTL* LOGIC LOCKING CONFIGURATIONS.

Circuit	#Keys	Percent	Conflicts	Decisions	CPU Time	Result	LCTL			CRTL		
							A (μm²)	P (μW)	D (ns)	A (μm²)	P (μW)	D (ns)
c1355 [25]	16	50	—	—	—	Timeout	380	150	60	280	120	25
c17 [25]	17	50	5	138	0.13	SAT	5	2.5	1.0	3	1.2	0.5
c1908 [25]	17	50	—	—	—	Timeout	300	140	55	220	110	20
c2670 [25]	17	50	0	0	16.7	UNSAT	475	240	90	350	190	35
c7552 [25]	17	50	—	—	—	Timeout	1100	550	210	850	420	80
s1494 [26]	16	20	—	—	—	Timeout	350	170	60	260	130	22
s386 [26]	18	80	104	10797	37.9	SAT	75	35	12	50	25	8
s526 [26]	18	80	109	19055	31.6	SAT	95	47	18	70	35	12
s5378 [26]	16	20	0	0	196.7	UNSAT	820	400	160	600	300	60
s713 [26]	18	70	309	35636	34.1	SAT	105	52	20	75	37	15
i10 [27]	15	50	—	—	—	Timeout	540	360	400	450	300	150
i8 [27]	15	80	—	—	—	Timeout	250	160	200	220	140	100
des⋆ [27]	10	5	—	—	—	Timeout	9000	6000	700	7600	5000	300
b15⋆ [28]	5	50	—	—	—	Timeout	2300	1500	180	2000	1300	100
b17⋆ [28]	10	10	—	—	—	Timeout	7200	5000	550	6100	4000	250

⋆ The comparator of the *LCTL* and *CRTL* was modified to have 10×width of PMOS and NMOS for a faster and functional operation.

with a moderate locking percentage, the charge recycling architecture consistently required less area and delay, reinforcing its suitability for high fan-in configurations. Modified circuits, such as b15, further emphasized this trend, where increased transistor widths for handling higher fan-in resulted in enhanced performance metrics in the recycling-based implementation.

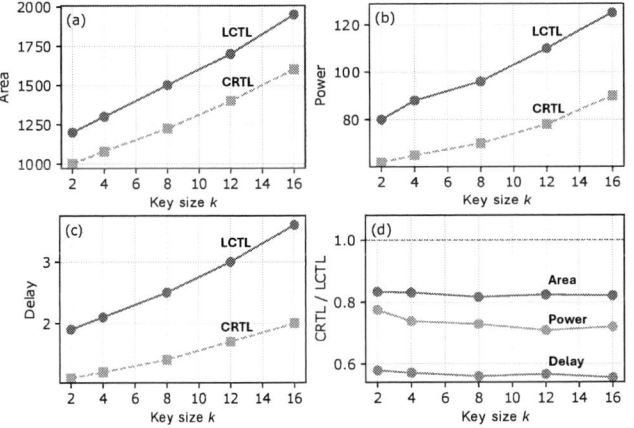

Fig. 5. CRTL vs. LCTL across key sizes k (example c1355; trend holds across suites). (a) CRTL/LCTL ratio (< 1 favors CRTL). (b) Area [μm²]. (c) Power [μW]. (d) Delay [ns]. LCTL has higher area; CRTL achieves lower power and delay.

B. Trade-Off Analysis of Key Weight Assignment

TLG-based locking treats keys as weighted inputs to the threshold, enabling tunable security–overhead trade-offs. We sweep the *total* key weight on a representative 4-input TLG and measure dynamic power, propagation delay, and the *corruption rate*—the fraction of primary-output mismatches averaged over all incorrect keys.

As shown in Fig. 4, corruption peaks at intermediate total key weights (≈ 2–3), where keys strongly influence the decision without over-biasing it; power and delay grow roughly linearly with weight. Thus, moderate key weights offer the best security per cost.

To contextualize the trade-off, Table II normalizes common schemes across SAT time (higher is better), area, power, and

delay. XOR incurs low overhead but weak resilience; SFLL improves resilience at higher cost; TLG locking reaches a favorable middle ground with a tunable knob (key weight).

C. Performance Analysis of TLG Architectures

Fig. 5 contrasts LCTL and CRTL using normalized area/power/delay versus key size k for c1355 (trend holds across benchmarks). LCTL incurs higher area due to latch-control overhead (e.g., s713, b17), while CRTL remains compact across k. CRTL also lowers power via charge recycling (e.g., s386, i10) and shortens delay, especially for high fan-in designs (e.g., s526, b15), making it preferable on timing-critical paths.

TABLE II
NORMALIZED LOCKING COMPARISON (SAT TIME: HIGHER IS BETTER).

Method	SAT Time	Area	Power	Delay
XOR Locking	0.1×	1.0×	1.0×	1.0×
SFLL-HD⁰	1.0×	2.2×	1.9×	1.8×
TLG (W=2–3)	**3.0×**	**1.5×**	**1.2×**	**1.3×**

V. CONCLUSION

We present a secure, efficient logic-locking scheme that embeds key-driven obfuscation directly into AIG-synthesized Threshold Logic Gates (TLGs). Exploiting the TLG weighted-sum decision, correctness holds only under the intended key, improving resistance to SAT and related attacks. On ISCAS'85/'89, ITC'99, and MCNC benchmarks, *CRTL* consistently outperforms *LCTL* in delay/power—up to 29% lower power and 58% faster; on b17 it reduces area by 26% while retaining security, and on s386/s526 it reaches 25 μW and 8 ns. Integrating locking within TLG synthesis reduces logic depth and further hardens designs; *CRTL* suits delay/power-constrained targets, whereas *LCTL* can favor area.

REFERENCES

[1] M. Tehranipoor and F. Koushanfar, "A survey of hardware trojan taxonomy and detection," *IEEE design & test of computers*, vol. 27, no. 1, pp. 10–25, 2010.

[2] M. Yasin, B. Mazumdar, and O. Sinanoglu, "Sarlock: Sat attack resistant logic locking," in *2016 IEEE International Symposium on Hardware Oriented Security and Trust (HOST)*. IEEE, 2016, pp. 236–241.

[3] Y. Xie and A. Srivastava, "Anti-sat: Mitigating sat attack on logic locking," *IEEE Transactions on Computer-Aided Design of Integrated Circuits and Systems*, vol. 38, no. 2, pp. 199–207, 2018.

[4] M. Yasin, A. Sengupta, M. T. Nabeel, M. Ashraf, J. Rajendran, and O. Sinanoglu, "Provably-secure logic locking: From theory to practice," in *ACM SIGSAC Conference on Computer and Communications Security (CCS)*, 2017, pp. 1601–1618.

[5] V. Beiu, J. M. Quintana, and M. J. Avedillo, "Threshold logic gates," *IEEE Potentials*, vol. 22, no. 1, pp. 36–40, 2003.

[6] R. Medina and J. G. Delgado-Frias, "Reconfigurable alus using threshold logic," in *2019 IEEE International Symposium on Circuits and Systems (ISCAS)*, 2019, pp. 1–5.

[7] S. Angizi, J. Sun, W. Zhang, and D. Fan, "Aligns: A processing-in-memory accelerator for dna short read alignment leveraging sot-mram," in *Proceedings of the 56th Annual Design Automation Conference 2019*, 2019, pp. 1–6.

[8] S. Angizi, W. Zhang, and D. Fan, "Exploring dna alignment-in-memory leveraging emerging sot-mram," in *Proceedings of the 2020 on Great Lakes Symposium on VLSI*, 2020, pp. 277–282.

[9] Y. He, D. Fan, and S. Bhunia, "Energy-efficient digital logic using threshold logic gates," *IEEE Transactions on Very Large Scale Integration (VLSI) Systems*, vol. 24, no. 10, pp. 3194–3204, 2016.

[10] G. Papandroulidakis and P. Kogge, "A practical implementation of threshold logic gates using emerging technologies," *IEEE Transactions on Nanotechnology*, vol. 18, pp. 299–310, 2019.

[11] A. Sahruri, M. Margala, and U. Cilingiroglu, "Hictl: High fan-in differential capacitive-threshold-logic gate implementation with an offset-compensated comparator," in *2024 25th International Symposium on Quality Electronic Design (ISQED)*. IEEE, 2024, pp. 1–7.

[12] H. Ozdemir, A. Kepkep, B. Pamir, Y. Leblebici, and U. Cilingiroglu, "A capacitive threshold-logic gate," *IEEE Journal of Solid-State Circuits*, vol. 31, no. 8, pp. 1141–1150, 1996.

[13] J. López-García, J. Fernández-Ramos, and A. Gago-Bohórquez, "A balanced capacitive threshold-logic gate," *Analog Integrated Circuits and Signal Processing*, vol. 40, pp. 61–69, 2004.

[14] Y. Leblebici, H. Ozdemir, A. Kepkep, and U. Cilingiroglu, "A compact high-speed (31, 5) parallel counter circuit based on capacitive threshold-logic gates," *IEEE Journal of Solid-State Circuits*, vol. 31, no. 8, pp. 1177–1183, 1996.

[15] M. Avedillo, J. Quintana, A. Rueda, and E. Jiménez, "Low-power cmos threshold-logic gate," *Electronics Letters*, vol. 31, no. 25, pp. 2157–2159, 1995.

[16] P. Celinski, J. López, S. Al-Sarawi, and D. Abbott, "Low power, high speed, charge recycling cmos threshold logic gate," *Electronics Letters*, vol. 37, no. 17, p. 1, 2001.

[17] S. Roy and I. L. Markov, "Epic: Ending piracy of integrated circuits," in *Design, Automation & Test in Europe Conference & Exhibition (DATE)*. IEEE, 2010.

[18] P. Subramanyan, D. Ray, and S. Malik, "Evaluating the security of logic encryption algorithms," *IEEE Transactions on Very Large Scale Integration (VLSI) Systems*, vol. 23, no. 11, pp. 2210–2223, 2015.

[19] N. Limaye, E. Kalligeros, N. Karousos, I. G. Karybali, and O. Sinanoglu, "Thwarting all logic locking attacks: Dishonest oracle with truly random logic locking," *IEEE Transactions on Computer-Aided Design of Integrated Circuits and Systems*, vol. 40, no. 9, pp. 1740–1753, 2021.

[20] H. M. Kamali, K. Z. Azar, F. Farahmandi, and M. Tehranipoor, "Advances in logic locking: Past, present, and prospects," *Cryptology ePrint Archive*, 2022.

[21] S. Roshanisefat, H. Mardani Kamali, and A. Sasan, "Srclock: Sat-resistant cyclic logic locking for protecting the hardware," in *Proceedings of the 2018 on Great Lakes Symposium on VLSI*, 2018, pp. 153–158.

[22] A. Rezaei, Y. Shen, and H. Zhou, "Cycsat-unresolvable cyclic logic encryption using unreachable states," *Asia and South Pacific Design Automation Conference (ASP-DAC)*, pp. 358–363, 2019.

[23] N.-Z. Lee, H.-Y. Kuo, Y.-H. Lai, and J.-H. R. Jiang, "Analytic approaches to the collapse operation and equivalence verification of threshold logic circuits," in *2016 IEEE/ACM International Conference on Computer-Aided Design (ICCAD)*. IEEE, 2016, pp. 1–8.

[24] A. Neutzling, J. M. Matos, A. Mishchenko, A. Reis, and R. P. Ribas, "Effective logic synthesis for threshold logic circuit design," *IEEE Transactions on Computer-Aided Design of Integrated Circuits and Systems*, vol. 38, no. 5, pp. 926–937, 2018.

[25] F. Brglez, D. Bryan, and K. Kozminski, "Combinational profiles of sequential benchmark circuits," in *Proceedings of the IEEE International Symposium on Circuits and Systems (ISCAS)*. IEEE, 1985, pp. 1929–1934.

[26] F. Brglez and H. Fujiwara, "A neutral netlist of 10 combinational benchmark circuits and a target translator in fortran," in *Proceedings of the IEEE International Symposium on Circuits and Systems (ISCAS)*. IEEE, 1989, pp. 695–698.

[27] S. Yang, "Logic synthesis and optimization benchmarks, version 3.0," *Tech. Report*, 1991.

[28] S. Davidson, "Itc'99 benchmark circuits-preliminary results," in *International Test Conference 1999. Proceedings (IEEE Cat. No. 99CH37034)*. IEEE, 1999, pp. 1125–1125.

Correlation Between Process Variability and Radiation Hardness in Digital Circuits

Elias Ramos[1,2], Augusto Weber[2], Wilian Padilha[2], Renan Carlos Gomes de Farias[2], João Baptista Martins[2], Ricardo Reis[1]

[1]Instituto de Informática, PGMICRO. Universidade Federal do Rio Grande do Sul (UFRGS)

[2]Santa Maria Design House (SMDH). Universidade Federal de Santa Maria (UFSM)

{elias.ramos, reis}@inf.ufrgs.br, {augusto.weber, wilian.padilha}@ecomp.ufsm.br, renan.farias@ufsm.br, batista@inf.ufsm.br

Abstract— **The objective of this work is to unify, with mathematical rigor, the analyses to estimate the effects of process variability and the effects caused by radiation in digital circuits using a single analysis. First, we proved that to estimate the effects of process variability on delay and power, it is enough to estimate at least one of the two. Later, a relationship between the effects of radiation and the effects of variability was obtained through a common element between the two: the entropy of the circuit. We applied the presented methods to compare circuits with the operations: matrix multiplication, graph searches, encryption algorithms, and neural networks. All of the circuits are designed using a 180nm technology. We were able to verify which circuits are more sensitive to the effects of variability and radiation simultaneously through the proposed method.**

Keywords— *Process variability, Radiation tolerance, Digital circuits, Microelectronics*

I. INTRODUCTION

Due to the design of components utilizing increasingly smaller dimensions, current integrated circuits (ICs) exhibit an increase in the number of functionalities, enabling the development of more complex systems, as well as an improvement in performance. On the other hand, reducing the size of devices can amplify the effects of variability in manufacturing processes and increase susceptibility to radiation effects [1] [2]. Regarding radiation effects, there are single-event effects (SEEs), classified as either destructive or non-destructive. This leads to the natural question: given two or more circuits with different topologies, which one is the most robust in relation to radiation effects and those caused by process variability? The primary objective of this work is to formally develop a method that estimates the effects of both variability and radiation in a single analysis. First, we prove that, to estimate the effects of process variability on power and delay in a circuit, it is sufficient to estimate one of the two, as the other follows as a consequence. Subsequently, we define a relationship between radiation tolerance and variability, using the relationship between Lyapunov exponents and the entropy of a dynamic system, providing a unified analysis to address both issues. Section II presents related work and an overview of other studies on the topic. Section III outlines the theoretical foundations, including definitions and techniques used. Section IV presents the results, and finally, conclusions and future work are discussed.

II. RELATED WORKS

Several studies have addressed the effects of process variability [3] [4], typically through extensive simulations ranging from 1,000 to 10,000 per case making the process time-consuming and technology-dependent. Transitioning between technologies (e.g., FinFET to FDSOI) requires new parameter sets and simulations. While reducing transistor size can enhance performance, nanometric technologies demand deeper analysis [5]. In [4] it is analyzed work function fluctuations in 14 nm FinFETs. Studies on analog circuits highlight the importance of variability in circuit stability [9]. [10] demonstrated that different transistor arrangements for the same logic function impact power, delay, and PDP. [1] found complex cells to be more sensitive than multilevel ones and showed topology-dependent variability effects in the C17 benchmark. It is well known that different transistor arrangements in a cell affect its stability, and the use of nanometric technologies has intensified process variations and radiation effects [6]. In [1] it is shown that, for the analyzed cells, multilevel topologies are less sensitive to SET effects despite their larger area using SET fault injection based on Messenger equations and SPICE simulations. In [7] it is demonstrated that transistor arrangements impact radiation sensitivity using different C17 benchmark variants. However, no general conclusion can be drawn about which arrangements or logic levels consistently improve robustness. Simulations are limited to a small set of cells and are computationally expensive for large circuits. [8] extended the discussion to analog circuits, showing SET effects in charge-redistribution SAR ADCs.

The effects of process variability and radiation tolerance were initially studied independently. Starting from the work [1], these phenomena began to be investigated in a correlated manner. From [11] and [12], dynamic systems theory was introduced to estimate the effects of process variability, and in [13], the same theory was applied to estimate the impact of radiation on digital circuits. However, no analysis has yet been conducted that addresses both issues simultaneously.

III. THEORETICAL BACKGROUND

Let $f\colon R^n \to R^n$ point $x \in R^n$, the Lyapunov exponent is defined by equation (1) [3] [14]:

$$\lambda\left(x_0\right) = \frac{1}{N} \sum_{i=0}^{N-1} ln\left|f'\left(x_i\right)\right| \qquad (1)$$

As seen in [11], [12], the Lyapunov exponent was used to estimate the effects of process variability. A bit flip occurs when the value of a bit is inverted, from 0 to 1 or from 1 to 0, typically due to radiation, interference, or hardware faults, potentially affecting the system's operation. Masked Faults: The fault occurs but does not impact the system's behavior or the final result, being "masked" or undetected. Detected Faults: The fault impacts the system's behavior and is identified by detection mechanisms, and can be corrected or signaled.

Consider $X = \{p_1, p_2, p_3, \dots p_n\}$ a set of probabilities corresponding to some experiments. A way of measuring the average degree of uncertainty about information sources, which consequently allows quantifying the information that flows in the system. The entropy of X will be defined by Equation (2) [13] [14].

$$H(X) = -\sum_{i=0}^{n} p_i(p_i) \tag{2}$$

Equation (3) presents a detailed description of the entropy of faults at each level. $H\big(X_i(DF)\big)$ shows the entropy of detected faults and $H\big(X_i(MF)\big)$ the entropy of masked faults.

$$H\big(X_i(F)\big) = H\big(X_i(DF)\big) + H\big(X_i(MF)\big) \tag{3}$$

Finally, equation (4) presents the main metric used in our work, that is the ratio of the entropy of detected faults to the total entropy of faults at each level. Therefore, the estimates will be calculated and to estimate circuit faults, simply calculate the arithmetic mean of all R_i called R. The higher the R value, the greater the sensitivity to circuit faults [13].

$$R_i = \frac{H\big(X_i(DF)\big)}{H\big(X_i(F)\big)} \tag{4}$$

To establish a relationship between the effects of process variability on dynamic power and delay, the following definitions are required: A diffeomorphism $f: R^n \to R^n$ is a smoothly differentiable, bijective function whose inverse is also smoothly differentiable, establishing a smooth equivalence between two differentiable manifolds. We say that two functions are diffeomorphic if there exists a diffeomorphism that conjugates them.

Theorem 1: If f and g are contraction mappings on $[0,1]$ (i.e., there exist constants $0<c<1$ such that for all $x, y \in [0,1]$, we have $|f(x)-f(y)| \leq c\,|x-y|$ and $|g(x)-g(y)| \leq c\,|x-y|$), and if f and g are differentiable, with continuous and invertible derivatives (i.e., $f'(x)\neq0$ and $g'(x)\neq0$ for all $x \in [0,1]$), then f is diffeomorphic to g [22].

Theorem 2: Let $f:[0,1]\to R$ and $g:[0,1]\to R$ be differentiable functions on $[0,1]$, with continuous and invertible derivatives, i.e., $f'(x)\neq0$ and $g'(x)\neq0$ for all $x \in [0,1]$. If f and g generate smooth dynamical systems, then the map $Lyap(f)\to Lyap(g)$, which maps the Lyapunov exponent of f to the Lyapunov exponent of g, is continuous [23] [24].

We present the calculation of dynamic power in a combinational circuit by equations (5) and (6). P is the dynamic power, C_i is the capacitance at node i, H_i is the entropy at node i, and t_i is the delay at node i [13] [25].

$$P_{din} = \sum_{i=1}^{N} V_{DD}^2 H_i C_i f \tag{5}$$

$$P_{din} = \sum_{i=1}^{N} V_{DD}^2 H_i C_i t_i^{-1} \tag{6}$$

The equations (5) and (6) show a relationship between dynamic power and delay. It is easy to see that power and delay satisfy the conditions of Theorem 1. Thus, the correspondence $Lyap(P)\to Lyap(D)$ is continuous and invertible by Theorem 2, ensuring a mapping from the

power exponent to the delay exponent. Therefore, we conclude that by observing the behavior of the first, we can estimate the behavior of the second, and vice versa. We now proceed to establish an analytical relationship between process variability and the effects of radiation.

Dynamic power can also be expressed using the transition probabilities (P_i) by equation (7) [26].

$$P_{din} = \frac{1}{2T_c} V_{dd}^2 \sum_{i=1}^{N} C_i P_i \tag{7}$$

High Transition Probability: With high transition probability, the bit frequently changes. A fault caused by a disturbance is more likely to be masked, as the system may already be in transition, potentially correcting or ignoring the error. **Low Transition Probability:** With low transition probability, the system stays in a fixed state longer, making faults more detectable since any bit change is more noticeable.

The probability of a detected fault should be inversely proportional to the transition probability, as in systems with rapid transitions, faults (like bit flips) are more likely to be corrected automatically before being noticed. Thus, the formula for the detected fault probability is (P_t is the probability transition) by equation (8):

$$P_d = \frac{1}{1 + P_t} \tag{8}$$

The probability of a masked fault should be directly proportional to the transition probability, since when the system is constantly transitioning, faults are more likely to be automatically corrected or hidden. Therefore, the formula, by equation (9) for the masked fault probability is:

$$P_m = \frac{P_t}{1 + P_t} \tag{9}$$

From the previous equations, we can deduce a new model of them using the concept of entropy introduced in [12], by equation (10).

$$H_d = \frac{H_T}{H_T + H_t} \tag{10}$$

Where H_d is the entropy of detected faults, H_T is the total entropy that can be reached by the logical level, and H_t is the entropy of the logical level.

Main Theorem: Consider two combinational circuits with the same number of inputs. The sensitivity to the variability of the manufacturing process is greater in the first circuit if and only if the first circuit is more sensitive to radiation under the action of single event transients.

Dem: Let A and B be two combinational circuits with the same number of inputs. Suppose that A is more robust to process variability than B. By definition, A has a lower Lyapunov exponent. Therefore, the average entropy across logic levels is higher in circuit A than in circuit B. This is because power and delay are strictly increasing contractions. Consequently, according to equations (8) and (10), the entropy associated with the detected faults will be lower in A. This proves the first part of the Theorem.

Now suppose that A is more robust to radiation. Then, the entropy of the detected faults will be lower than that of B. This implies that the average entropy across the logic levels in A will be higher than in B. Consequently, A will have a lower Lyapunov exponent than B, which proves the second part of the Theorem.

IV RESULTS

In this work, Genus logical synthesis was used to convert high-level descriptions, such as RTL code, into optimized logic gate networks, focusing on area, power, and performance. Innovus physical synthesis defined the chip layout, placing and routing logic cells to minimize power consumption and ensure signal integrity. These steps are crucial for integrated circuit efficiency. For the CMOS inverter in 180nm technology, the following parameters were used: supply voltage (VDD) ranged from 0.6V to 2.4V, load capacitance (C) was 1 pF, operating frequency (f) was 100 MHz, and the activity factor (α) was 0.5. PMOS and NMOS transistor widths were 2 μm and 1 μm, respectively, with threshold voltages of 0.4V and 0.5V.

These parameters were used to estimate dynamic power and delay across different VDD values. For the presentation of the results, the following units are used: power is expressed in watts (W), delay in picoseconds (ps), and area in square micrometers (μm^2). In the tables that follow, the Lyapunov exponent with respect to power is denoted as LP, with respect to delay as LD, and Hd represents the entropy of the detected faults.

Matrix multiplication can be performed using various algorithms, each with distinct computational characteristics. The classical method is straightforward and has a computational complexity of O(n^3). The Strassen[15] algorithm enhances this performance by employing a recursive strategy, reducing the complexity to O($n^2.81$). The Karatsuba [16] algorithm, which is based on the divide-and-conquer paradigm, achieves a lower complexity of O($n^{1.585}$). The Winograd [17] algorithm, a variant of Strassen's method, further optimizes the number of addition and subtraction operations, maintaining a similar asymptotic complexity to Strassen. The efficiency of each algorithm depends on factors such as matrix size and structure. In this study, we focus on the multiplication of 4×4 matrices as shown in Table 1.

TABLE 1 SENSITIVITY TO PROCESS VARIABILITY AND RADIATION TOLERANCE IN MATRIX MULTIPLICATION CIRCUITS

Algorithm	Power	Delay	Area	#Cells	LP	LD	H
Classic	4.44E-02	17668	853705.01	28260	-5.88	-20.42	0.034
Strassen	4.07E-03	12868	94384.581	3762	-7.90	-20.56	0.212
Karatsuba	2.82E-02	31793	70865.619	2304	-6.74	-20.44	0.078
Winograd	8.10E-04	22765	34929.092	1170	-9.82	-20.30	0.649

The Classic algorithm exhibits the lowest variability (with LP, LD, and H values closest to zero), particularly in entropy (H = 0.034), which indicates high stability, strong radiation tolerance, and consistent predictions—though it comes at the highest cost in terms of power, delay, area, and number of cells, making it the most hardware-intensive. The Strassen algorithm provides a favorable balance, offering moderate variability and a substantial reduction in hardware cost compared to the Classic algorithm, making it a more efficient alternative. Karatsuba demonstrates intermediate behavior, with even lower hardware demands than Strassen and slightly higher variability; however, its low entropy (H = 0.078) still suggests good radiation tolerance and prediction reliability. Winograd, although the most

lightweight in terms of power consumption and cell count, exhibits the highest variability, with LP and H values farthest from zero, implying reduced stability and lower resistance to radiation effects. In summary, the Classic algorithm ensures the highest stability and radiation tolerance at a high hardware cost; Strassen offers the best trade-off between robustness and efficiency; and Winograd, despite its minimal hardware footprint, is the most vulnerable to variability and radiation-induced errors.

Depth-First Search (DFS) and Breadth-First Search (BFS) are fundamental graph traversal algorithms. DFS explores a graph by traversing as deeply as possible along each branch before backtracking, typically implemented using recursion or a stack. In contrast, BFS operates by visiting nodes level by level, starting from a root node and exploring all neighboring nodes before progressing to the next level. DFS is particularly well-suited for tasks such as cycle detection, while BFS is commonly used for finding the shortest path in unweighted graphs [18]. In this study, both algorithms are applied to search operations on an 8×8 graph, as shown in Table 2.

TABLE 2 SENSITIVITY TO PROCESS VARIABILITY AND RADIATION TOLERANCE IN SEARCH ALGORITHMS CIRCUITS

Algorithm	Power	Delay	Area	#Cells	LP	LD	H
DFS	1.29E-04	6782	4843.005	155	-11.151	-21.006	0.777
BFS	1.39E-04	8227	5396.064	173	-11.282	-21.013	0.798

The analysis of the table shows that the DFS and BFS algorithms exhibit very similar performance in both hardware metrics and output variability. BFS consumes slightly more power and has higher delay, area, and cell count than DFS, indicating a marginally higher implementation cost. Regarding variability, both algorithms have closely aligned LP and LD values, reflecting comparable stability. However, DFS demonstrates a slight advantage in entropy (H = 0.777 vs. H = 0.798), implying marginally better radiation tolerance and slightly more reliable predictions. Therefore, DFS emerges as the more efficient option, with lower hardware cost and improved resilience to radiation-induced variability, even though the overall differences between the two are minimal.

Cryptographic algorithms employ diverse approaches to ensure data confidentiality and integrity. RSA is a public-key encryption scheme that relies on the mathematical properties of large prime numbers. XOR encryption is a simple symmetric technique that applies the XOR logical operation between the plaintext and a key. The Feistel structure, exemplified by the Data Encryption Standard (DES) [19], is an iterative framework that partitions data into blocks and processes them through a series of transformations. Substitution and permutation techniques, as used in the Advanced Encryption Standard (AES), involve systematically replacing and rearranging data to achieve diffusion and confusion. The Shift Cipher, including the well-known Caesar cipher [19][20], encrypts text by shifting characters within the alphabet. Each algorithm provides distinct trade-offs in terms of security and computational efficiency. In this study, the encryption circuits were applied to 16-bit data words, as shown in Table 3.

979-8-3315-9813-6/25 $31.00 © 2025 IEEE

TABLE 3 SENSITIVITY TO PROCESS VARIABILITY AND
RADIATION TOLERANCE IN ENCRYPTION CIRCUITS

Algorithm	Power	Delay	Area	#Cells	LP	LD	H
RSA	1.74E-05	4087	814.868	29	-13.035	-21.394	0.478
XOR	4.17E-05	12684	1716.724	40	-11.875	-19.974	0.223
Feistel	**1.10E-04**	**2336**	**3965.656**	**65**	**-10.217**	**-19.903**	**0.151**
S and P	3.94E-05	10520	1454.132	24	-12.087	-20.315	0.262
Shift Cipher	6.42E-05	14470	3659.965	161	-12.137	-20.536	0.270

The analysis of the table indicates that among the evaluated algorithms, the variability metrics (LP, LD, and H) suggest that the Feistel algorithm exhibits the best stability performance, with the lowest absolute values—particularly for LP and H indicating reduced uncertainty in the network outputs and higher radiation tolerance. In contrast, RSA, while showing the lowest power consumption and area usage, displays the highest variability, with significantly large negative LP and LD values, reflecting instability in predictions. The XOR, S and P, and Shift Cipher algorithms show intermediate behavior with moderate variability; among them, XOR stands out for its relatively low entropy (H) and area, suggesting higher radiation tolerance, though it suffers from higher delay. In terms of hardware cost, RSA is the most lightweight, whereas Shift Cipher incurs the highest delay and number of cells, making it the most expensive to implement. Overall, the Feistel algorithm offers the most favorable trade-off between prediction stability, radiation tolerance, and hardware efficiency, while RSA, despite its low hardware footprint, is hindered by high output variability.

Neural networks vary in complexity according to the number of hidden layers. The Perceptron, a single-layer architecture, is limited to solving linearly separable problems. Networks with one hidden layer are capable of addressing more complex tasks, such as the XOR problem [21]. According to the Universal Approximation Theorem, a feedforward neural network with a single hidden layer containing a finite number of neurons can approximate any continuous function on a compact domain, given appropriate activation functions. However, in practice, deep networks with multiple hidden layers—ranging from two to eight or more—are often preferred, as they can represent complex functions more efficiently and with fewer neurons per layer. Deeper architectures promote hierarchical feature learning, improve generalization in high-dimensional tasks, and reduce the computational burden per layer, despite being more difficult to train and more susceptible to overfitting. Networks with six to eight hidden layers are commonly employed in advanced applications such as pattern recognition and deep feature extraction, as shown in Table 4.

TABLE 4 SENSITIVITY TO PROCESS VARIABILITY AND
RADIATION TOLERANCE IN NEURAL NETWORKS
CIRCUITS

N. network	Power	Delay	Area	#Cells	LP	LD	H
Perceptron	5.06E-04	13751	10132.175	344	-10.153	-21.667	0.562
1 hide layer	2.58E-03	24545	51808.237	1770	-9.216	-20.780	0.334
2 hide layers	5.57E-03	52374	68990.421	2520	-9.160	-20.735	0.322
3 hide layers	7.27E-03	50587	142162.851	4795	-8.835	-20.711	0.255
4 hide layers	8.79E-03	63360	168596.472	5665	-8.811	-20.651	0.251
5 hide layers	1.04E-02	76176	195030.094	6535	-8.810	-20.638	0.250
6 hide layers	1.20E-02	89034	221463.717	7405	-8.804	-19.705	0.249
7 hide layers	1.34E-02	102167	247897.339	8257	-8.802	-19.698	0.245
8 hide layers	**1.49E-02**	**114983**	**274330.962**	**9145**	**-8.769**	**-19.673**	**0.243**

The analysis of the table indicates that, as the number of hidden layers increases, the variability metrics (LP, LD, and H) improve, meaning they approach zero, which reflects greater stability, reduced uncertainty in the network's predictions, and higher radiation tolerance. However, the most substantial improvements occur with approximately 2 to 3 hidden layers; beyond this point, further enhancements in the metrics become marginal, while the hardware cost in terms of power consumption, delay, area, and number of cells escalates significantly. The Perceptron, which lacks hidden layers, exhibits the highest variability and thus the poorest performance in terms of stability and radiation resistance. Networks with more than 4 hidden layers provide only marginal gains in variability metrics but incur considerably higher implementation costs. Therefore, the optimal trade-off between prediction stability, radiation tolerance, and hardware efficiency lies in architectures with 2 to 3 hidden layers.

V CONCLUSIONS

The main goal of this work was to develop a comprehensive and unified methodology capable of simultaneously analyzing the effects of process variability and radiation-induced phenomena in digital circuits, within a single analytical approach. Through an analytical perspective, we demonstrated that estimating either circuit delay or power consumption individually provides sufficient information regarding the impact of process variability on overall circuit performance. Furthermore, we identified and established a fundamental correlation between radiation effects and process variability, with circuit entropy serving as a key unifying parameter that quantitatively links these two domains. This connection offers a novel perspective for understanding and modeling the interaction between manufacturing variations and environmental perturbations in advanced integrated circuit technologies.

This relationship proved essential for understanding the interaction between these two sources of disturbance in digital systems. The proposed methodology was applied to evaluate and compare a range of circuit implementations, including designs for matrix multiplication, graph traversal algorithms, cryptographic functions, and artificial neural networks, all synthesized using a 180 nm CMOS technology. By applying this unified approach, we identified the circuits most susceptible to the combined impact of process variability and radiation, thereby offering valuable insights into their reliability and performance under diverse operating conditions.

For future work, we plan to conduct experiments using more advanced, smaller technology nodes. This direction is motivated by the fact that process variations become significantly more pronounced and complex at reduced feature sizes, offering a richer and more challenging environment for analysis and validation of the proposed methodologies.

ACKNOWLEDGMENT - This paper is related to the project "DECISARR - Development of Integrated Circuit for Satellite using Radiation-Robust Cell Library", number 407642/2022-6, supported by the CNPq, as well the research work developed at UFRGS.

This work is dedicated to the Brazilian mathematician Jacob Palis, for all his work and inspiration.

REFERENCES

[1] L. H. Brendler, A. L. Zimpeck, C. Meinhardt and R. Reis, "Multi-Level Design Influences on Robustness Evaluation of 7nm FinFET Technology," in IEEE Transactions on Circuits and Systems I: Regular Papers, vol. 67, no. 2, pp. 553-564, Feb. 2020, doi: 10.1109/TCSI.2019.2927374.

[2] A. L. Zimpeck, Y. Aguiar, C. Meinhardt and R. Reis, "Robustness of Sub-22nm multigate devices against physical variability," 2017 IEEE International Symposium on Circuits and Systems (ISCAS), 2017, pp. 1-4, doi: 10.1109/ISCAS.2017.8050441.

[3] A. L. Zimpeck, C. Meinhardt And R. Reis. "Evaluating the impact of environment and physical variability on the ION current of 20nm FinFET devices". 2014 24th International Workshop on Power and Timing Modeling, Optimization and Simulation (PATMOS), 2014, pp. 1-8, DOI: 10.1109/PATMOS.2014.6951859

[4] Gupta, N. Chauhan, O. Prakash, And H. Amrouch. "Variability Effects in FinFET Transistors and Emerging NC-FinFET". 2021 International Conference on IC Design and Technology (ICICDT), 2021, pp. 1-4, DOI: 10.1109/ICICDT51558.2021.9626531.

[5] M. Orshansky, S. R. Nassif, And D. Boning, "Design for Manufacturability and Statistical Design: A Constructive Approach". Springer. 2008. doi 10.1007/978-0-387-69011-7.

[6] A. L. Zimpeck, Y. Aguiar, C. Meinhardt And R. Reis, "Robustness of Sub-22nm multigate devices against physical variability," 2017 IEEE International Symposium on Circuits and Systems (ISCAS), 2017, pp. 1-4, doi: 10.1109/ISCAS.2017.8050441.

[7] B. B. Sandoval, L. H. Brendler, A. L. Zimpeck, F. L. Kastensmidt, R. Reis and C. Meinhardt, "Exploring Gate Mapping and Transistor Sizing to Improve Radiation Robustness: A C17 Benchmark Case-study," 2021 IEEE 22nd Latin American Test Symposium (LATS), 2021, doi: 10.1109/LATS53581.2021.9651

[8] Thales E. Becker, Alisson J.C. Lanot, Guilherme S. Cardoso, Tiago R. Balen, "Single event transient effects on charge redistribution SAR ADCs", Microelectronics Reliability, Volume 73,2017. DOI: 10.1016/j.microrel.2017.04.002

[9] F.J. Alonso, D. Maldonado, A.M. Aguilera, J.B. Roldán, "Memristor variability and stochastic physical properties modeling from a multivariate time series approach", Chaos, Solitons & Fractals, Volume 143, 2021, doi.org/10.1016/j.chaos.2020.110461

[10] A. L. Zimpeck, C. Meinhardt, L. Artola, G. Hubert, F. L. Kastensmidt And R. Reis. "Impact of different transistor arrangements on gate variability". Microelectron. Rel., vol. 88, pp. 111-115, Sep. 2018. doi.org/10.1016/j.microrel.2018.06.090

[11] E. de Almeida Ramos and R. Reis, "Using Lyapunov Exponents to Estimate Sensitivity to Process Variability," 2023 IEEE 14th Latin America Symposium on Circuits and Systems (LASCAS), Quito, Ecuador, 2023, pp. 1-4, doi: 10.1109/LASCAS56464.2023.10108159.

[12] E. de Almeida Ramos and R. Reis, "Using Lyapunov Exponents and Entropy to Estimate Sensitivity to Process Variability," 2023 IEEE Computer Society Annual Symposium on VLSI (ISVLSI), Foz do Iguacu, Brazil, 2023, pp. 1-6, doi: 10.1109/ISVLSI59464.2023.10238486.

[13] E. De Almeida Ramos and R. Reis, "The Use of Entropy to Estimate the Effects of Radiation in Digital Circuits," 2024 31st IEEE International Conference on Electronics, Circuits and Systems (ICECS), Nancy, France, 2024, pp. 1-4, doi: 10.1109/ICECS61496.2024.10849271.

[14] Ricardo Mane. "Ergodic Theory and Differentiable Dynamics." Ergebnisse der Mathematik und ihrer Grenzgebiete. 3. Folge / A Series of Modern Surveys in Mathematics. Springer. 1987: https://link.springer.com/book/10.1007/978-3-642-70335-5.

[15] Nurhayati, A. Meizar, F. Tambunan and E. Ginting, "Optimizing the Complexity of Time in the Process of Multiplying Matrices in the Hill Cipher Algorithm Using the Strassen Algorithm," 2019 7th International Conference on Cyber and IT Service Management (CITSM), Jakarta, Indonesia, 2019, pp. 1-4, doi: 10.1109/CITSM47753.2019.8965416.

[16] M. Nursalman, A. Sasongko, Y. Kurniawan and Kuspriyanto, "Improved generalizations of the Karatsuba algorithm in GF(2n)," 2014 International Conference of Advanced Informatics: Concept, Theory and Application (ICAICTA), Bandung, Indonesia, 2014, pp. 185-190, doi: 10.1109/ICAICTA.2014.7005938.

[17] Z. Zhang, Z. Li and H. Chen, "A cache structure and corresponding data access method for Winograd algorithm," IET International Radar Conference (IET IRC 2020), Online Conference, 2020, pp. 634-639, doi: 10.1049/icp.2021.0723.

[18] Richard J. Trudeau . Introduction to Graph Theory. Dover Publications; 2nd Revised ed. ISBN-13 : 978-0486678702.

[19] Jeffrey Hoffstein, Jill Pipher, J. H. Silverman. An Introduction to Mathematical Cryptography. Springer; Softcover reprint of hardcover 1st ed. 2008. ISBN-13 : 978-1441926746.

[20] Jonathan Katz, Yehuda Lindell. Introduction to Modern Cryptography, Second Edition. CRC Press. ISBN-10 : 9781466570269. ISBN-13 : 978-1466570269.

[21] Simon Haykin. Neural Networks and Learning Machines. Prentice Hall; 3rd ed. 2008. ISBN-10 : 0131471392. ISBN-13 : 978-0131471399

[22] Rudin, Walter (1976) [1953]. Principles of Mathematical Analysis (3rd ed.). New York: McGraw-Hill. ISBN 007054235X.

[23] Ricardo Mane. "Ergodic Theory and Differentiable Dynamics." Ergebnisse der Mathematik und ihrer Grenzgebiete. 3. Folge / A Series of Modern Surveys in Mathematics. Springer. 1987. : https://link.springer.com/book/10.1007/978-3-642-70335-5.

[24] Viana, M., Yang, J. Continuity of Lyapunov exponents in the C0 topology. Isr. J. Math. 229, 461–485 (2019). https://doi.org/10.1007/s11856-018-1809-7.

[25] Rabaey, Jan. "Digital Integrated Circuits: a Design Perspective". Prentice Hall, 178609-7. 1996. ISBN:978-0-13.

[26] M. Nemani And F. N. Najm, Towards a high-Nível Lógic power estimation capability [digital ICs], in IEEE Transactions on Computer-Aided Design of Integrated Circuits and Systems, vol. 15, no. 6, pp. 588-598, June 1996, doi: 10.1109/43.503929.

CMOS Time Register With High Dynamic Range

Johnatan Felipe Silva Garcia
Graduate Program in Electrical Engineering (PPGEE)
Federal University of Minas Gerais (UFMG)
Belo Horizonte, Brazil
https://orcid.org/0009-0006-8679-2185

Dalton Martini Colombo
Graduate Program in Electrical Engineering (PPGEE)
Federal University of Minas Gerais (UFMG)
Belo Horizonte, Brazil
https://orcid.org/0000-0002-6781-9673

Kamal El-Sankary
Department of Electrical and Computer Engineering
Dalhousie University
Halifax, Canada
https://orcid.org/0000-0001-8104-6913

Mahsa Zareie
Department of Electrical and Computer Engineering
Dalhousie University
Halifax, Canada
https://orcid.org/0000-0001-6892-227X

Abstract—**This paper introduces a novel time register for time-domain signal processing, utilizing a delay line within a ring configuration topology and a digital counter to achieve high dynamic range. The proposed design was implemented and simulated using the AMS 350 nm CMOS process and tested under nominal supply voltage of 3.3 V. Corner simulation, considering supply voltage, temperature, and process variations, demonstrated an error rate below 2% for inputs exceeding 50 ns. A key advantage of the proposed solution is that its range can be easily scaled by increasing the number of counter bits, while the layout area overhead introduced by the counter remains minimal. This is the first CMOS time register capable of efficiently processing signals from picoseconds to milliseconds, providing an output pulse equal to the input pulse, and being feasible for physical implementation.**

Index Terms—**time register, time domain, delay line.**

I. INTRODUCTION

The miniaturization of semiconductor devices has advanced efficiency and processing speed but poses challenges for analog circuits due to reduced supply voltage, leading to lower signal-to-noise ratios and dynamic ranges. Time-domain signaling addresses these issues by encoding signal amplitude in pulse width, aligning with digital operations, and enhancing scalability. In this way, time, which is an analog quantity, is encoded in a digital format, resulting in several advantages. This type of data processing is called Time-Mode Signal Processing (TMSP) [1]–[3]. One of the most important blocks in TMSP systems is the time register. It is responsible for storing and retrieving time-domain signals. Moreover, this building block can be used to perform addition, subtraction, amplification, and filtering of time signals [3]–[6].

A very interesting CMOS time register composed of a chain of delay cells is proposed and clearly explained in [7]. This architecture was also successfully employed in [8]. Using a different approach, studies [9] and [10] use capacitors instead of an inverter chain to implement the time register. In [11], a feedback inverter is included in the chain of delay cells, making the time register topology more efficient in terms of energy consumption and silicon area. This topology performs

the signal recovery operation in a single step, instead of two steps as in traditional circuits. The limitation of the previous implementations is the dynamic range. To increase this range, a large number of delay cells or a large capacitor must be used. This characteristic results in a large silicon area and becomes a severe limiting factor for several applications.

Finally, in [12], a counter was recently included in the design to extend the dynamic range of the time register. However, the output pulse of the time register in [12] has a width equal to the clock period minus the input pulse, meaning it does not provide the actual value of the input pulse, limiting its applicability. This paper proposes a new topology for the time register with a high dynamic range, feasible for physical implementation, and capable of generating an output pulse equal to the input pulse.

II. TRADITIONAL TIME REGISTER AND ITS LIMITATIONS

The time register can be implemented simply using a series of controlled delay cells, referred to as a Controlled Delay Line [8], along with an XOR gate, as shown in Fig. 1.

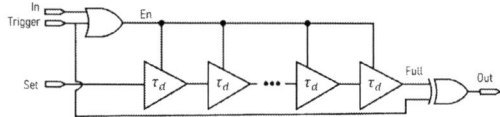

Fig. 1: Implementation of time register [8].

Each delay cell is responsible for generating a delay time τ_d. Therefore, the complete scale of the time register is defined by the number of delay cells, N, which produce the total delay time of the delay line. The full-scale time T_{FS} is defined as shown in Equation (1):

$$T_{FS} = N * \tau_d \qquad (1)$$

Therefore, the time storage capacity can be increased simply by increasing the number of delay cells. The operation of the time register is shown in Fig. 2. A good analogy for understanding the circuit is to see the time-register as a water

979-8-3315-9813-6/25 $31.00 © 2025 IEEE

tank. The tank fills when the Enable signal is high, triggered by either the *In* or *Trigger* signals.

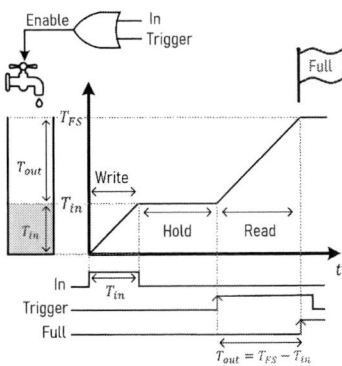

Fig. 2: Time register concept diagram [8].

Observe that when the input *In* is high, enabling *EN*, the *SET* signal propagates through the delay line. When neither *In* nor the *Trigger* signal are present, the propagation of the *SET* signal is halted. However, when the *Trigger* signal is applied, the propagation of the *SET* signal is reactivated. Once the *SET* signal reaches the end of the Controlled Delay Line, a *Full* signal is generated. We can conclude that when *SET* is high, the system can be in Write (*In* high), Read (*Trigger* high), or Hold (*In* low, *Trigger* low) mode of operation.

By measuring the time difference between the *Trigger* signal and the *Full* signal, we obtain the time information that was stored. The *Full* signal is generated by the XOR logic gate, so its output will remain high after the *Trigger* signal is released, until the full scale is reached. At the end of the conversion, the *SET* signal goes low and clears the time register.

Equation (2) defines the pulse width at the output of the time register.

$$T_{\text{out}} = T_{\text{FS}} - T_{\text{in}} \qquad (2)$$

To obtain an output signal ideally identical to the input signal T_{in}, two time registers with the same full scale capacity T_{FS} can be used in cascade. The traditional time register can be adapted to carry out addition and subtraction operations with minor modifications [8].

III. PROPOSED TIME REGISTER

The proposed time register is designed with two distinct inputs, T_{in} and *Trigger*, and just one output pin. The first input, T_{in}, is used to apply the pulse width of a signal in the time domain for storage. The second input, *Trigger*, serves as a control signal that retrieval of the stored input time signal. The stored signal is subsequently observed at the output, denoted as T_{out}.

The proposed novel Time Register is depicted in Fig. 3. The delay line block, is configured in a ring topology. This ring delay line is based on a fundamental delay line structure, as shown in Fig. 4, where the *SET* input is feedback with from the circuit output via an inverter [11].

The concept of the proposed CMOS Time Register with a High Dynamic Range can be explained in two modes: write

and read. These modes will be discussed in detail in the following subsections.

A. Write Mode

During the writing process, the T_{in} signal is applied at the input with a specific pulse width, causing first ring delay line to propagate the SET signal proportionally to the duration for which T_{in} remains active. Throughout this process, the *Flag* output oscillates as the ring delay line completes each propagating of SET signal cycle (full scale T_{FS}).

The T_{in} signal can be divided into two components. The first component, T_{complete}, represents the number of complete propagation cycles of the ring delay line, N_{cycles}, which is a integer number, multiplied by the full capacity of the delay line (T_{FS}). The second component, $T_{\text{incomplete}}$, corresponds to the residual portion from the last incomplete propagation, m, which is rational number and less than T_{FS}. This relationship can be mathematically expressed by equations (3)–(4):

$$T_{\text{in}} = T_{\text{complete}} + T_{\text{incomplete}} \qquad (3)$$
$$T_{\text{in}} = N_{\text{cycles}} \cdot T_{\text{FS}} + m \cdot T_{\text{FS}} \qquad (4)$$

In this operation, only the edge detector connected to the output of the first ring delay line is used. This edge detector monitors transitions in the *Flag* signal, which indicate that the ring delay has completed a propagation cycle.

When the input signal T_{in} is active, the *UP* input of the counter is enabled, allowing the counter to increment its value with each transition in the *Flag* signal from the first ring delay line. For every transition of the *Flag* signal, the edge detector generates a pulse that is connected to the *Clock* input of the counter. The residual time of the first ring delay line, $T_{\text{residual1}}$, is mathematically expressed by equations (5)–(7):

$$T_{\text{residual1}} = T_{\text{FS}} - T_{\text{incomplete}} \qquad (5)$$
$$T_{\text{residual1}} = T_{\text{FS}} - m \cdot T_{\text{FS}} \qquad (6)$$
$$T_{\text{residual1}} = (1 - m) \cdot T_{\text{FS}} \qquad (7)$$

The timing diagram illustrating the operation of the Time Register during the write mode is presented in Fig. 5.

B. Read Mode

To initiate the read mode, $T_{\text{residual1}}$ is first retrieved from the first ring delay line and applied to the input of the second ring delay line. Subsequently, the *Read* input of the second ring delay line is activated.

Once $T_{\text{residual1}}$ is applied to the second ring delay line, the residual time at the output of the second ring delay line is $T_{\text{residual2}}$, which is given by equations (8)–(10):

$$T_{\text{residual2}} = T_{\text{FS}} - T_{\text{residual1}} \qquad (8)$$
$$T_{\text{residual2}} = T_{\text{FS}} - (T_{\text{FS}} - m \cdot T_{\text{FS}}) \qquad (9)$$
$$T_{\text{residual2}} = m \cdot T_{\text{FS}} = T_{\text{incomplete}} \qquad (10)$$

At this point, only the edge detector connected to the output of the second ring delay line is active. Simultaneously, the *DOWN* input of the counter is enabled, allowing the counter to decrement its value with each transition of the *Flag* signal from the second ring delay line. When the *Read* input of second ring delay line is active, the output of the time register,

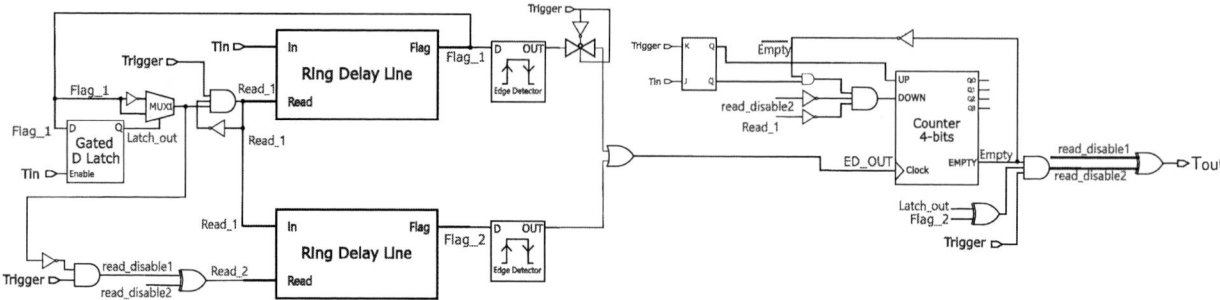

Fig. 3: Proposed Time Register.

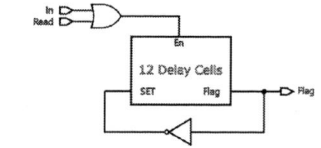

Fig. 4: Ring delay line.

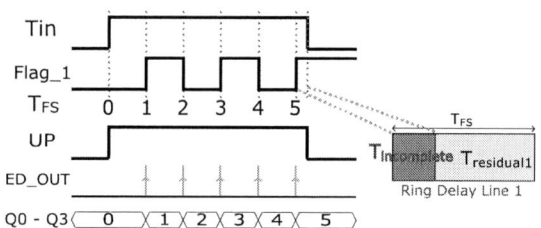

Fig. 5: Time Diagram in the write mode of the Time Register.

T_{out}, remains high until the counter is decremented to zero. The pulse width of the T_{out} signal is determined by equations (11)–(13):

$$T_{\text{out}} = T_{\text{residual2}} + T_{\text{complete}} \tag{11}$$
$$T_{\text{out}} = m \cdot T_{\text{FS}} + N_{\text{cycles}} \cdot T_{\text{FS}} \tag{12}$$
$$T_{\text{out}} = N_{\text{cycles}} \cdot T_{\text{FS}} + m \cdot T_{\text{FS}} = T_{\text{in}} \tag{13}$$

Fig. 6 presents the timing diagram that illustrates the operation of the Time Register in read mode.

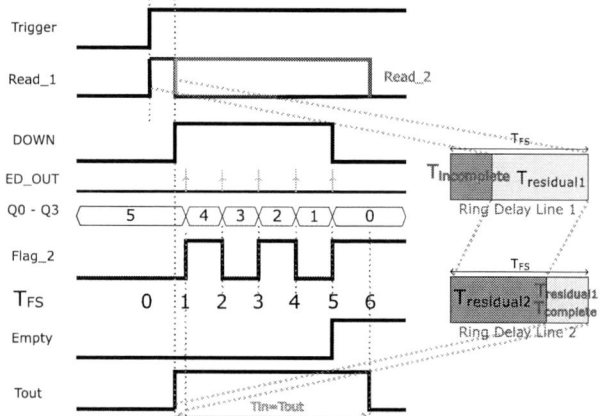

Fig. 6: Time Diagram in the read mode of the Time Register.

IV. SIMULATIONS RESULTS AND ANALYSIS

The proposed circuit was implemented in a full-custom design using the AMS 350nm technology process, with a supply voltage of 3.3 V, and was simulated in Cadence Virtuoso. To demonstrate the operation of the proposed topology, the ring delay line was implemented with 12 stages and the counter with 4 bits. These values, however, are fully parameterizable to enable a configurable dynamic range. Each delay line produce a total delay of 23.44 ns. Consequently, the effective total delay of the system was calculated as $T_{\text{eff}} = 16 \cdot 23.44 ns = 375.1 ns$.

A. Corners Simulations and Monte Carlos Analysis

In this subsection, we present the results of a comprehensive set of robustness tests conducted on the proposed time register. The Fig. 7 exposed the percentual error throughout the range of the proposed topology for temperature variation, distinct supply voltage and distinct of process conditions.

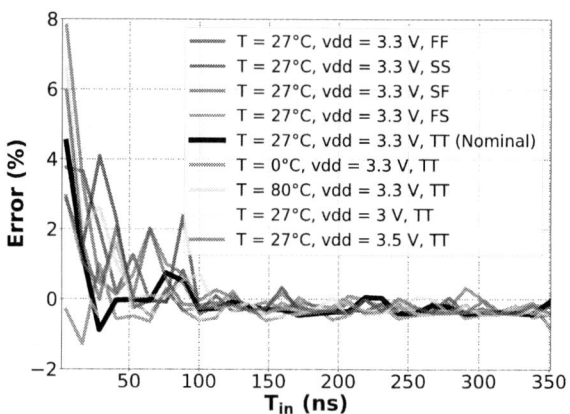

Fig. 7: Corners Simulations.

This demonstrates that the error remains below 0.5% for most of the full range of the proposed time register. However, for T_{in} values below 7 ns, a higher error is observed compared to the rest of the time register's range. This elevated error occurs because, for very narrow pulses, any leakage current in the transistors while the time register is active becomes comparable in magnitude to the input pulse width. Leakage can be mitigated by enhancing the time register design — for

979-8-3315-9813-6/25 $31.00 © 2025 IEEE

instance, through the inclusion of small capacitors between its stages — at the expense of a slight area overhead.

In the Monte Carlo simulation, shown in Fig. 8, the value of T_{in} is fixed at 108 ns. The percentage error is measured at the circuit output T_{out} relative to the input T_{in}. As can be seen, the average error is about -2.5%, and the worst case scenario considering 99.7% of the samples, is still less than 10%.

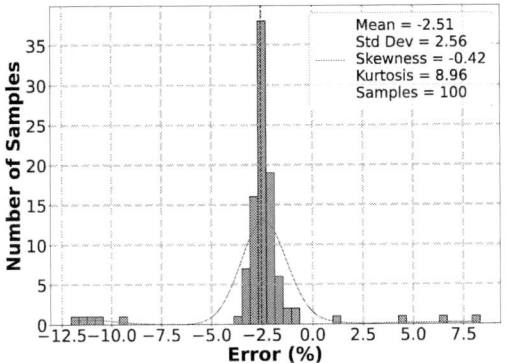

Fig. 8: Monte Carlo analysis for 100 iterations.

B. Linearity

Fig. 9 illustrates the linearity of the circuit.

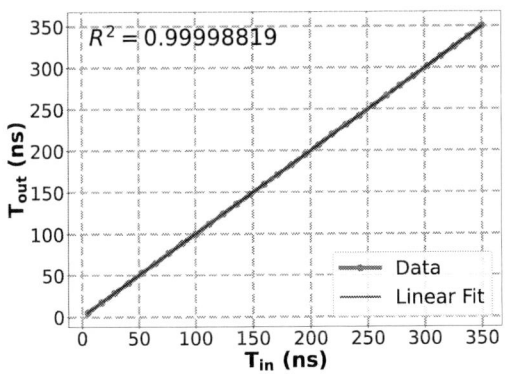

Fig. 9: Linearity of the proposed time register.

For the normal conditions, the coefficient of determination(R^2) is ~ 0.99, as noted in Fig. 9, indicates a excellent fit between the T_{in} and T_{out}. This demonstrates that the proposed time register exhibits great linearity, as T_{out} closely matches T_{in}.

C. Comparative Analysis

Table I and Table II presents a comparative analysis between existing time register topologies and the proposed design.

As observed in Table II, the proposed time register achieves a significantly higher dynamic range without incurring a substantial increase in layout area. Notably, the dynamic range of the proposed architecture can be doubled by incrementing the counter width by a single bit, with negligible impact on area and power consumption. In contrast, traditional time register

Table I: Average power consumption for different topologies.

Topology	Delay Cells	Average Power Consumption (μW)[a]	
		Write Mode	Read Mode
Based on [8]	24	337.3	431.2
Based on [8]	384	565.9	1335.7
Based on [11]	12	269.9	182.1
Based on [11]	24	286.5	163.1
Proposed	24 + 4 bits	169.2	172.7
Proposed	24 + 5 bits	163.4	144.8

[a] Average power measured with input $T_{in} = T_{FS}/2$.

Table II: Area, time range, and dynamic range for different topologies.

Topology	Delay Cells	Area (μm^2)[a]	Range (ns)	Dynamic Range (dB)[b]
Based on [8]	24	167.5	23.5	15.0
Based on [8]	384	2559.1	375.1	39.0
Based on [11]	12	81.3	23.2	14.9
Based on [11]	24	160.2	46.4	20.9
Proposed	24 + 4 bits	232.1	375.1	39.0
Proposed	24 + 5 bits	241.8	750.2	45.0

[a]Estimated minimal area (W*L), excluding spacing and routing.
[b]Dynamic range: $20 \log(T_{FS}/T_{in\_min})$, with $T_{in\_min} = 4ns$.

designs require an increase in the number of delay cells to extend the dynamic range, leading to substantial growth in area and an exponential increase in power usage. It is worth mentioning that all the circuits presented in Table I and Table II were redesigned, and all simulation results were generated from the design view. The impact of post-layout extraction is not severe for this type of circuit and would affect all the circuits in Table I and Table II similarly. The key innovation here is the proper integration of the digital counter.

V. CONCLUSION

In conclusion, the proposed time register offers superior performance over conventional designs by maintaining a constant layout area, with only the counter size adjusting to extend the dynamic range. By using a 5-bit counter, the operation range increases by approximately 30 times (or 30 dB), with a small area penalty—less than 50%—corresponding to the area occupied by the digital counter. This solution is more efficient and scalable, and by increasing the counter bits, it can reach millisecond-level performance, making it unique in the literature.

VI. ACKNOWLEDGMENTS

The authors thank Danilo Garcia Mariano for his support during the proposed time register testing. This study was financed in part by Instituto SerraPilheira (grant number Serra – process 2211-42117), in part by Fundação de Amparo à Pesquisa de Minas Gerais (FAPEMIG, APQ-05837-23), in part by "National Council for Scientific and Technological Development – CNPq (process 404467/2024-5), and in part by Coordenação de Aperfeiçoamento de Pessoal de Nível Superior - Brasil (CAPES) - Finance Code 001. Also, the authors thank the APCI Program from the Brazilian Microelectronics Society (SBMICRO).

REFERENCES

[1] G. W. Roberts and M. Ali-Bakhshian, "A Brief Introduction to Time-to-Digital and Digital-to-Time Converters," in IEEE Transactions on Circuits and Systems II: Express Briefs, vol. 57, no. 3, pp. 153-157, March 2010, doi: 10.1109/TCSII.2010.2043382.

[2] Zhu G, Yuan F, Khan G (2013) Time-Mode Approach for Mixed Analog-Digital Signal Processing. J Elec Electron 2: e109. doi:10.4172/2167-101X.1000e109.

[3] Asada, Kunihiro & Nakura, Toru & Iizuka, Tetsuya & Ikeda, Makoto. (2018). Time-domain approach for analog circuits in deep sub-micron LSI. IEICE Electronics Express. 15. 20182001-20182001. 10.1587/elex.15.20182001.

[4] Yuan, F. (2018), Design techniques of all-digital arithmetic units for time-mode signal processing. IET Circuits Devices Syst., 12: 753-763. https://doi.org/10.1049/iet-cds.2017.0327.

[5] Ali-Bakhshian, M.. (2011). Digital storage, addition and subtraction of time-mode variables. Electronics Letters. 47. 910-911. 10.1049/el.2011.1406.

[6] Panetas-Felouris, O., & Vlassis, S. (2022). A 3rd-Order FIR Filter Implementation Based on Time-Mode Signal Processing. Electronics, 11(6), 902. https://doi.org/10.3390/electronics11060902.

[7] K. Kim, W. Yu, and S. Cho, "A 9 bit, 1.12 ps resolution 2.5 b/stage pipelined time to- digital converter in 65 nm cmos using time-register,"IEEE Journal of Solid-State Circuits, vol. 49, no. 4, pp. 1007–1016, 2014

[8] P. S. Locatelli, D. M. Colombo and K. El-Sankary, "Time-Domain Multiply–Accumulate Unit," in IEEE Transactions on Very Large Scale Integration (VLSI) Systems, vol. 31, no. 6, pp. 762-775, June 2023.

[9] A. Karmakar, V. De Smedt and P. Leroux, "Pseudo-Differential Time-Domain Integrator Using Charge-Based Time-Domain Circuits," 2021 IEEE 12th Latin America Symposium on Circuits and System (LASCAS), Arequipa, Peru, 2021, pp. 1-4, doi: 10.1109/LAS-CAS51355.2021.9459120.

[10] P. -F. Orfeas and S. Vlassis, "A novel time register with process and temperature calibration," 2021 10th International Conference on Modern Circuits and Systems Technologies (MOCAST), Thessaloniki, Greece, 2021, pp. 1-4, doi: 10.1109/MOCAST52088.2021.9493414.

[11] M. Zareie, K. El-Sankary, D. M. Colombo and E. El-Masry, "A Time-Domain Frequency Analyzer Based on Goertzel Algorithm," in IEEE Transactions on Very Large Scale Integration (VLSI) Systems, doi: 10.1109/TVLSI.2025.3541539.

[12] R. Cela, N. Spanos, K. P. Pagkalos and S. Vlassis, "All-Digital Time Register Based On Recursive Delay-Line," 2024 Panhellenic Conference on Electronics & Telecommunications (PACET), Thessaloniki, Greece, 2024, pp. 1-4, doi: 10.1109/PACET60398.2024.10497083.

Energy-Efficient Computation of TensorFloat32 Numbers on an FP32 Multiplier

Per Larsson-Edefors
Chalmers University of Technology, Gothenburg, Sweden
Email: perla@chalmers.se

Abstract—Several new shorter floating-point formats have been proposed to match requirements of emerging application workloads. To simplify hardware development in the presence of an increasing number of formats, one practical design option is to use as much as possible preexisting hardware, such as standard 32-bit IEEE-754 (FP32) floating-point units, to handle emerging, less complex formats. We evaluate the case where we use an FP32 multiplier to run Nvidia TensorFloat32 data. While the FP32 multiplier area is not as small as a dedicated TensorFloat32 multiplier, we show that energy per operation scales well with the mantissa width reduction and that smart pin assignment can leverage uneven input vector switching activities to significantly decrease energy for reduced precisions.

I. INTRODUCTION

Machine learning is but one example of important applications that fuel a trend towards floating-point formats with reduced computing precision. The key rationale behind this trend is that shorter formats lead to less complex compute and memory circuits and less interconnects linking them. Choice of format and precision of data representations has a direct impact on resource and energy usage of floating-point units. A case in point is the integer multiplier—a core component of floating-point units—whose complexity decreases quadratically as the precision is reduced.

A number of floating-point formats with reduced precision and/or dynamic range have recently been proposed: IEEE-754 half-precision format (binary16 or FP16), IBM's DLFloat, AMD's fp24, Google's bfloat16, and Nvidia's TensorFloat32 (TF32) are examples of formats which relax the demands on the hardware in comparison to a IEEE-754 single-precision (binary32 or FP32) solution. Requirements on precision and dynamic range vary not only across application domains, but also between workload phases of some applications. Thus, processor floating-point units that can handle mixed-precision workloads, which can be found in machine learning and signal processing, have been proposed [1]–[3].

A circuit implementation customized to one reduced-precision format clearly will perform more efficiently than standard FP32 hardware, which is bound to have higher circuit complexity and area. But the latter one-size-fits-all hardware solution is practical as it can handle many types of formats, as long as they are not more complex than FP32. As we will show in this work, the energy dissipation of FP32 when executing shorter formats can be reduced substantially, to levels closer to FP16 operation than to FP32. We will specifically investigate how TF32 [4], [5] multiplications can be efficiently run on a standard FP32 multiplier.

II. BACKGROUND

The IEEE-754 standard for floating-point arithmetic [6] defines a number as $(-1)^S 2^{exponent-bias} 1.mantissa$, where S (the sign bit) indicates the sign, $exponent$ represents the w-bit exponent, and $mantissa$ represents the t-bit mantissa (significand). For normalized numbers, the resulting exponent is defined as the subtraction of the coded $exponent$ and a $bias$ which depends on the format: $2^{(w-1)} - 1$. Additionally, there exists an implicit 1 which is leading the mantissa bits and makes the resulting p-bit mantissa, where $p = t + 1$, take on numbers in an interval of $[1, 2)$. The IEEE-754 standard supports also denormalized numbers, in which case an implicit 0 leads the mantissa bits and the exponent is set to 0.

As shown in Fig. 1, IEEE-754's FP32 format uses three different fields with 23 mantissa bits, 8 exponent bits, and one sign bit. It turns out that the half-precision FP16 format, with its 10 mantissa bits and 5 exponent bits, provides sufficient precision for many workloads. The TF32 format proposed by Nvidia is a compromise between FP16 and FP32 [4], [5]: It has the same 10-bit mantissa field as FP16, but shares the 8-bit exponent field of FP32, thus offering the same dynamic range as FP32, with the precision of FP16.

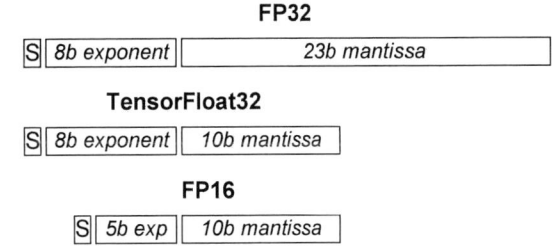

Fig. 1: Floating-point formats under consideration.

As a baseline we also consider fixed-point arithmetic, which is often based on two's complement numbers. Here, an n-bit number is defined as $-x_{n-1}2^{n-1} + \sum_{i=0}^{n-2} x_i 2^i$. We can scale this number any way we prefer, using an implicit binary point. This is in contrast to floating point, where the mantissa bits are directly trailing an implicit 0 and 1 for denormalized and normalized numbers, respectively.

III. EVALUATION METHOD

Floating-point multipliers are implemented for different formats: FP32, FP16, and TF32. We also implement 24-bit and 11-bit fixed-point multipliers as comparison baselines for FP32 and FP16. Their precisions consequently should mirror those of their floating-point counterparts; thus, we let $n = p = t + 1$.

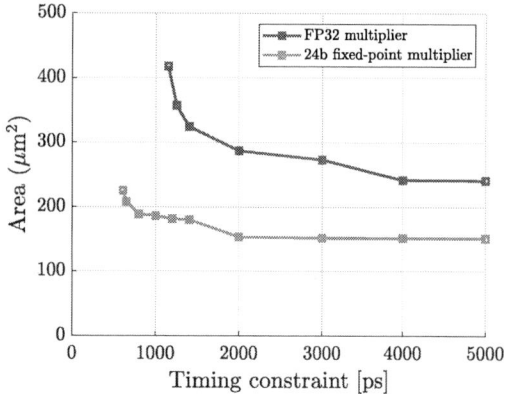

Fig. 2: Area of an FP32 multiplier vs a 24-bit fixed-point multiplier, as function of timing constraint.

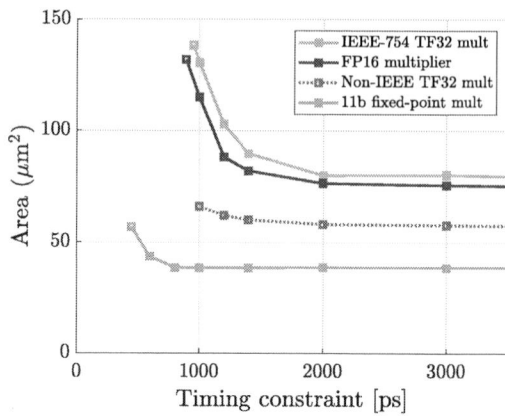

Fig. 3: Area of IEEE754-compliant TF32 and FP16 floating-point multipliers vs non-IEEE-compliant TF32 and 11-bit fixed-point multipliers, as function of timing constraint.

We use Cadence Genus [7] to synthesize the HDL code under different timing constraints. We use regular-VT cells from the open-source ASAP7 library [8], which was developed by Arizona State University and ARM Ltd. to represent a predictive 7-nm FinFET process technology. In general, ASIC logic synthesis strives to optimize area under a timing constraint. Thus, as the timing target is gradually reduced, the area starts to grow when the synthesizer struggles to handle the longest logic paths of the circuit. To make the results as general as possible, we choose to make all circuits combinational.

The implementation metric of area is retrieved after synthesis, but the metric of energy per operation requires more evaluation steps: We generate in Matlab input vector sets for all number formats. For floating point, each set is made up of 10,000 random IEEE-754-compliant numbers, representing normals and denormals, but not NaNs. The three different floating-point fields are created independently and concatenated at the bit level. The fixed-point inputs are more straightforward to generate as there is only one single n-bit field. Additionally, the results of reference multiplications performed in Matlab are stored in output vector sets, for fixed-point and IEEE-754-compliant floating-point multiplications, in all formats. All gate netlists are verified for logic functionality using Cadence Xcelium [9], which compares the netlist outputs to the ground-truth reference of the stored output vector sets.

During the above netlist verification, we save switching data for each circuit node and backannotate this to the netlist. Using the system clock rate f as reference, we define α_i as the per node switching activity. This is the fraction of clock cycles when a circuit node i with capacitance C_i switches from 0 to 1. Assuming N nodes, the switching power is defined as $P_{sw} = f V_{DD}^2 \sum_{i=1}^{N}(C_i \alpha_i)$, where $V_{DD} = 0.7\,\mathrm{V}$ is the supply voltage of the cell library. For all energy evaluations, we assume a constant $f = 200\,\mathrm{MHz}$ which corresponds to a relaxed system operating point that satisfies every timing constraint used in our evaluations.

While the use of random input data will overestimate the power dissipation that we can expect in most practical scenarios, in which signal switching is significantly lower, this assumption works here: We focus on comparing energy per operation for multipliers that have the same input switching

activity profile. Energy per operation is calculated as P_{tot}/f, in which P_{tot} includes an insignificant level of static power.

IV. IMPLEMENTATION AREA

Fig. 2 contrasts the area of an FP32 multiplier with that of a 24-bit fixed-point multiplier, which has the same precision as FP32. As expected, the fixed-point multiplier has a smaller area and it can meet tighter timing constraints. Fig. 3 shows the area of the multiplier implementations of the TensorFloat32 and IEEE's FP16 formats. As comparison we include the area of an 11-bit fixed-point multiplier. The lowest timing constraint of the TF32 multiplier is longer than that of the FP16 multiplier. This is due to TF32's exponent field being wider than FP16. The wider exponent field is also the reason the TF32 multiplier is slightly larger than the FP16 multiplier. The graphs show that the delay overhead of using floating-point over fixed-point for multipliers is close to 2X.

We consider multipliers which are compliant with IEEE-754. But as shorter floating-point formats are being pursued, some features in conventional formats may be removed to reduce hardware complexity. For example, in bfloat16 [10] denormals are not supported but input and output numbers are flushed to zero. To illustrate this option, we include in Fig. 3 also a variant of TF32 which is not compliant with IEEE-754, but lacks leading zero anticipation and normalization features.

V. ENERGY PER OPERATION

The previous section showed area results for several multiplier implementations. A widespread assumption is that power dissipation scales linearly with area, but there are several caveats to this simplified view. For example, glitching power, due to unwanted spurious signal transitions in gates with unbalanced input arrival times, will make total switching power depend also on logic functions and computing workload.

XOR-intensive arithmetic circuits are known to have excessive glitching power dissipation. As shown in [11], the glitching power can constitute more than half the total power of a 16-bit fixed-point complex-valued multiplier. Adding pipeline registers to an arithmetic circuit is a known remedy to glitches as this stops them from propagating [12]. However,

979-8-3315-9813-6/25 $31.00 © 2025 IEEE

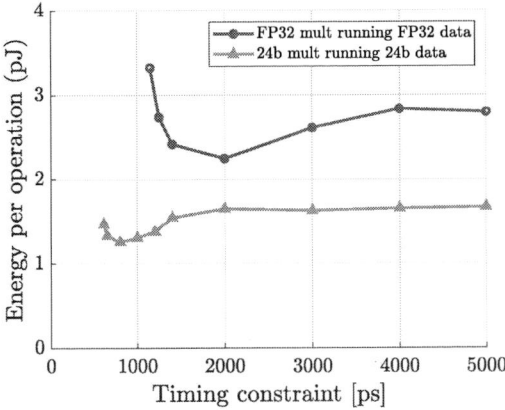

Fig. 4: Energy per operation of FP32 and 24-bit fixed-point multipliers, as function of timing constraint.

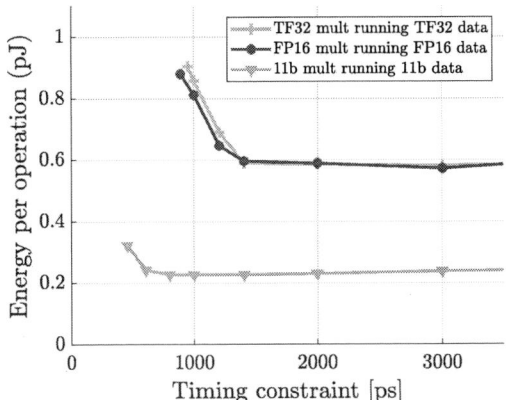

Fig. 5: Energy per operation of TF32 and FP16 floating-point multipliers and an 11-bit fixed-point multiplier.

using an extra pipeline stage is not straightforward in latency-restricted architectures and systems using feedback loops.

Fig. 4 shows the energy per operation for an FP32 multiplier operating on FP32 numbers and a 24-bit fixed-point multiplier operating on 24-bit two's complement numbers. As shown in this figure, the energy per operation decreases for tighter timing constraints, before the area begins to grow fast (Fig. 2). The reason we observe the decreasing energy dissipation is because for tighter timing constraints, the logic paths become more balanced in terms of timing: In logic gates where inputs arrive simultaneously, the generation of glitches is inhibited.

Fig. 5 shows the energy per operation for two floating-point multipliers (TF32 and FP16) and an 11-bit fixed-point multiplier. Here, TF32 operates on TensorFloat32 numbers, the FP16 multiplier on FP16 data, and the fixed-point multiplier on 11-bit two's complement numbers. There is a small but clear area difference between TF32 and FP16 in Fig. 3. However, except for the tightest timing constraints, their energy per operation is almost identical. This is because in floating-point multipliers, the exponent computation is less complex and less power dissipating than the mantissa computation [13].

VI. USING REDUCED-PRECISION WORKLOAD DATA

We will again perform an energy evaluation, but now we will include cases where we apply data with shorter floating-point formats to an FP32 multiplier. Fig. 6 shows the energy per operation for an FP32 multiplier operating on TF32 and FP16 data, respectively. We have also included two curves from previous graphs, viz. FP32 running FP32 data from Fig. 4 and TF32 running TF32 data from Fig. 5.

Clearly input data with lower precision (mantissa) and dynamic range (exponent) leads to significantly lower energy dissipation. As shown in Fig. 1, TF32 numbers have a narrower mantissa field than FP32 numbers, whereas FP16 numbers have both narrower mantissa and exponent fields. Since the two cases where FP32 is running shorter formats yield almost the same energy per operation, we can make a useful observation when using workloads with shorter floating-point formats: Reducing the number of mantissa bits in the input data has a greater effect on energy dissipation than reducing the number

of exponent bits. This is because the exponent bias changes, which impacts the bit information in this field.

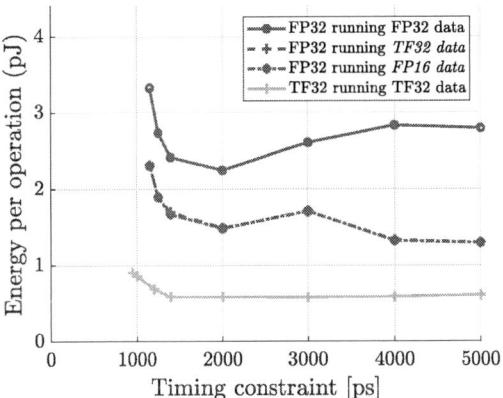

Fig. 6: Energy per operation strongly depends on input data precision.

A. Pin Assignment for Reduced-Activity Data

It is well known that when input data to fixed-point multipliers have different dynamic ranges, the power dissipation can be reduced if we assign data to the 'right' input pins. As shown previously [14], [15], a lower dynamic integer range (which leads to longer strings of consecutive '0' or '1' in the most significant bit positions) can be exploited in fixed-point multipliers where Booth recoding of bit patterns like '000' and '111' lead to partial products which are evaluated to zero. Since they are implemented in a different way, floating-point multipliers cannot exploit dynamic range reductions the same way fixed-point multipliers can. But as shown in Fig. 6, a reduced mantissa precision, which implies consecutive '0' in the *least* significant bit positions of the mantissa integer, leads to reduced energy.

In many applications, multiplications are performed on data input vectors which are different in how frequently their bits switch. This is often the case where weights or coefficients are changing more slowly than the data they operate on. This difference in switching activities has been previously used for low-power design of multipliers [16], FIR filters [17]

979-8-3315-9813-6/25 $31.00 © 2025 IEEE

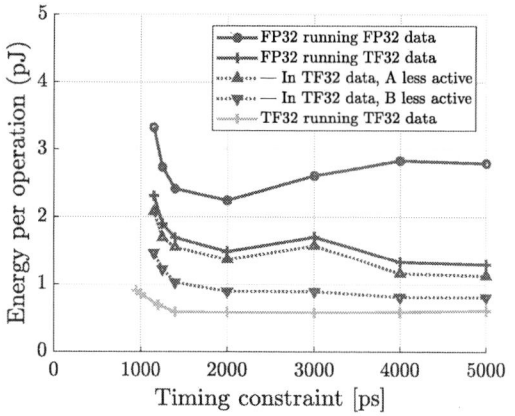

Fig. 7: Energy per operation depends on input switching activity.

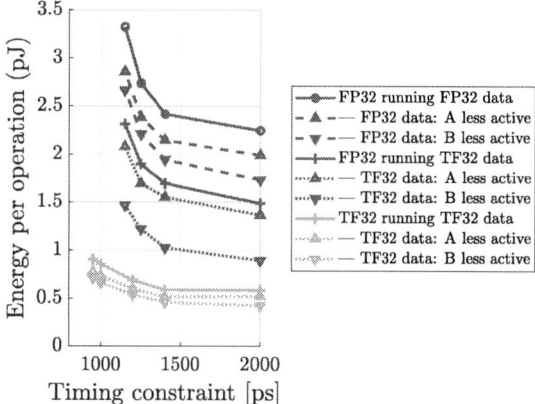

Fig. 8: Impact of pin assignment on energy per operation for performance-oriented implementations with strict timing targets.

and FFTs [18]. In [19], we unify the two approaches of pin assignment—using dynamic range and switching activity—for fixed-point multipliers for complex numbers.

The effect of pin assignment on an FP32 multiplier running TF32 data is shown in Fig. 7. We show as references (top and bottom) the previous curves (Fig. 6) for the FP32 multiplier running FP32 data and the TF32 multiplier running TF32 data. In addition, we show as baseline the energy per operation for the FP32 multiplier as it operates on TF32 (Fig. 6). Then we add two curves which show the effect of reducing the switching activity of one of the input data vectors by 4X: We assign this reduced-activity data to, in the first simulation, multiplier pin A. In the subsequent simulation run, we swap the inputs so that the reduced-activity data goes to pin B. As shown, we substantially reduce energy per operation when the slowly changing data is assigned to pin B.

Pin assignment can also be applied to the reference cases. Fig. 8 includes the result of pin assignments also for the FP32 multiplier running FP32 data and the TF32 multiplier running TF32 data. Optimal pin assignment has a favorable impact on energy, but the gain is the largest when we reduce precision, making an FP32 multiplier running TF32 data approach the energy efficiency of a dedicated TF32 multiplier.

VII. CONCLUSION

We evaluate using standard FP32 floating-point multipliers for TensorFloat32 workloads which use less precision than the FP32 format. While using a one-size-fits-all FP32 multiplier also for numbers with less complex floating-point representations comes with an area overhead, we show that energy per operation substantially decreases with a reduced mantissa precision in the workload. Additionally, we show that optimally assigning input data that have different switching activities to the input pins has a large impact on energy dissipation when the precision of the workload is reduced.

ACKNOWLEDGEMENT

This project is financially supported by the Swedish Foundation for Strategic Research (SSF).

REFERENCES

[1] J. Lee, J. Lee, D. Han, J. Lee, G. Park, and H.-J. Yoo, "LNPU: a 25.3TFLOPS/W sparse deep-neural-network learning processor with fine-grained mixed precision of FP8-FP16," in *IEEE Int. Solid-State Circuits Conf. (ISSCC)*, 2019, pp. 142–144.

[2] A. Nannarelli, "Variable precision 16-bit floating-point vector unit for embedded processors," in *IEEE 27th Symp. Computer Arithmetic (ARITH)*, 2020, pp. 96–102.

[3] H. Tan, G. Tong, L. Huang, L. Xiao, and N. Xiao, "Multiple-mode-supporting floating-point FMA unit for deep learning processors," *IEEE Trans. Very Large Scale Integration (VLSI) Systems*, vol. 31, no. 2, pp. 253–266, 2023.

[4] J. Choquette, W. Gandhi, O. Giroux, N. Stam, and R. Krashinsky, "NVIDIA A100 tensor core GPU: Performance and innovation," *IEEE Micro*, vol. 41, no. 2, pp. 29–35, 2021.

[5] D. Stosic and P. Micikevicius, "Accelerating AI training with TF32 tensor cores," https://developer.nvidia.com/blog/accelerating-ai-training-with-tf32-tensor-cores/, Nvidia.com, 2021, Accessed Aug. 27, 2025.

[6] "IEEE Standard for Floating-Point Arithmetic," *IEEE Std 754-2019 (Revision of IEEE 754-2008)*, pp. 1–84, 2019.

[7] *Cadence® Genus®, v. 18.14*, Cadence Design Systems, Inc., 2019.

[8] V. Vashishtha, M. Vangala, and L. T. Clark, "ASAP7 predictive design kit development and cell design technology co-optimization," in *IEEE/ACM Int. Conf. Computer-Aided Design*, Nov. 2017, pp. 992–998.

[9] *Cadence® Xcelium®, v. 22.09*, Cadence Design Systems, Inc., 2023.

[10] *BFLOAT16 — Hardware Numerics Definition White Paper*, https://software.intel.com/sites/default/files/managed/40/8b/bf16-hardware-numerics-definition-white-paper.pdf, Intel Corp., 2018.

[11] P. Larsson-Edefors and E. Börjeson, "Implementation evaluation of fixed-point multipliers for complex numbers," in *IEEE 32nd Symp. Computer Arithmetic (ARITH)*, 2025, pp. 81–84.

[12] A. Chandrakasan and R. Brodersen, "Minimizing power consumption in digital CMOS circuits," *Proc. IEEE*, vol. 83, no. 4, pp. 498–523, 1995.

[13] L. Bertaccini, G. Paulin, T. Fischer, S. Mach, and L. Benini, "MiniFloat-NN and ExSdotp: An ISA extension and a modular open hardware unit for low-precision training on RISC-V cores," in *IEEE 29th Symp. Computer Arithmetic (ARITH)*, 2022, pp. 1–8.

[14] P.-M. Seidel, "Dynamic operand modification for reduced power multiplication," in *Asilomar Conf. on Signals, Systems and Computers*, vol. 1, 2002, pp. 52–56.

[15] N.-Y. Shen and O.-C. Chen, "Low-power multipliers by minimizing switching activities of partial products," in *IEEE Int. Symp. Circuits and Systems*, vol. 4, 2002, pp. 93–96.

[16] P. Larsson-Edefors and E. Börjeson, "Activity-based input operand assignment for reduced multiplier power dissipation," in *IEEE Latin American Symp. Circuits and Systems (LASCAS)*, 2025.

[17] C. J. Nicol and P. Larsson, "Low power multiplication for FIR filters," in *Int. Symp. Low Power Electronics and Design*, 1997, pp. 76–79.

[18] O. Meteer and M. J. G. Bekooij, "Low-power Booth multiplication without dynamic range detection in FFTs for FMCW radar signal processing," in *Asia-Pacific Signal and Information Processing Association Annual Summit and Conf.*, 2021, pp. 44–48.

[19] P. Larsson-Edefors and E. Börjeson, "Low-power complex multiplier pin assignment based on spatial and temporal signal properties," in *IEEE Int. Symp. Circuits and Systems (ISCAS)*, 2025.

Delay Mismatch Optimization in Routing Dominated Multi-Path Systems: A Case Study on an IR-UWB Edge-Combiner Transmitter Front End

Kyla Marie H. Juruena, Maria Ena R. Rosales, Trisha Renee G. Capulong, Louis P. Alarcón

Microelectronics and Microprocessors Laboratory, Electrical and Electronics Engineering Institute

University of the Philippines Diliman, Quezon City 1101, Philippines

{kyla.juruena, maria.ena.rosales, trisha.renee.capulong, louis.alarcon}@eee.upd.edu.ph

Abstract—Routing-dominated multi-path architectures play a critical role in enhancing data throughput and resource efficiency in computing, communication, and RF systems. Hence, ensuring matched point-to-point signal arrival is critical to minimizing distortion, latency, and data corruption while managing increased routing density and interconnect complexity. In this paper, delay mismatch in routing-dominated systems is investigated while assuming negligible mismatch within circuit blocks. A variance-based framework is proposed to quantify and optimize the delay mismatch components across multiple levels of abstraction: the system, block, and layout. To validate this model, systematic routing optimization techniques are applied to an impulse-radio ultra-wideband (IR-UWB) edge-combiner transmitter front end (TFE) operating at 6.5-8.0 GHz center frequency. The TFE requires precise pulse widths in the picosecond range, making it highly susceptible to interconnect-induced delay variability. After optimization, the total variance of propagation delay was decreased by 94.37% based on post-layout simulations.

Index Terms—propagation delay mismatch, interconnects, multi-level abstraction, ultra-wideband transmitter

I. INTRODUCTION

Mitigating delay mismatch across multiple parallel signal paths is a critical challenge in multi-path architectures. This challenge is even more pronounced in routing-dominated systems, where signal behavior and timing consistency are significantly affected by physical wiring on a chip, making precise layout strategies essential for optimal functionality. Additionally, ensuring the matched arrival of point-to-point signals while managing increased routing density and interconnect complexity is particularly crucial in high-frequency systems, where even a sub-picosecond mismatch can lead to significant performance degradation.

Extensive reviews of placement and routing tools for physical implementation of analog/mixed-signal designs are provided in [1] and [2]. A major class of these routing frameworks is constraint-driven, focusing on finding optimal solutions to satisfy constraints such as symmetry, wirelength matching, and parasitics [3]. However, delay matching as a primary objective is rarely addressed in the literature and is usually treated as equivalent to length matching. Wirelength alone, however, is insufficient to guarantee matched delays, as factors such as routing topology, layer assignments, via locations,

and coupling effects significantly influence the electrical characteristics of interconnects [4]. The routing approach in [5] considers symmetry, common-centroid, topology, and length-matching constraints simultaneously, while [6] targets the exact matching of nets in terms of wirelength, number of bends, and routing layers. These routing frameworks have proven effective, but only for a limited number of analog routing paths, and the broader challenge of routing high-frequency multi-path architectures remains unaddressed.

To address this gap, this work introduces a systematic framework for analyzing delay mismatch in routing-dominated systems at multiple levels of abstraction: system, block, and layout. To validate the framework in a practical application, optimization techniques were applied to a full-custom impulse radio ultra-wideband (IR-UWB) edge-combiner transmitter front end (TFE) operating at 6.5–8 GHz [7]. Because this system relies on precise pulse shaping and signal timing, it serves as an ideal case study for assessing delay mismatch effects in high-frequency applications.

II. SOURCES OF DELAY MISMATCH

The variance-based multi-level abstraction framework in this paper focuses solely on interconnect mismatch, assuming negligible mismatch within circuit blocks.

A. System and Block-level Mismatch

Systematic mismatch in propagation delay between parallel nets is predominantly attributed to inconsistencies in interconnect length and routing topology. At the system level, the primary contributor to this mismatch is the variance of the Euclidean distance, σ_d^2, between the driver (source) and the load (sink) of the parallel nets. A higher σ_d^2 indicates a system with a large spread in port-to-port distances, leading to significant variations in signal propagation time. Hence, the initial approach to reducing delay mismatch at the system level is to optimize the spatial arrangement of blocks and ports.

At the block level, variations in the internal routing between blocks, such as differences in routing strategies, introduce additional delay mismatch. Partitioning helps mitigate this mismatch by creating identical segments within the floor plan, significantly reducing the number of unique routing shapes

---

979-8-3315-9813-6/25 $31.00 © 2025 IEEE

required. This approach enhances uniformity, enables layout cell reuse, and streamlines implementation by allowing a single representative segment to serve as the focus for further internal routing optimization.

B. Layout-level Mismatch

Delay mismatch persists at the physical wiring level due to the disparities in the interconnect parasitics of parallel nets. Fig. 1 illustrates the first-order capacitance and resistance components of a wire. Capacitive components include the area capacitance to the substrate (C_a), lateral coupling capacitance between adjacent interconnects on the same metal layer (C_x), and inter-layer capacitance between vertically stacked metal layers (C_z). Fringing field effects are ignored in this simplified model, allowing the use of parallel-plate capacitor expressions as the following, where $\eta = \frac{\epsilon_0 \epsilon_r}{d}$ for C_a and C_z and $\gamma = \epsilon_0 \epsilon_r h$ for C_x are process-dependent constants.

$$C_{tot} = C_a + C_z + C_x \tag{1}$$

$$C_a = \eta_a w L, \quad C_x = \gamma \frac{L}{s}, \quad C_z = \eta_z w L, \tag{2}$$

Meanwhile, the resistive components include the line resistance (R_w), additional corner resistance introduced at bends (R_c), and the via resistance (R_v). Wire resistance is proportional to its length, and is inversely proportional to its cross section, as expressed by the following equations, where $\beta = \frac{\rho}{h}$ is the sheet resistance, k is the correction factor for wire bends and N is the number of identical vias in parallel:

$$R_w = \beta \frac{L}{w}, \quad R_c = k R_w, \quad R_v = \frac{\beta \frac{L_v}{w_v}}{N} \tag{3}$$

$$R_{tot} = R_w + R_c + R_v \tag{4}$$

Given the equations for total capacitance and resistance, the first-order approximation of the nominal propagation delay t is represented by (5) and the contribution of small changes in R and C to the total change in t is expressed in (6). This equation estimates how sensitive t is to changes in R and C, assuming they are independent random variables. Through the propagation of uncertainty [8], the variance of Δt is estimated in (7).

$$t \propto RC \tag{5}$$

$$\Delta t \approx R \Delta C + C \Delta R \tag{6}$$

$$\sigma_{\Delta t}^2 \approx R^2 \sigma_{\Delta C}^2 + C^2 \sigma_{\Delta R}^2 \tag{7}$$

Equation (8) shows the total variance in capacitance $\sigma_{\Delta C_{tot}}^2$ that captures the variations in the length of the wire, the laterally coupled lengths and the number of overlaps. Meanwhile, (9) expresses the total variance in resistance $\sigma_{\Delta R_{tot}}^2$ that accounts for variances associated with wirelengths, number of corners and number of vias. Wire widths, corner and via characteristics, and loads are assumed to be the same across all paths to simplify the analysis. Based on this model, the total delay mismatch depends not only on matching wirelengths but also on disparities in bend and via counts, as well as coupling

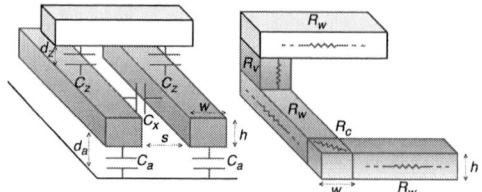

Fig. 1: Capacitive and resistive parasitics of a wire

and overlap capacitance. Therefore, the topology of each path must be carefully designed to minimize these factors.

$$\sigma_{\Delta C_{tot}}^2 \approx \sigma_{\Delta C_a}^2 + \sigma_{\Delta C_x}^2 + \sigma_{\Delta C_z}^2 \tag{8}$$

$$\sigma_{\Delta C_{tot}}^2 \approx (\eta_a w)^2 \sigma_{\Delta L}^2 + (\eta_z w L_z)^2 \sigma_{\Delta N_{overlap}}^2 + \left(\frac{\gamma}{s}\right)^2 \sigma_{\Delta L}^2$$

$$\sigma_{\Delta R_{tot}}^2 \approx \sigma_{\Delta R_w}^2 + \sigma_{\Delta R_c}^2 + \sigma_{\Delta R_v}^2 \tag{9}$$

$$\sigma_{\Delta R_{tot}}^2 \approx \left(\frac{\beta}{w}\right)^2 \sigma_{\Delta L}^2 + (k R_w)^2 \sigma_{\Delta N_{corner}}^2 + (R_v)^2 \sigma_{\Delta N_{via}}^2$$

III. ROUTING ANALYSIS IN IR-UWB TRANSMITTERS

An IR-UWB edge-combiner TFE operating at a 6.5–8.0 GHz center frequency, adapted from [7], was analyzed at the system, block, and layout levels. The architecture, shown in Fig. 2(a), generates a 2 ns Gaussian pulse envelope with picosecond-range sub-pulses, requiring minimized delay mismatch to ensure precise signal alignment, achieve the target frequency, and maintain spectral quality.

The TFE consists of 16 delay units (D) that alternate odd (α) and even (β) impulses. Each impulse is directed by a programmable router (R) to four outputs based on a 4-bit tuning word, then serialized by combiners (C) α and β. The resulting pulse trains drive the power amplifier (PA), producing binary-weighted contributions to the output pulse envelope, which is then processed by the matching network (MN). This interleaved structure, where signals are divided into odd and even components, introduces additional complexity in routing, highlighting the need for optimization techniques, which will be discussed in the following sections.

A. System and Block-level Optimization

Given the signal path, a sequential arrangement of the blocks at each stage is the most straightforward approach. As shown in Fig. 2(b), interleaving connections create numerous intersections and convoluted wiring configurations, especially between the pulse shaper blocks. In multi-stage systems, prioritizing the placement of stages with the highest number of ports and parallel wires is more effective, as variations introduced at this stage accumulate and become increasingly difficult to compensate at lower levels of abstraction.

Dividing the system into odd and even components and partitioning the pulse shaper into four identical sections, as shown in Fig. 2(c), enables convenient merging of the parallel signals at the power amplifier. It also reduces the number of paths that require manual routing and allows faster implementation through layout cell replication.

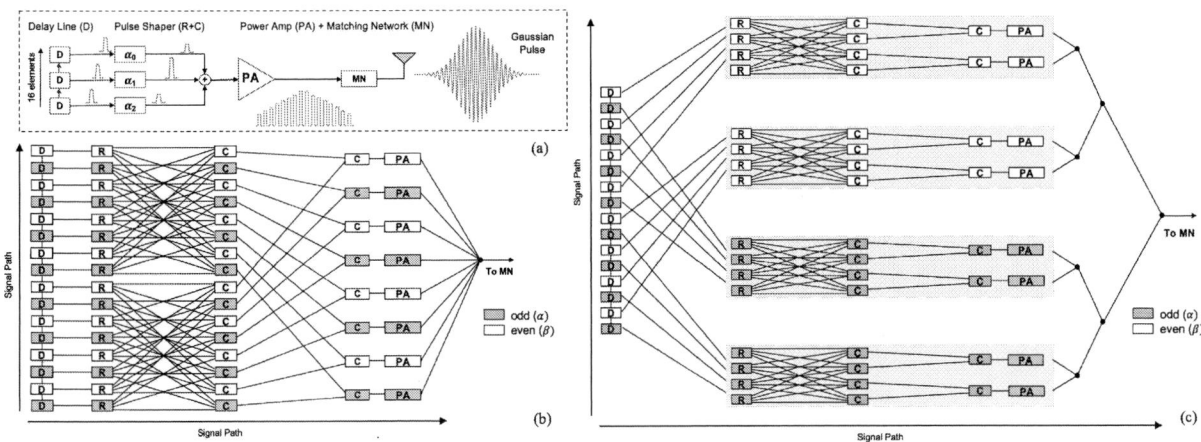

Fig. 2: (a) IR-UWB edge-combiner TFE architecture adapted from [7] (b) Unoptimized floorplan (c) Optimized floorplan

B. Layout-level Optimization

Based on (9), mismatch in corner resistance and via resistance can be avoided by ensuring that all routing paths have the same number of bends and vias. This constraint is enforced while maintaining equal wire lengths along the vertical and horizontal axes by dividing the paths into an equal number of segments. The vertical segments introduced at each direction change are kept equal in length, while the horizontal segments may vary individually to prevent path intersections, with their total length balanced across all paths to preserve overall length matching. Additionally, consistent layer assignments, wire widths, and spacings are used for the corresponding wire segments to ensure a uniform capacitance profile relative to the substrate, as well as to the adjacent and overlapping wires. Another technique used is symmetric Y-tree routing, which distributes signals from one source to multiple ports (e.g., D to R) and merges multiple sources into one (e.g., PA to MN) while maintaining uniform delay. Using 45° tree bends shortens wires and avoids sharp corners that cause reflections and impedance mismatches. Fig. 3 illustrates the different configurations of topology matching and symmetric tree routing applied at different stages of the system.

IV. RESULTS AND DISCUSSION

The TFE layout was implemented in Cadence Virtuoso using a 22 nm FDSOI process. To isolate interconnect-induced delay variability, post-layout simulations were performed with parasitics extracted from the interconnects using Calibre xACT, while keeping the internal circuit blocks at the schematic level. This hybrid approach ensures that delay variations arise solely from routing and layout factors, enabling accurate evaluation of the implemented optimization.

System and block-level mismatch was quantified by measuring the variance of Euclidean distances between connected ports at each stage, before and after floorplan optimization. As presented in Table I, σ_d^2 decreased by 51%, while the mean distance μ_d increased, reflecting a larger vertical area

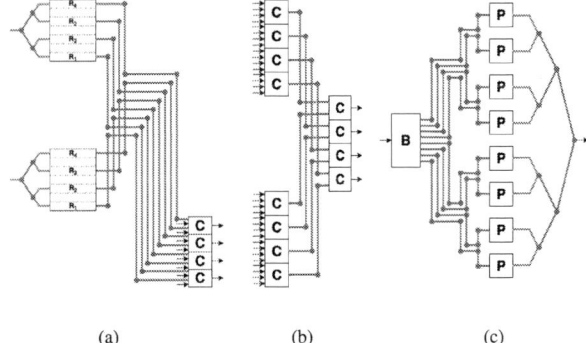

Fig. 3: Topology matching between (a) router and 4-input combiner (b) 4-input combiner to 2-input combiner (c) buffer to power amplifier

to facilitate wirelength matching in the layout level. This tradeoff is acceptable since, in an edge-combining system, matched delays are more critical than absolute delay for preserving the Gaussian output shape and spectral quality. This contrasts with other performance-driven placement methods for analog and mixed-signal systems, such as the "place-like-schematic" framework with signal flow considerations in [9] and pin-clustering with symmetry constraints in [10]. These frameworks prioritize minimizing area and wirelength to improve the signal-to-noise-and-distortion ratio for an ADC.

Meanwhile, Table II shows a 53.66% reduction in the variance of total capacitance after optimization. Equalizing the area of all parallel wires reduced $\sigma_{\Delta C_a}^2$ to zero, while $\sigma_{\Delta C_z}^2$ increased by 34% because overlap uniformity per path was relaxed to achieve equal length, segments, and vias across all parallel paths. Consequently, the variance in total resistance was completely eliminated after optimization. These results show that matching wirelengths and routing topologies across all paths effectively reduces parasitic mismatch. A similar concept was presented in [3], where interconnect electromigration and parasitic matching were considered. In contrast, their

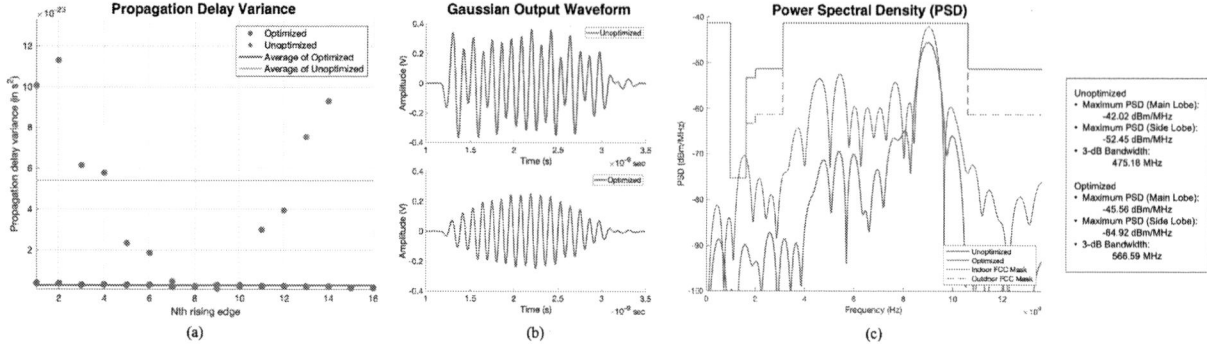

Fig. 4: Post-layout simulation results before and after optimization: (a) Propagation delay variance at each rising edge (b) Gaussian output waveform and (c) Output power spectral density

methodology seeks an optimal path with minimum wirelength and adapts interconnect widths to match parasitics among nets.

Finally, after minimizing the variances of the parasitic components, the variance in total propagation delay from the delay unit to the 2-input combiner was computed for each sub-pulse using eight parallel rising edges. As shown in Fig. 4(a), post-layout simulations indicate that the optimized design reduced this variance by 94.37% compared to the unoptimized design. Fig. 4(b) further illustrates a significant improvement in the Gaussian pulse envelope transient response after optimization, enabling its spectral quality to comply with FCC indoor and outdoor masks as seen in Fig. 4(c). FCC compliance is essential for UWB transmitters to avoid interference with other existing wireless systems, highlighting the practical viability of the proposed optimization.

TABLE I: System and block-level mismatch

	Unoptimized	Optimized
Mean $\mu_d (\mu m)$	346	413
Var $\sigma_d^2 (\mu m)$	7077	3452

TABLE II: Layout-level mismatch for capacitance and resistance

(F^2)	$\sigma_{\Delta C_a}^2$	$\sigma_{\Delta C_x}^2$	$\sigma_{\Delta C_z}^2$	$\sigma_{\Delta C_{tot}}^2$
Unopt.	5.60×10^{-28}	3.51×10^{-17}	4.23×10^{-33}	3.51×10^{-17}
Opt.	0	1.63×10^{-17}	5.65×10^{-33}	1.63×10^{-17}
(Ω^2)	$\sigma_{\Delta R_w}^2$	$\sigma_{\Delta R_c}^2$	$\sigma_{\Delta R_v}^2$	$\sigma_{\Delta R_{tot}}^2$
Unopt.	1.41×10^4	0	0	1.41×10^4
Opt.	0	0	0	0

V. CONCLUSION

This paper demonstrates the effectiveness of a multi-level abstraction approach in mitigating delay mismatch in routing-dominated multi-path architectures such as an IR-UWB edge-combiner transmitter. Post-layout simulations confirmed a 94.37% reduction in propagation delay variance and achieved the desired Gaussian shape and spectral quality, validating the effectiveness of the methodology. These results highlight the critical role of coordinated optimization across multiple levels

of abstraction in realizing high-performance, delay-sensitive systems. Given the limited literature on delay mismatch optimization in routing-dominated high frequency analog/mixed-signal systems, the presented framework provides a good foundation for advancing automated methodologies in this domain.

ACKNOWLEDGMENT

This study is done under the Center for Integrated Circuits and Devices Research (CIDR) program, and the authors would like to acknowledge DOST-PCIEERD for the project funding and support.

REFERENCES

[1] H. E. Graeb, Ed., *Analog Layout Synthesis: A Survey of Topological Approaches*. Boston, MA: Springer US, 2011.

[2] R. M. F. Martins and N. C. C. Lourenço, "Analog Integrated Circuit Routing Techniques: An Extensive Review," *IEEE Access*, vol. 11, pp. 35 965–35 983, 2023.

[3] M. Torabi and L. Zhang, "Electromigration- and Parasitic-Aware ILP-Based Analog Router," *IEEE Transactions on Very Large Scale Integration (VLSI) Systems*, vol. 26, no. 10, pp. 1854–1867, Oct. 2018.

[4] M. M. Ozdal and R. F. Hentschke, "Exact route matching algorithms for analog and mixed signal integrated circuits," in *2009 IEEE/ACM International Conference on Computer-Aided Design - Digest of Technical Papers*, Nov. 2009, pp. 231–238.

[5] H.-C. Ou, H.-C. C. Chien, and Y.-W. Chang, "Nonuniform Multilevel Analog Routing With Matching Constraints," *IEEE Transactions on Computer-Aided Design of Integrated Circuits and Systems*, vol. 33, no. 12, pp. 1942–1954, Dec. 2014.

[6] M. M. Ozdal and R. F. Hentschke, "Maze routing algorithms with exact matching constraints for analog and mixed signal designs," in *2012 IEEE/ACM International Conference on Computer-Aided Design (ICCAD)*, Nov. 2012, pp. 130–136.

[7] T. Haapala, T. Rantataro, and K. A. I. Halonen, "A Fully Integrated Programmable 6.0–8.5-GHz UWB IR Transmitter Front-End for Energy-Harvesting Devices," *IEEE Journal of Solid-State Circuits*, vol. 55, no. 7, pp. 1922–1934, Jul. 2020.

[8] J. Soch, "Variance of the linear combination of two random variables," https://statproofbook.github.io/P/var-lincomb.html, Jul. 2020.

[9] K. Zhu, H. Chen, M. Liu, X. Tang, N. Sun, and D. Z. Pan, "Effective Analog/Mixed-Signal Circuit Placement Considering System Signal Flow," in *2020 IEEE/ACM International Conference On Computer Aided Design (ICCAD)*, pp. 1–9.

[10] H. Chen, K. Zhu, M. Liu, X. Tang, N. Sun, and D. Z. Pan, "Toward silicon-proven detailed routing for analog and mixed-signal circuits," in *2020 IEEE/ACM International Conference On Computer Aided Design (ICCAD)*, 2020, pp. 1–8.

979-8-3315-9813-6/25 $31.00 © 2025 IEEE

Conjunctive Merge Instruction to Accelerate Sparse Matrix - Dense Vector Multiplication

Manuel Osterno*, César Marcon*, Jarbas Silveira†, Fernando Moraes*, Jardel Silveira†

*School of Technology – Pontifical Catholic University of Rio Grande do Sul (PUCRS), Porto Alegre, Brazil
†UFC - Universidade Federal do Ceará

adahilmuniz@hotmail.com, jardel@ufc.br, jarbas@lesc.ufc.br, fernando.moraes@pucrs.br, cesar.marcon@pucrs.br

Abstract—Sparse linear algebra is essential in many domains due to reduced computation and efficient memory usage. However, the irregularity of sparse data poses challenges for conventional software and hardware. While specialized accelerators offer performance gains, they lack general-purpose flexibility and rely on processor communication, creating bottlenecks. This work addresses these issues by proposing a tiling strategy to improve vector register usage and extending the RISC-V Vector (RVV) ISA with a custom merge instruction. Experiments using the gem5 simulator show that the tiled vector version achieved speedups of up to 1.30x (95% sparsity) and 1.72× (65%). In contrast, the version with merge instructions reached up to 1.81× and 6.04×, respectively, over a baseline implementation.

Index Terms—Sparse Algebra, Risc-V, Vector ISA, Acceleration, MAC.

I. INTRODUCTION

Sparse algebra operations are essential in applications such as linear solvers [1], graph mining, and machine learning [2]. In neural networks, sparsity arises from zero activations and pruned weights, often reaching over 70% [3]. To reduce memory usage, sparse matrices are typically stored in formats like Compressed Sparse Row (CSR) or Compressed Sparse Fiber (CSF).

By skipping zero computations, sparse operations reduce execution time, energy, and memory bandwidth. These formats avoid unnecessary activations of arithmetic units and minimize data flow, improving performance and lowering power consumption. However, their irregular structure leads to scattered memory access and frequent cache misses, which remain challenging for general-purpose processors. While some accelerators target sparse operations [4], they lack generality and rely on external cores. In contrast, Instruction Set Architecture (ISA) level support for sparse data remains limited [5], [6].

This work addresses this gap by extending the RISC-V Vector (RVV) ISA with a custom instruction, vconv, to perform conjunctive merge operations over integer data. We integrate this extension into the gem5 simulator and demonstrate its application in Sparse Matrix–Dense Vector Multiplication (SpMxDV), leveraging a tiling strategy to improve vector register utilization.

II. RELATED WORKS AND MOTIVATION

To boost the performance of applications such as linear solvers [1], GPM [6], and Deep Learning [2], some efforts focus on designing Application-Specific Integrated Circuits (ASICs) with multiple Processing Elements (PEs) and sparse acceleration units [7], [8], or using FPGAs [9]. While ASICs deliver high performance, they lack flexibility and depend on external general-purpose processors. To address this, other projects explore Application-Specific Instruction Processors (ASIPs) [10].

This chapter reviews representative ASIC and ASIP solutions to assess their strengths and limitations, and to distinguish this work's contribution.

A. ASIC Approach

The Tensor Marshaling Unit (TMU) [7] integrates with the Central Processing Unit (CPU) to handle sparse matrix index operations, enabling acceleration of various algorithms such as Gustavson, SpAdd, and SpMxDV.

NeCTAr [8] is a System-on-chip (SoC) that places sparse accelerators near the cache, handling sparse data loading, decoding, and computation. These units are tightly coupled to caches and scratchpads to optimize data flow.

Both projects design dedicated hardware for sparsity, requiring CPU integration for general-purpose tasks. In contrast, our work proposes extending the processor ISA itself.

B. ASIP Approach

Scheffler et al. [11] to optimize indirect memory accesses for sparse linear algebra within a specialized SoC [12].

IndexMAC [13], [14] extends RVV with a vindexmac instruction for SpMxDV acceleration. However, its focus on structured sparsity limits general applicability.

VIA [15] introduces a co-designed vector architecture with Smart Scratchpad Memory (SSPM), a Content Addressable Memory (CAM) that stores fiber values and indices to improve data handling. While effective, this adds area overhead.

SparseCore [6] defines a custom ISA for "streams"—structured key-value data—supported by 14 specialized instructions. Though innovative, it results in a highly domain-specific core.

To conclude, some of the works chose the skip strategy, that implements hardware to avoid computation by skipping the zero operands like EIE SAVE and SparCE, this strategy, however, seems to be going out of use, since it consumes a considerable amount of area only to do the zero identification on fetch stage and the current formats used to represent sparse data already omit the zero values. Other promising approach is the indirection access optimization architectures (UVE and ISSR); these architectures still add an area and time overhead to the processor datapath. In the end, we have the merge optimizati onstrategy, the merge operations (conjunctive

979-8-3315-9813-6/25 $31.00 © 2025 IEEE

and disjunctive) are presented on the basic sparse algorithms (SpMxDV, SpMxDM, SpMxSpM, and SpAdd), as shown in Chapter III, so, optimizing this process may result in good performance results [4] [16] [7] [10].

To conclude, our work contrasts from the cited ones in some points, aiming to accelerate SpMxDV by extending the RVV ISA to add a conjunctive merge instruction, maintaining CPU generality, reusing existing hardware, and minimizing compiler and SoC changes.

III. BACKGROUND

Sparse matrices, vectors, and tensors are characterized by many zero elements, enabling performance and memory optimizations. The sparsity level, defined in Equation 1, quantifies this characteristic.

$$Sparsity = ZeroElements/TotalElements \quad (1)$$

Higher sparsity favors formats that omit zero values, such as Compressed Sparse Row (CSR) and Compressed Sparse Fiber (CSF), the latter being an extension of CSR [17].

The CSR format compresses row information to reduce memory usage, avoiding the redundancy found in the CO-Ordinate (COO) format. It uses three arrays: IA, JA, and A. The IA array stores the number of non-zero (NZ) elements until that row, therefore, to find the number of NZ elements in the i-th row, use Equation 2.

$$NZ[i] = IA[i+1] - IA[i] \quad (2)$$

The *JA* array contains the column indices where to find NZ elements and the *A* array contains the actual NZ value.

Figure 1 illustrates a CSR-encoded sparse matrix.

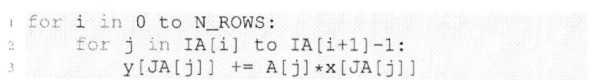

Fig. 1. CSR Sparse Matrix Representation.

In that example, the 0-th row contains two NZ elements (2-0, see Equation 2) and those are α and β placed within columns 2 and 4.

A key sparse operation is Sparse Matrix–Dense Vector Multiplication (SpMxDV), widely used in deep learning [2]. Code 1 shows its algorithm, which iterates over each matrix row and performs element-wise multiplication for non-zero entries. The algorithm performs a conjunctive merge of indices, followed by multiplication and accumulation into a dense output vector.

```
1 for i in 0 to N_ROWS:
2     for j in IA[i] to IA[i+1]-1:
3         y[JA[j]] += A[j]*x[JA[j]]
```

Listing 1. SpMxDV algorithm.

IV. METHODOLOGY

To evaluate our approach, we developed two versions of the SpMxDV algorithm: one based on the RVV ISA and another

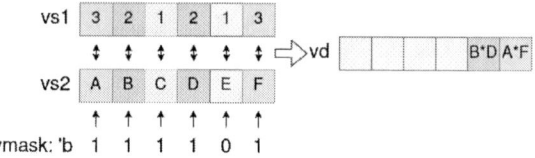

Fig. 2. vconv usage example.

incorporating a custom conjunctive merge instruction. We also implemented a tiling strategy to optimize vector register usage and memory access.

A. Conjunctive Merge Instruction

The SpMxDV operation involves matching column indices and multiplying the corresponding values, followed by a reduction. We extended the RVV ISA in gem5 to accelerate this with the vconv instruction. This instruction accepts index and data vectors (vs1, vs2), a mask (vmask), and stores the result in vd.

```
1 # Syntax pattern for conjunctive merge inst.
2 vconv.vv vd, vs1, vs2, vmask.t
```

Listing 2. Proposed conjunctive merge instruction pattern.

The vconv instruction compares the values present in vs1 (index vector) and, when any of these values match, the values in vs2 (data vector) relative to the values in vs1 are multiplied and the result is stacked at the bottom of vd (result vector). Figure 2 shows two input vectors (vs1 and vs2), where each pair of elements of each vector is colored the same color and the result elements are colored the same color as the match that resulted in these values. So, reading from right to left (0 to N) the first index to match is three (for values A and F), so, the result A*F is stored at the position zero of vd, then the second index which match is two (for values B and D) and the result B*D is stored at position one of vd. Finally, the third index that might have matched is one, however, position four (where the second 1 is found) of the vectors is masked, and then this result is ignored.

B. vconv Hardware Instruction

We used the existing comparators employed by vmseq implementation in a multicycle scheme to minimize hardware overhead, avoiding dedicated hardware. The number of comparisons and cycles required for a vector of size n is given by:

After comparisons, the results are passed to existing multipliers used by the vmul instruction.

$$nb\_comparisons = \frac{n^2 - n}{2}, \quad nb\_rounds = \left\lceil \frac{n-1}{2} \right\rceil \quad (3)$$

C. SpMxDV Vectorized and Merge Versions

The classic C-based SpMxDV code was not vectorized automatically due to indirect memory access. Thus, we created:

- **SpMxDV Vectorized**: operates with IA, JA, A, and x in vector registers and explicitly performs scalar operations within loops (Code 3).

- **SpMxDV Merge**: replaces the inner loop with the `vconv` instruction and performs reduction via `vredsum` (Code 4).

```
1  vload IA, JA, A, x
2  for i in 0 to N_ROWS:
3      # BLOCK 0: Get IA[i] and IA[i+1]
4      j_init = IA[i] # Move IA[i] to scalar
5      j_final = IA[i+1] # Move IA[i+1] to scalar
6
7      for j in j_init to j_final:
8          # BLOCK 1: Get A[j], x[JA[j]] and
9          # A[j] * x[JA[j]]
10         A_j = A[j] # Move A[j] to scalar reg
11         JA_j = JA[j] # Move JA[j] to scalar reg
12         x_JA_j = x[JA[j]] # x[JA[j]] to scalar
13         res = A_j * x_JA_j # Mult. the scalars
14
15         # BLOCK 2: Positioning Result
16         v_aux = 0 # Zero the entire vector
17         v_aux[i] = res # Set mult. result
18
19         # BLOCK 3: Storing result into y vector
20         y += v_aux # Vector accum. the result
21 vstore y
```

Listing 3. SpMxDV Vectorized

```
1  vload IA, JA, A, x
2  create idx_x # Create the dense index vector
3  # BLOCK 0: Vectors Initialization
4  v_val_concat = {x, A} # Concatenation of values
5  v_idx_concat = {idx_x, JA} # Concat. index vect.
6  # END OF BLOCK 0
7  for i in 0 to N_ROWS:
8      # BLOCK 1: Get number of non-zeroes on row i
9      j_init = IA[i] # Move IA[i] to scalar
10     j_final = IA[i+1] # Move IA[i+1] to scalar
11     nnz = j_final - j_init # Number of non-zeroes
12
13     # BLOCK 2: Build the mask
14     v0 = build_mask(nnz)
15
16     # BLOCK 3: Calculate A[:], x[JA[:]] for all
17     JA and A elements of the row and sum them up
       #Conjunction
18     "vmconv.vv v_merge_res, v_val_concat,
       v_idx_concat, v0.t \n"
19     #Sum Reduction
20     res = redsum(v_merge_res)
21
22     # BLOCK 4: Store the value in the result
       vector
23     y[i] = res
```

Listing 4. SpMxDV Merge - using vconv instruction.

The merged version improves performance by simplifying the inner loop using a masked conjunctive merge.

The SpMxDV Vectorized algorithm is divided into four blocks, considering that the vectors are already instantiated. In "BLOCK 0" the IA vector is manipulated to get the IA[i] and IA[j] values, the boundaries of the row, in "BLOCK 1" these values are used to manipulate A, JA and x vector to get the elements to be multiplied and then multiply them (A[j] * x[JA[j]]). In "BLOCK 2" and "BLOCK 3", the result is put in the correct place and accumulated into the result vector; creating and manipulating the masks omitted in the pseudocode is necessary.

The SpMxDV Merge version erases the inner loop of the SpMxDV Vectorized, since that loop was used to "merge" the vector, which is played by the instruction vconv. As in

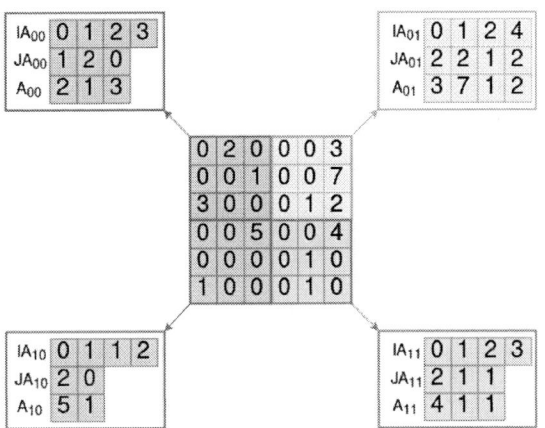

Fig. 3. Proposed tiling strategy.

the prior code, this one also considers that the vectors have already been initialized.

The BLOCK 0 is responsible for concatenating the sparse and dense data vectors (A and x) and the sparse and dense index vectors (JA and idx_x) because only one vector register is shared by data and only one is shared by indices. The code BLOCK 1 gets the number of non-zeroes elements from row i (A[i+1]-A[i]) and BLOCK 2 builds the mask that will be used by vconv instruction, this mask enables all the elements from the dense part and enables only the elements of the row being used by the iteration. For example, if we have a vector of sixteen elements (eight for dense and eight for sparse), but in the row zero, only two elements are valid, the mask of this iteration should assume the following binary value: 'b1111_1111_0000_0011.

In BLOCK 3, the conjunction merge instruction uses the concatenated registers and the mask built previously. The results (considering the example, A[0]*x[JA[0]] and A[1]*x[JA[1]]) are then reduced by the sum reduction instruction (vredsum). The final result is placed in the correct position of the final vector (v20) in BLOCK 4.

D. Tiling Sparse Matrices

Since CSR arrays often exceed vector register capacity, we proposed a tiling strategy that splits the original dense matrix into uniform tiles, ensuring the number of rows and non-zero elements per tile satisfy $NZ \leq VL$, where NZ is the number of Non-Zero elements and VL is the vector length. Figure 3 illustrates this strategy.

We implemented two methods:

- **Optimized**: iteratively refines tile sizes to meet $NZ \leq VL$.
- **Fast**: uses fixed $\sqrt{VL} \times \sqrt{VL}$ tiles for guaranteed compliance.

The optimized method was adopted for evaluation. After tiling, matrix-vector multiplications are performed tile-wise using the proposed SpMxDV versions, and results are accumulated using the vadd instruction.

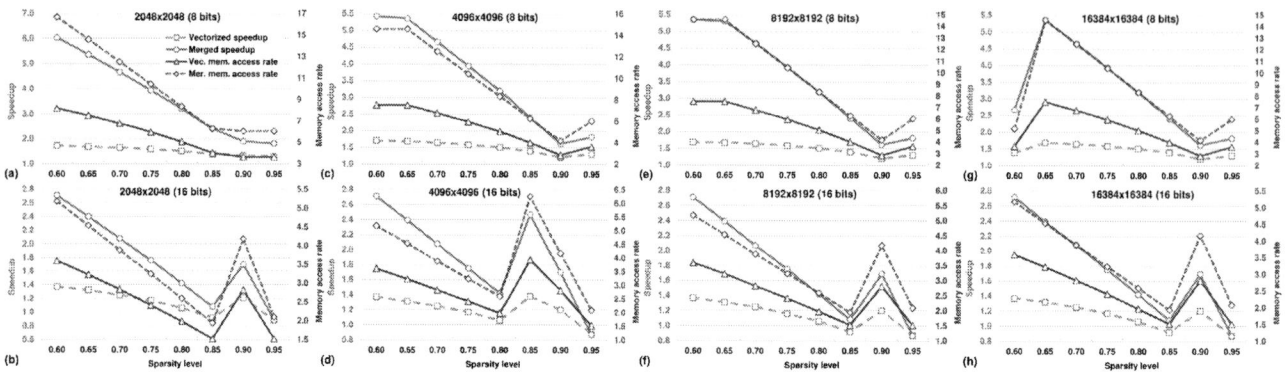

Fig. 4. Speedup (left axis, red lines) and Memory Access Rate (right axis, blue lines) for vectorized and merged SpMxDV versions |(a), (c), (e), (g) - 8 bit data type |(b), (d), (f), (h) - 16 bit data type.

V. RESULTS

A. Experimental Setup

We developed a Python-based infrastructure to generate sparse matrices (via *Scipy Sparse*), tile them using the optimized strategy, and convert them into CSR format. Parameters include matrix size (from 2048×2048 to 16384×16384), sparsity level (60%–95%), and data type (8-/16-bit integers). The tiled and classic CSR matrices are compiled into ELF binaries for execution on gem5.

The target architecture is a RISC-V RV64GCV system (Table I), featuring a 256-bit vector engine, DDR4 memory, and L1/L2 caches. Emulation results include execution time (ticks), memory accesses, and cache statistics. Theoretical latency per vector operation was 16 cycles (8-bit) and 8 cycles (16-bit), based on Equation 3.

TABLE I
EMULATION SYSTEM PROPERTIES.

	RISC-V 64GCV
Processor core	256-bit vector engine L1I: 16kB, 2 cycle latency L1D: 64kB, 2 cycle latency
L2 cache	512kB 8 cycle latency
Main memory	DDR4-2400

B. Experimental Results

Figure 4 shows speedups and memory access rates across eight sparsity levels (60% to 95%). Performance improves with lower sparsity (higher density) due to greater row occupancy, which enhances vector register utilization. An exception occurs in Figure 4 (d) for a 4096x4096 matrix, where at 85% sparsity the tile size is larger than at 80%, yielding better performance despite lower density.

Maximum speedups for 8-bit data at 60% sparsity (2048x2048) were 1.72× (Vectorized) and 6.04× (Merged), tied to a tile size of 16. For 16-bit data, both versions had the lowest speedups (0.88×) due to tile management overhead that outweighed performance gains.

Both approaches enhance data locality via vector registers. Memory access is reduced by over 8× (Vectorized) and 16× (Merged) compared to the classic method for 8-bit data.

SpMxDV Merged outperforms the Vectorized version due to fewer memory accesses, especially at lower sparsity and 8-bit data. At 65% sparsity, it requires half the memory accesses and executes 1.88× fewer instructions, leading to faster execution. Merged version also spends more time (46.18%) on arithmetic operations than Vectorized (26.66%), indicating better computational efficiency.

Performance improves with lower sparsity due to better use of vector registers. However, this trend may reverse if tile sizes are smaller for denser matrices, as seen in Figure 4(b), comparing 85% vs. 90% sparsity.

TABLE II
ALGORITHM EXECUTION PROFILE (MEAN VALUES)

	SpMxDV Vectorized	SpMxDV Merged
Number of executed instructions	825,367,204.23	438,035,436.31
Arithmetic instruction rate	26.66%	46.18%
Memory accesses	115,141,942.35	79,653,370.09
Speedup over classic	1.31	2.60

VI. CONCLUSION

This work proposed a tiling strategy and a conjunctive merge instruction to accelerate Sparse Matrix–Dense Vector Multiplication (SpMxDV). Experimental results show that, depending on data type and sparsity, the Vectorized and Merged versions achieved speedups of up to 1.68× and 5.90×, respectively, with average gains of 1.30× and 2.47×. Significant reductions in memory accesses support the improvements, up to 8× (Vectorized) and 16× (Merged), compared to the classic approach. Additionally, the Merged version executed 1.81× fewer instructions than the Vectorized version, contributing to its superior performance. The proposed strategy effectively boosts SpMxDV performance, offering promising benefits for machine learning workloads, particularly in deep neural networks and graph processing.

ACKNOWLEDGMENTS

This work was financed in part by Coordenação de Aperfeiçoamento de Pessoal de Nível Superior (CAPES), Finance Code 001; Conselho Nacional de Desenvolvimento Científico e Tecnológico (CNPq), 308182/2023-5, 305621/2024-6; National Institute of Science and Technology (INCT-Signals) under grant 406517/2022-3; Fundação de Amparo à Pesquisa do Estado do Rio Grande do Sul (FAPERGS), 23/2551-0002200-1; HP Brasil Indústria e Comércio de Equipamentos Eletrônicos LTDA.

REFERENCES

[1] H. Anzt, S. Tomov, and J. Dongarra, "Accelerating the lobpcg method on gpus using a blocked sparse matrix vector product," in *Symposium on High Performance Computing (HPC)*, 2015, pp. 75–82.

[2] T. Hoefler, D. Alistarh, T. Ben-Nun, N. Dryden, and A. Peste, "Sparsity in deep learning: pruning and growth for efficient inference and training in neural networks," *The Journal of Machine Learning Research*, vol. 22, no. 1, pp. 10 882–11 005, 2021.

[3] A. Parashar, M. Rhu, A. Mukkara, A. Puglielli, R. Venkatesan, B. Khailany, J. Emer, S. Keckler, and W. Dally, "SCNN: An accelerator for compressed-sparse convolutional neural networks," 2017. [Online]. Available: https://arxiv.org/abs/1708.04485

[4] G. Zhang, N. Attaluri, J. Emer, and D. Sanchez, "Gamma: leveraging gustavson's algorithm to accelerate sparse matrix multiplication," in *Proceedings of the ACM International Conference on Architectural Support for Programming Languages and Operating Systems (ASPLOS)*, 2021, pp. 687–701.

[5] P. Scheffler, F. Zaruba, F. Schuiki, T. Hoefler, and L. Benini, "Indirection stream semantic register architecture for efficient sparse-dense linear algebra," in *Design, Automation and Test in Europe Conference and Exhibition (DATE)*, 2021, pp. 1787–1792.

[6] G. Rao, J. Chen, J. Yik, and X. Qian, "SparseCore: stream ISA and processor specialization for sparse computation," in *Proceedings of the ACM International Conference on Architectural Support for Programming Languages and Operating Systems (ASPLOS)*, 2022, pp. 186–199.

[7] M. Siracusa, V. Soria-Pardos, F. Sgherzi, J. Randall, D. J. Joseph, M. Moretó Planas, and A. Armejach, "A tensor marshaling unit for sparse tensor algebra on general-purpose processors," in *Proceedings of the 56th Annual IEEE/ACM International Symposium on Microarchitecture*, ser. MICRO '23. Association for Computing Machinery, pp. 1332–1346. [Online]. Available: https://doi.org/10.1145/3613424.3614284

[8] V. Schmulbach, J. Kim, E. Gao, L. Revina, N. Jha, E. Wu, and B. Nikolic, "NeCTAr: A Heterogeneous RISC-V SoC for Language Model Inference in Intel 16," Mar. 2025, arXiv:2503.14708 [cs]. [Online]. Available: http://arxiv.org/abs/2503.14708

[9] J. Oliver, C. Álvarez, T. Cervero, X. Martorell, J. D. Davis, and E. Ayguadé, "b8c: SpMV accelerator implementation leveraging high memory bandwidth," in *2023 IEEE 31st Annual International Symposium on Field-Programmable Custom Computing Machines (FCCM)*, May 2023, pp. 216–216, iSSN: 2576-2621. [Online]. Available: https://ieeexplore.ieee.org/document/10171501

[10] P. Scheffler, F. Zaruba, F. Schuiki, T. Hoefler, and L. Benini, "Sparse stream semantic registers: A lightweight ISA extension accelerating general sparse linear algebra," vol. 34, no. 12, pp. 3147–3161. [Online]. Available: http://arxiv.org/abs/2305.05559

[11] F. Schuiki, F. Zaruba, T. Hoefler, and L. Benini, "Stream Semantic Registers: A Lightweight RISC-V ISA Extension Achieving Full Compute Utilization in Single-Issue Cores," Apr. 2020.

[12] G. Paulin, P. Scheffler, T. Benz, M. Cavalcante, T. Fischer, M. Eggimann, Y. Zhang, N. Wistoff, L. Bertaccini, L. Colagrande, G. Ottavi, F. K. Gürkaynak, D. Rossi, and L. Benini, "Occamy: A 432-Core 28.1 DP-GFLOP/s/W 83% FPU Utilization Dual-Chiplet, Dual-HBM2E RISC-V-Based Accelerator for Stencil and Sparse Linear Algebra Computations with 8-to-64-bit Floating-Point Support in 12nm FinFET," in *2024 IEEE Symposium on VLSI Technology and Circuits (VLSI Technology and Circuits)*, Jun. 2024, pp. 1–2, iSSN: 2158-9682. [Online]. Available: https://ieeexplore.ieee.org/document/10631529

[13] V. Titopoulos, K. Alexandridis, C. Peltekis, C. Nicopoulos, and G. Dimitrakopoulos, "IndexMAC: A custom RISC-v vector instruction to accelerate structured-sparse matrix multiplications." [Online]. Available: http://arxiv.org/abs/2311.07241

[14] ——, "Optimizing Structured-Sparse Matrix Multiplication in RISC-V Vector Processors," Jan. 2025, arXiv:2501.10189 [cs]. [Online]. Available: http://arxiv.org/abs/2501.10189

[15] J. Pavon, I. V. Valdivieso, A. Barredo, J. Marimon, M. Moreto, F. Moll, O. Unsal, M. Valero, and A. Cristal, "VIA: A Smart Scratchpad for Vector Units with Application to Sparse Matrix Computations." IEEE Computer Society, Feb. 2021, pp. 921–934. [Online]. Available: https://www.computer.org/csdl/proceedings-article/hpca/2021/223500a921/1t0HUpXQTQs

[16] N. Srivastava, H. Jin, J. Liu, D. Albonesi, and Z. Zhang, "MatRaptor: A sparse-sparse matrix multiplication accelerator based on row-wise product," in *2020 53rd Annual IEEE/ACM International Symposium on Microarchitecture (MICRO)*, Oct. 2020, pp. 766–780. [Online]. Available: https://ieeexplore.ieee.org/document/9251978

[17] B. Xie, J. Zhan, X. Liu, W. Gao, Z. Jia, X. He, and L. Zhang, "CVR: efficient vectorization of SpMV on x86 processors," in *Proceedings of the International Symposium on Code Generation and Optimization (CGO)*, 2018, pp. 149–162.

Marker-Based Recognition for Autonomous Micro-Drone Flight: An FPGA-Optimized Feasibility Study

Diego Marcelo Ramirez Jove
Grad. School of Sci. and Tech.
University of Tsukuba
Tsukuba, Japan
0009-0003-9169-8431

Keisuke Sugiura
Inst. of Systems and Info. Eng.
University of Tsukuba
Tsukuba, Japan
0000-0001-8534-2381

Yoshiki Yamaguchi
Inst. of Systems and Info. Eng.
University of Tsukuba
Tsukuba, Japan
0000-0001-9744-8271

Abstract—Drones are nowadays essential in construction and structural maintenance, providing high-resolution images for structural assessments and precise interventions. However, complex maneuvering in confined spaces and reduced communication with external positioning systems, such as GPS, pose challenges for conventional drones. Micro-drones, with their small size that minimizes failure impact and enhances adaptability in restricted or hazardous environments, have gained increasing research interest for expanding drone applications across various fields. This study presents an energy-efficient design that incorporates an optimized combination of basic color separation and a custom downsized version of the Sobel operator to detect markers placed along the trajectory of the drone in real time. The flight instructions stem from the marker identification to fine-tune altitude, horizontal position, and orientation of the micro-drone before moving forward to the next marker. The proposed strategy demonstrated accurate marker detection under favorable lighting conditions 97% of the time, while significantly reducing BlockRAM usage.

Index Terms—FPGA, micro-drone, autonomous navigation, real-time, energy efficiency

I. INTRODUCTION

Unmanned aerial vehicles (UAVs), also known as drones, have been predominantly adopted in the military, healthcare, and agricultural domains [1], while civilian applications are still in development. Among these, infrastructure inspection and maintenance have emerged as key areas of growth [2]. Autonomous drones have proven effective for inspecting large-scale structures, demonstrating their utility in fields such as energy, telecommunications, insurance, and construction [3], [4]. These platforms offer safe and efficient monitoring, particularly in hazardous or hard-to-reach environments.

Despite this progress, operating drones in GPS-denied or electromagnetically noisy environments, such as tunnels, high-rise interiors, and underground facilities, remains technically challenging. Conventional systems rely heavily on remote control or global positioning, limiting their applicability in such scenarios. Autonomous navigation methods like SLAM [5], reinforcement learning [6], and marker-based navigation using ArUco tags [7] have addressed some of these challenges, but often at the cost of high computational and energy demands. This makes them unsuitable for micro-drones, which are constrained by limited payloads and short battery life [8].

To overcome these limitations, we propose a power-efficient, FPGA-based marker recognition system for autonomous flight of 30-gram micro-drones. By leveraging FPGAs' parallel processing, low-latency, and flexibility, our system enables real-time visual navigation without requiring GPS, external computing, or high-power processors. The platform integrates a CMOS camera and detects predefined color-coded markers to guide flight along a target trajectory.

The proposed method combines color separation and binary marker recognition in hardware, allowing accurate symbol detection with minimal computational overhead. Designed for scalability, the system can adapt to different marker types and colors. It enables low-power, autonomous operation in environments where conventional methods are impractical.

The key contributions of this work are as follows:

1) We demonstrate real-time autonomous navigation on a 30-gram micro-drone by utilizing the low-latency, low-power characteristics of FPGAs.
2) We propose a lightweight, embedded-friendly algorithm for accurate detection of color-coded geometric markers using minimal resources.
3) We achieve navigation without GPS or external communication, enabling fully self-contained flight even in constrained environments.
4) We contribute to the standardization of lightweight navigation marker systems, facilitating broader deployment of autonomous drones in indoor and infrastructure inspection tasks.

II. BASELINE APPROACHES

A. Experimental Environment

The experimental setup was built using an OmniVision OV7670 camera [9] connected to a Xilinx Spartan-6 XC6SLX9 FPGA [10] on a custom board (Fig. 1). Images captured by the drone were recorded at 640×480 resolution and 30 fps in RGB565 format using the SCCB protocol. The FPGA communicated with a Crazyflie 2.0 micro-drone [11] via SPI and a Crazyradio PA dongle [12], with onboard velocity sensing supported by FlowDeck v2 [13].

To validate the system, captured images were used to confirm both marker recognition precision and the correctness of flight commands, under well-lit conditions.

979-8-3315-9813-6/25 $31.00 © 2025 IEEE

Fig. 1: (Left) 30-gram micro drone equipped with FPGA, (Right) FPGA and CMOS sensor board (front and back).

B. Color Separation

Previous work by Chen [14] demonstrated the feasibility of real-time autonomous flight by integrating a small CMOS camera with an FPGA to control a micro-drone. Using a simple color separation strategy, base RGB colors were assigned to flight commands: green for forward, blue for right turn, red for left turn, and white for landing. This system successfully issued real-time commands, with the drone acting as an actuator for the FPGA as shown on the left of Fig.2.

While the study validated the potential of FPGA-based control, it also revealed two key limitations: limited precision in positional control over markers, and slight desynchronization between processing and data transmission, affecting movement accuracy.

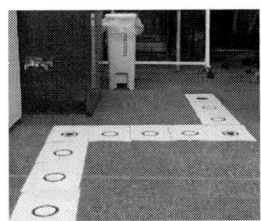

Fig. 2: Autonomous flight by color (left) and markers (right).

C. Concentric Circle Markers

Chenguang et al. [15] proposed a marker recognition strategy using monochromatic concentric circles, offering improved drone positioning over symbols. While employing the same hardware setup described in Sec.II-A, their method relies on a five-stage image processing pipeline. It uses a sliding window register strategy [16] that stores two consecutive image rows in blockRAM (Fig.3).

The process begins with RGB565 image input, followed by conversion to YCbCr color space to extract the luminance component. Binarization amplifies contrast, then median and Sobel filters extract marker edges, from which marker size and position are computed. This enables the drone to adjust its altitude and alignment before proceeding (Fig. 2 right). A key challenge reported was directional instability caused by external factors, such as wind or load imbalance.

III. PROPOSED METHODOLOGY

To enable robust and efficient symbol recognition for autonomous micro-drone flight, we propose a marker-based navigation system leveraging FPGA hardware. Unlike previous approaches using monochromatic markers [15], which are vulnerable to lighting variations, our design uses color

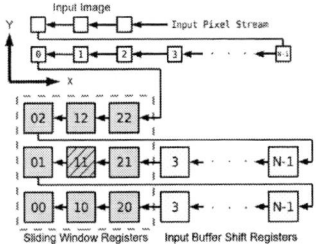

Fig. 3: Sliding window register [16].

separation in the RGB565 space. A blue triangle with a red line at the front vertex serves as the marker. This shape enables direction estimation and allows for smooth turns without compromising angular resolution.

A. System Overview and Core Algorithm

As shown in Fig. 4, the proposed system consists of three core stages: (1) color separation (truncate stage), (2) edge detection via a simplified Sobel filter, and (3) symbol recognition and command generation. The use of a triangle marker and the encoding of directional information allow precise adjustments of position, altitude, and yaw angle.

a) Color Separation: The FPGA receives 16-bit RGB565 pixel data. Red and blue components are compared using a fixed threshold (empirically set to 3, 10% of the 5-bit scale) to classify pixels into red, blue, or background. This replaces binarization and median filtering in [15], significantly reducing logic resource usage. The classified pixels are stored in two parallel layers (Fig.5) using shift registers to maintain local spatial information for edge detection.

b) Modified Sobel Filter: For efficient edge detection, we introduce a 1D Sobel filter that utilizes only three horizontal pixels, thereby eliminating the need for block RAM or whole 3×3 kernels. The filter detects left/right edges by encoding each pixel with 3-bit patterns: color (000), white (010), and background (001). Filter output patterns (e.g., 011 or 111) correspond to edge types (Table I), enabling continuous and lightweight edge detection as new pixel data arrives.

c) Symbol Recognition and Adjustment Strategy: From detected edges, the marker's three triangle vertices (A, B, C) are identified based on geometric criteria (Fig. 6). The free vertex indicates the direction of movement. Once the marker is detected, the drone performs the following adjustments:

- Border Correction: If the marker is near a frame edge (within 5 pixels), the drone shifts accordingly.
- Altitude Control: Based on the marker's length (ideal: 125–160 pixels), the drone adjusts its height (Fig. 7a).
- Centering: If the marker is off-center (outside a ±32 pixel tolerance), lateral correction is applied (Fig. 7b).
- Landing Signal: A blue circle is recognized when two of the triangle's vertices align vertically (Fig. 8), triggering a landing command.
- Yaw Direction: The angle between the triangle's tip and red line center is computed using a pipelined CORDIC algorithm [17] with 10 iterations and quadrant correction (Table III). A ±5° tolerance ensures stable alignment (Fig. 7c).

This architecture enables real-time navigation on a 30-gram micro-drone without GPS, off-chip memory, or external computing.

B. Core Algorithm Execution

The proposed algorithm (Fig. 4) enables real-time symbol recognition for autonomous micro-drone navigation. It consists of three stages: (1) color separation, (2) edge detection using a simplified Sobel filter, and (3) symbol recognition with flight command generation.

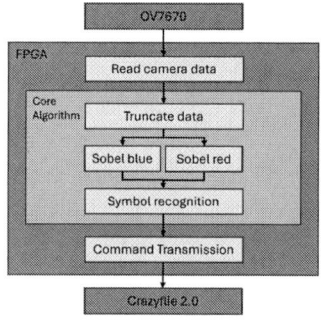

Fig. 4: Core algorithm execution.

a) Color Separation (Truncate Data): Incoming pixel data in RGB565 format is compared to classify each pixel as red, blue, or background using a threshold of 3. This lightweight operation replaces filtering and binarization used in [15], reducing hardware complexity. Pixels are stored in red and blue layers using FIFO-style shift registers (Fig. 5), enabling edge detection across adjacent pixels.

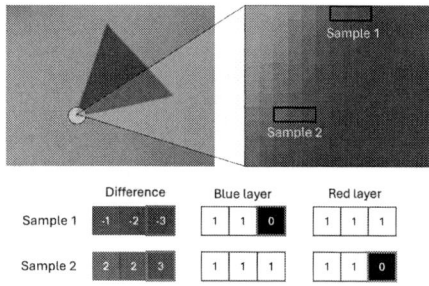

Fig. 5: Layer generation.

b) Edge Detection (Reduced Sobel Filter): To minimize resource usage, a 1D Sobel operator is applied using only horizontal neighbors:

$$G = \begin{bmatrix} -1 & 0 & +1 \end{bmatrix} * A_{mod} \qquad (1)$$

Each pixel is encoded with 3 bits—000 (color), 010 (white), 001 (background)—and output patterns such as 011 and 111 are used to classify left and right edges, respectively (Table I).

Detected edges are used to identify the triangle's three vertices (A, B, C), where A lies above C and B is left of A (Fig. 6). The front-facing vertex defines the flight direction.

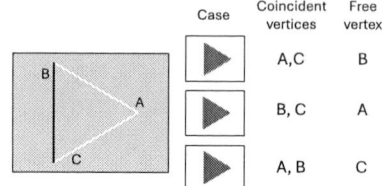

Fig. 6: Triangle vertex labeling.

c) Symbol Recognition and Flight Adjustment: Based on marker geometry, the following control decisions are made:

- Border: If the marker is within 5 pixels of any edge, the drone shifts direction.
- Altitude: If marker width deviates from 125–160 pixels, the drone adjusts its height (Fig. 7a).
- Position: Lateral corrections are triggered when the horizontal midpoint deviates from center by more than ±32 pixels (Fig. 7b).
- Stop Command: A blue circle is detected when two triangle vertices align at the vertical midpoint (Y_{midB}), prompting a landing (Fig. 8).
- Yaw Direction: Direction is computed using the triangle tip and red line center via a 10-iteration pipelined CORDIC algorithm [17], with quadrant correction applied (Eqs.(2)–(3); Fig.7c).

$$\begin{bmatrix} x_{i+1} \\ y_{i+1} \end{bmatrix} = \begin{bmatrix} 1 & -\tan(\gamma_i) \\ \tan(\gamma_i) & 1 \end{bmatrix} \begin{bmatrix} x_i \\ y_i \end{bmatrix}, \qquad (2)$$

$$\beta_{i+1} = \beta_i - \sigma_i \gamma_i, \quad \gamma_i = \arctan(2^{-i}). \qquad (3)$$

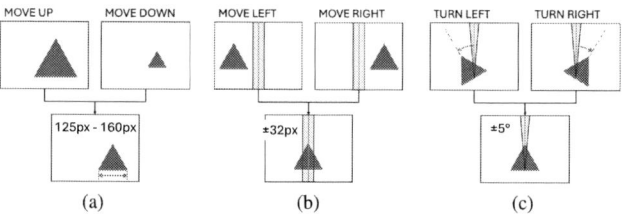

Fig. 7: (a) Altitude adjustment. (b) Position adjustment. (c) Direction adjustment.

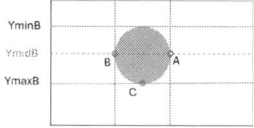

Fig. 8: Circle recognition.

TABLE I: Meaning of data in reduced Sobel filter

Pixel a_1	Pixel a_3	Sobel detection	Meaning
Color	Color	001 - 000 + 000 = 001	No change
Color	White	001 - 000 + 010 = 011	Left edge
White	Color	001 - 010 + 000 = 111	Right edge
White	White	001 - 010 + 010 = 001	No change

C. Instruction Encoding and FPGA Implementation

Flight commands are transmitted to the micro-drone via SPI using a compact 2-byte format: the first byte encodes the instruction, and the second encodes the magnitude (in centimeters or degrees), as shown in Table II.

To accommodate rotation commands within a single byte, rotational angles are scaled down before transmission. Specifically, the calculated yaw angle is halved, ensuring that 180° turns fit within the 8-bit range. The drone then multiplies the received value by two to recover the original angle. This avoids protocol overhead while maintaining accuracy.

TABLE II: Instruction syntax

Op Code (base16)	Instruction	Multiplier
00	Start	-
01	Stop	-
02	Up	x0.01
03	Down	x0.01
04	Left	x0.01
05	Right	x0.01
06	Front	x0.01
07	Back	x0.01
08	Turn Left	x2°
09	Turn Right	x2°
0A	Forward	-

Quadrant-specific corrections for yaw control are handled using CORDIC-based logic (Table III), ensuring correct turning direction and angle in all coordinate spaces. The entire system, including camera input, color processing, and SPI-based instruction transmission, is implemented on an FPGA, as shown in Fig. 9.

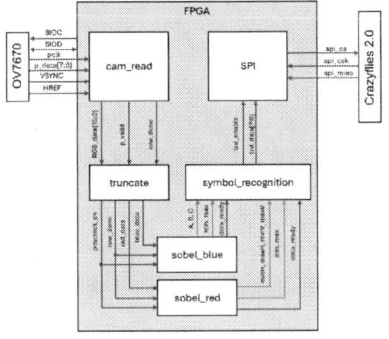

Fig. 9: Overview of FPGA-based system implementation.

TABLE III: CORDIC quadrant corrections

Quadrant	Conditions	x_0	y_0	Turn	Angle β_n
I	$x > 0, y \geq 0$	-	-	Left	-
II	$x < 0, y > 0$	$-x_0$	-	Left	$180° - \beta_n$
III	$x < 0, y < 0$	$-x_0$	$-y_0$	Right	$180° - \beta_n$
IV	$x > 0, y \leq 0$	-	$-y_0$	Right	-

IV. RESULTS

The proposed system was evaluated through FPGA-based simulations using real images captured by the onboard CMOS camera. Intermediate results such as edge detection and coordinate updates were visually confirmed (Fig. 10) and validated using Icarus Verilog. The FPGA performed real-time updates during pixel reception, with the min/max coordinate computation completing in 140 ns per frame, and directional

angle calculation requiring 460 ns. These results confirm that all processes operate within real-time constraints and do not overlap with data transmission.

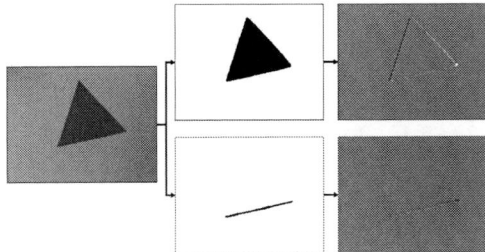

Fig. 10: Visual simulation of Sobel operation.

Robustness was further evaluated under varying illumination and symbol positioning. As shown in Table IV, symbol recognition exceeded 90% under adequate lighting (intensity ≥ 102), but declined sharply in dark or cropped scenarios, confirming the importance of controlled input conditions.

TABLE IV: Effect of illumination in marker recognition

Illumination	Accuracy by Image Type			
	Centered	Big	Small	Cropped
204 - 255	96%	97%	94%	77%
153 - 203	97%	97%	92%	76%
102 - 152	91%	94%	86%	72%
51 - 151	72%	79%	68%	45%
0 - 50	17%	33%	17%	7%

Hardware utilization results (Table V) demonstrate that the design requires only minimal logic resources—1% of registers, 4% of LUTs, and no block RAM—making it well suited for energy-constrained micro-drone applications.

TABLE V: Slice utilization summary

Slice	Used	Available	Utilization
Register	172	11440	1%
LUTs	237	5720	4%
Block RAM	0	32	0%
I/O Block	26	102	3%

V. CONCLUSION

This paper presented a lightweight, real-time visual marker recognition system for autonomous micro-drone flight, prototyped entirely on an FPGA with minimal hardware cost. By combining an optimized color separation stage and a custom 1D Sobel filter, the system achieves robust symbol detection and flight instruction generation under strict latency and power constraints. The architecture avoids the use of block RAM, external memory, or CPU intervention, making it highly suitable for deeply embedded control.

While developed on an FPGA for validation, the design intentionally reflects the constraints and expectations of ASIC-based VLSI systems, focusing on parallelism, low resource footprint, and deterministic timing. The results demonstrate that the core processing pipeline can be efficiently mapped to ASIC logic, confirming the feasibility of low-power, marker-based autonomous control module for future System-on-Chip (SoC) integration. Thus, this work serves not only as a functional prototype but also as a validated design exploration path toward VLSI implementation.

REFERENCES

[1] M. Emimi, M. Khaleel, and A. Alkrash, "The Current Opportunities and Challenges in Drone Technology," *International Journal of Electrical Engineering and Sustainability*, vol. 1, no. 3, pp. 74–89, Jul. 2023.

[2] A. World, *Global Drone Utilization: Trends, Applications, and Market Dynamics*, Accessed: 2025-06-10, 2023. [Online]. Available: https://www.adwworld.com/news/urban-air-operations/global-drone-utilization.

[3] C. Kanellakis, E. Fresk, S. S. Mansouri, D. Kominiak, and G. Nikolakopoulos, "Autonomous visual inspection of large-scale infrastructures using aerial robots," *arXiv preprint arXiv:1901.05510*, 2019.

[4] Z. Lotfi, *Commercial Drone Use in 2024*, Accessed: 2025-06-10, 2024. [Online]. Available: https://droneii.com/commercial-drone-use-in-2024.

[5] S. Yang, S. A. Scherer, X. Yi, and A. Zell, "Multi-camera visual SLAM for autonomous navigation of micro aerial vehicles," *Robotics and Autonomous Systems*, vol. 93, pp. 116–134, 2017.

[6] A. Romero, A. Shenai, I. Geles, E. Aljalbout, and D. Scaramuzza, *Dream to Fly: Model-Based Reinforcement Learning for Vision-Based Drone Flight*, 2025.

[7] A. Cyba, H. Szolc, and T. Kryjak, "A simple vision-based navigation and control strategy for autonomous drone racing," in *Procdings of the International Conference on Methods and Models in Automation and Robotics (MMAR)*, 2021, pp. 185–190.

[8] B. Boroujerdian, H. Genc, S. Krishnan, *et al.*, "The Role of Compute in Autonomous Micro Aerial Vehicles: Optimizing for Mission Time and Energy Efficiency," *ACM Transactions on Computer Systems*, vol. 39, no. 1–4, pp. 1–44, Jul. 2022.

[9] I. OmniVision Technologies, *OV7670/OV7171 CMOS VGA (640x480) CAMERACHIP Sensor with OmniPixel technology*, Accessed: 2025-03-07, 2006. [Online]. Available: https://web.mit.edu/6.111/www/f2016/tools/OV7670_2006.pdf.

[10] Xilinx, *Spartan-6 family overview*, Accessed: 2025-06-10, 2011. [Online]. Available: https://docs.amd.com/v/u/en-US/ds160.

[11] Bitcraze, *Crazyflie 2.0*, Accessed: 2025-06-10, 2020. [Online]. Available: https://store.bitcraze.io/products/crazyflie-2-0.

[12] B. AB, *Crazyradio PA*, Accessed: 2025-06-10, 2024. [Online]. Available: https://www.bitcraze.io/products/crazyradio-pa/.

[13] B. AB, *Flow Deck v2*, Accessed: 2025-06-10, 2024. [Online]. Available: https://www.bitcraze.io/products/flow-deck-v2/.

[14] J. Chen, *FPGA image processing to support the autonomous flight of microdrones*, Master's thesis, University of Tsukuba, Mar. 2021.

[15] Z. Chenguang, *Self-flying drone system by marker recognition*, Master's thesis, University of Tsukuba, Mar. 2022.

[16] C. Moore, H. Devos, and D. Stroobandt, "Optimizing the FPGA memory design for a Sobel edge detector.," in *Proceedings of the International Conference on Engineering of Reconfigurable Systems & Algorithms*, Jan. 2009, pp. 299–300.

[17] J. E. Volder, "The CORDIC Trigonometric Computing Technique," *IRE Transactions on Electronic Computers*, vol. EC-8, no. 3, pp. 330–334, 1959.

Cultivating Security: Debug Authentication for Ensuring the Security of SoC's Root of Trust

Arash Vafaei, Sujan Kumar Saha, Mark Tehranipoor, and Farimah Farahmandi

Dept of Electrical and Computer Engineering, University of Florida

Gainesville, Florida

Email: {arash.vafaei, sujansaha}@ufl.edu, and {tehranipoor, farimah}@ece.ufl.edu

Abstract—Hardware-assisted debugging provides the necessary infrastructure for developers to closely monitor program behaviors at the microarchitectural level in a system-on-chip (SoC). However, debug infrastructure jeopardizes the security of the system by providing a backdoor for accessing crucial assets embedded in the system because of the inevitable increase in observability. While trusted execution environments (TEE) provide an extra level of security and isolate design assets, the security implication of hardware debug integration on TEEs has not been investigated. In this paper, we introduce a multi-level bidirectional access authentication mechanism over the debug module that defines the minimum number of privilege levels needed and the access details at each level so that debug users are authorized and blocked from accessing assets private to other entities. Trust is established by exchanging certificates both from the debugger and SoC sides to implement a bidirectional authorization platform in order to restrict the debugger's access to SoC assets as well as prevent the debugger's test data from being accessed by an SoC impersonator through emulation. We provide a prototype of the debug authentication platform on RISC-V architecture that proves the small overhead of the approach while staying compatible with traditional debug efforts.

I. INTRODUCTION

With the increasing complexity of modern System-on-Chip (SoC) designs—featuring multi-core processors, deep memory hierarchies, numerous on-chip peripherals, and Trusted Execution Environments (TEEs)—debugging these systems has become more difficult. Hardware-assisted debugging is a vital tool that allows engineers to observe internal behaviors in real time without modifying software execution, making it more effective than traditional software debugging, which often distorts system behavior by inserting payloads. However, while hardware debugging improves visibility into the system, it also introduces new attack surfaces that can compromise the confidentiality and integrity of sensitive data [14, 8].

Hardware debug interfaces such as JTAG and scan chains allow direct access to internal signals of IP blocks, enable arbitrary instruction execution, and permit full memory extraction [18]. These powerful features can be misused by attackers to leak encryption keys, tamper with memory contents, or even inject malicious payloads. For instance, AES and RSA modules have been shown to be vulnerable to scan-chain-based attacks that expose round keys during operation [1, 19]. Debug interfaces also allow attackers to pause execution and inject instructions that can result in privilege escalation or unauthorized memory access [11]. Firmware extraction is another threat where adversaries can read or patch firmware through debug ports for reverse engineering or bypassing security checks [18, 16].

These risks are especially concerning in manufacturing and field-deployed environments, where sensitive assets such as encryption keys, secure firmware, and test seeds may be exposed over debugging paths. To mitigate such threats, various countermeasures have been proposed. One set of approaches involves encrypting the debug data path using stream or block ciphers to protect information in transit [17, 16]. Others enforce authentication using passwords or challenge-response mechanisms. Password-based debug control, while simple, lacks strong identity binding [10], whereas PUF-based authentication methods offer stronger guarantees by binding debugger access to unique hardware fingerprints [7].

Another strategy is to permanently disable debug interfaces after manufacturing using e-fuses [11]. While this prevents unauthorized access, it also eliminates in-field debugging capabilities. Overall, most solutions fall into three categories: encrypting data, authenticating debuggers, and enforcing access control through monitoring [12]. These methods often come with trade-offs in area, performance, or flexibility, and many remain vulnerable to reverse engineering.

To address these challenges, this paper proposes a certificate-based secure debugging framework for SoCs that enables controlled and authenticated access across the product lifecycle. As illustrated in Figure 1, different stakeholders—including test engineers, field service providers, and security analysts—require varying levels of debug access. Our proposed solution enables mutual authentication between the SoC and debugger using certificates and provides role-based privilege levels to enforce access policies dynamically.

The contributions of this work are as follows:

- Introducing a mutual authentication protocol based on certificates to ensure only authorized debuggers can interact with SoC debug infrastructure.

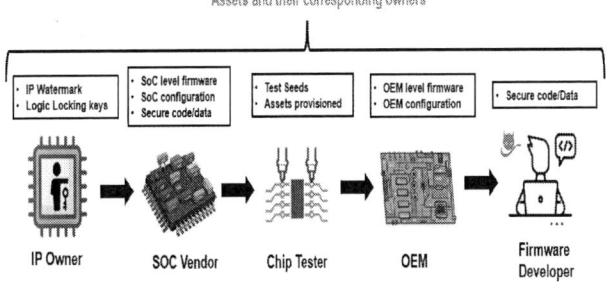

Fig. 1. Various entities using debug infrastructure and the attack targets (assets) at each stage.

979-8-3315-9813-6/25 $31.00 © 2025 IEEE

- Defining multiple privilege levels to allow differentiated access for users based on operational domain (e.g., secure vs. non-secure world).
- Enforcing access control policies through runtime monitors and evaluating the proposed framework on a RISC-V SoC, including hardware overhead analysis.

The rest of this article is organized as follows. Section II explains the concepts required to understand the solution, while Section III explains the access levels in general secure SoC architectures. The last Section will introduce the results for a RISC-V example.

II. BACKGROUND AND PRELIMINARY CONCEPTS

In this section, the main concepts that provide the essence for understanding the architectures, procedures, and authentication protocols involved in the authorization of an external debugger are explained in detail.

A. Threat Model

To understand debug-related threats, it's crucial to identify the various users of the debug interface throughout the SoC lifecycle. Entities such as testers, RMA operators, and firmware developers require different levels of debug access. However, this access can be misused to bypass physical memory protection and compromise confidential assets, such as Boot ROM code or OEM-proprietary data. Non-secure developers must never access secure programs or data, making strict isolation between secure and non-secure worlds essential. Any debug access violating these isolation boundaries is a critical security vulnerability. Additionally, with the rise of remote debugging, the debugger itself can become a threat—transmitting private code and configurations to potentially unauthorized SoCs. Our threat model therefore addresses vulnerabilities from both SoC and debugger perspectives to ensure robust and granular access control across trusted and untrusted domains. Figure 1 illustrates stakeholders, assets, and their provisioning timeline throughout the SoC lifecycle.

B. Certificate-Based Authentication

Certificates, based on Public Key Infrastructure (PKI), are widely used for authentication due to their flexibility in dynamic validation and resistance to reverse engineering—advantages over hardware-based methods like PUFs. A certificate contains the public key of the issuing entity along with other identifying information, enabling secure communication. To prevent impersonation, certificates are digitally signed using the private key of a trusted authority. This signature is generated by first hashing the certificate content and then encrypting the hash using the authority's private signing key. This ensures the certificate's integrity—any modification changes the hash and invalidates the signature.

As an example, X509 certificate includes the issuer ID, public key, digital signature, and custom fields such as SoC ID and debugger privilege levels used in our secure debug framework. During validation, the system recomputes the hash from the certificate content (excluding the signature), then decrypts the provided signature using the corresponding public validation key. If the decrypted hash matches the computed

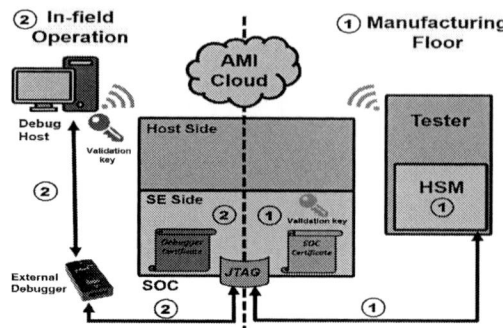

Fig. 2. The communications and assets are marked with number 1 for the manufacturing floor and number 2 for in-field operations. Each certificate and its validation key have the same color and AMI has a table that saves all the keys for both signing and validating as well as the whole certificate issued.

one, the certificate is deemed authentic and unaltered. This mechanism ensures trusted authentication of both debugger and SoC in the proposed secure debug infrastructure. The following section explains how keys and certificates are provisioned during manufacturing and deployment.

C. Secure SoC Architecture

To counter growing cyber threats, modern SoCs have adopted process isolation since 2018, confining attacks to vulnerable components and protecting critical assets. Central to this effort is the Trusted Execution Environment (TEE) or Security Engine (SE), an isolated hardware-supported environment that ensures confidentiality, integrity, and authenticity of sensitive data—even if the OS or hypervisor is compromised [6]. SEs either retain assets within their boundary or securely provide services to the host, such as encrypted data transfers [13]. TEE implementations include Intel SGX with enclave-based isolation and attestation [5], ARM TrustZone's secure/normal world partitioning [2], and Keystone for RISC-V, which supports memory encryption and remote attestation [9]. Recent extensions also protect against supply chain threats like IP theft through features such as watermarking [3], logic locking, and secure boot [13]. However, the integration of SEs introduces new vulnerabilities via the debug port, which often connects both secure and non-secure domains. This shared access can expose high-risk services to attackers. Existing defenses—such as disabling the debug interface or applying weak authentication—are either impractical or poorly managed by OEMs [11]. This paper investigates secure debugging in SE-equipped SoCs, focusing on protecting sensitive assets from unauthorized debug access.

D. Asset provisioning and Security Engine

The certificate signing and validation process involves two key phases. In the manufacturing phase, SoC-specific assets are provisioned, including the debugger certificate validation key and the SoC certificate. Since the tester is untrusted, a Hardware Secure Module (HSM) facilitates secure provisioning and establishes trust between the Asset Management Infrastructure (AMI) and the tester. The AMI, acting as the Certificate Authority (CA), generates key pairs and certificates, maintains a database indexed by SoC IDs, and issues certificates tailored to each chip [13]. Each SoC securely stores its

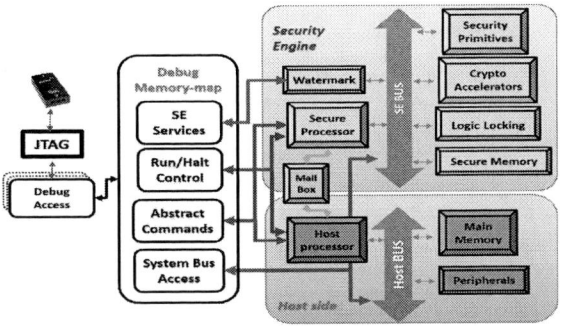

Fig. 3. Debug platform services and connections in the presence of a Security Engine in the SoC.

validation key and SoC certificate in a tamper-proof memory. In the in-field debug phase, the external debug host contacts the AMI and requests a certificate for a specific SoC ID, authenticating itself through standard internet protocols. The AMI then signs a debugger certificate with the private key linked to that SoC ID, embedding necessary debug parameters such as privilege level and a public encryption key for secure communication. The Security Engine (SE) inside the SoC uses the pre-provisioned validation key to verify the debugger certificate's authenticity. Figure 2 shows both phases, highlighting the secure certificate flow and validation path for authenticated debug access.

III. SECURE DEBUG

With the foundational concepts of secure debug architecture and certificate validation established, we now examine common debug architectures to define a privilege-based access model that is independent of the underlying instruction set architecture (ISA). Understanding how a Security Engine (SE) integrates with the host-side SoC and shares debug infrastructure enables us to enforce isolation with minimal architectural changes.

In a secure SoC, the SE provides critical services such as cryptographic operations, watermarking, IP unlocking (for both SE and host-side IPs), and security primitives like True Random Number Generators (TRNG) and odometers [13]. The SE includes a secure processor hardened against attacks like cache timing and power side-channels. Certain services must be accessed via debug to maintain trust, especially when the host processor is untrusted. For instance, watermark extraction cannot safely be issued from an untrusted host processor. Furthermore, secure firmware development necessitates debug access to the SE processor for functionality verification and bug tracing. On the host side, multiple cores typically require hardware-assisted debugging. Host-side buses and memory hierarchies are also exposed via debug, but SE-provided services such as memory encryption and protection units ensure the security of sensitive assets.

To generalize this architecture, Figure 3 presents a high-level view of a secure SoC. Debug services are accessed through a memory-mapped interface and routed to modules within the SE via specific debug port registers. Connections to processors and buses—illustrated using color-coded pathways—highlight the scope of debug interactions.

To secure debug access, a multi-level privilege model is introduced. This model governs read/write operations on debug services and restricts access based on entity roles within the supply chain. The minimum required privilege levels are derived from the threat model and connection paths shown in Figure 3. With the big picture of the SoC in mind, we can now define a multi-level access paradigm and how and which entities can utilize each of these levels for their debug purposes. These privilege levels differ by blocking specific accesses to certain entities on the supply chain and our discussion will be based on the connections presented in various colors in Figure 3.

The debug memory-map services have connections to different entities inside SoC architecture and the minimum number of privilege levels necessary to address the threat model mentioned so far consists of:

Level 0: Level zero is a special privilege access dedicated to IP owners. In case the IP owner wants to extract its IP watermark in a court of law or has provided a logically locked IP and demands to check the required key, or any IP-specific service that SE provides, this level guarantees secure access without host-side involvement through debug infrastructure.

Level 1: This level has full permission to access the connections on the SE side. It is considered a high-privilege level of access used by secure firmware developers, SoC owners, and any entity with private assets inside SE boundaries. When AMI grants this level of access to a debugger they can run code on the secure processor and access its registers and allocated memory. The only restriction at this level is special accesses reserved for IP owners such as watermark extraction services.

Level 2: In this level all the services including run/halt control, abstract command, and SE bus access are disabled while the mailbox configuration still allows for the host side code to issue requests to SE through secure channels already implemented inside mailbox IP. This will make debugging host-side code with calls to the security function possible while preventing access to assets inside SE by debug channels. The SE services shown with the red arrow are also disabled for this level.

Level 3: The only difference between this level and with previous level is the complete isolation of SE via disabling the mailbox. This level exists so that the AMI can disable access to SE in case a vulnerability exists that can be exploited by offloading tasks through the mailbox to SE. General developers who should be prohibited from accessing assets within SE boundaries will be issued a certificate with this privilege level of access.

Level 4: These levels are defined per host-side requirements and are based on the host SoC architecture. These levels are similar to Level 3 because they are also blocked from accessing assets within the SE boundary but are different based on the specific restriction imposed on the host-side SoC architecture. For example, one level can block direct access to the system bus that can bypass physical memory protection units while another level allows such access.

These privilege levels present the minimum number of zones required for safe access to debug infrastructure in a secure SoC that has an integrated security engine. Based on

979-8-3315-9813-6/25 $31.00 © 2025 IEEE

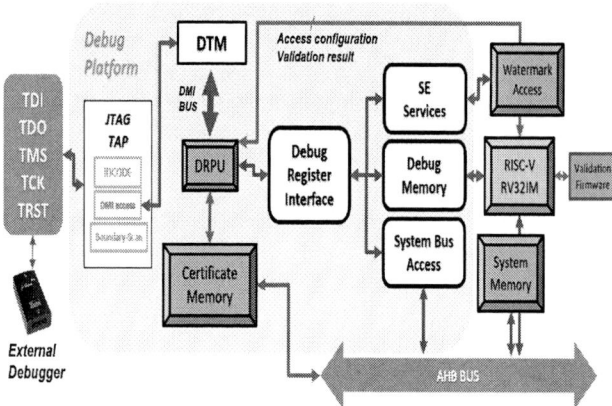

Fig. 4. RISC-V debug platform and with its connection to a simplified RISC-V SoC with a 32-bit core and a small memory for storing firmware.

TABLE I
POWER AND AREA OVERHEAD OF SECURE DEBUG IPS

	Secure Debug	RISCV CVA6 Core	RISCV Debug	Overhead for debug	overall overhead
Area (nmm)	2608.24	245871.38	49440.84	5%	0.8%
Dynamic Power (uW)	311.19	8861.1	2311.8	13%	2%

the implementation, each level can be fragmented so we have a finer grain of policy enforcement.

IV. RESULTS AND CONTRIBUTIONS

RISC-V, as a rapidly adopted open-source architecture, exposes significant security risks through its conventional debug interface. Unlike proprietary solutions like ARM's SDC-600, RISC-V lacks built-in secure debug, prompting the need for additional IPs to enhance its security without altering the existing specification.

A. RISC-V Debug

The RISC-V debug architecture provides a flexible and feature-rich interface for hardware-assisted debugging of RISC-V processors. At full functionality, it enables reading and writing of all hardware threads' (harts) registers, including control and status registers (CSRs). Memory access can be performed either from the hart's perspective or directly via a master port on the system bus. The architecture supports independent debugging of each hart and allows automatic discovery of system configuration, requiring minimal manual setup. Debugging can begin from the first instruction using the debug reset interface, with support for breakpoints, single-stepping, and other granular control mechanisms. Optional features include halting/resuming subsets of harts, executing custom instructions on halted harts, non-intrusive register access, and programmable triggers based on PC, memory address, or data [15].

A simplified RISC-V SoC with its debug architecture is illustrated in Fig.4, excluding the green-shaded contributions of this paper. External debuggers typically connect via JTAG (IEEE 1149)[4], which interfaces with the Debug Transport Module (DTM). The DTM drives the Debug Module Interface (DMI), connecting to the Debug Module (DM), which exposes the full debug capability via a register-based interface. The DM supports three main categories: (1) Run/Halt control for pausing/resuming harts, (2) Abstract commands and program buffers for executing arbitrary instructions and accessing registers, and (3) System bus access for interacting with SoC peripherals. Access to these features is managed through read/write operations on specific debug module registers, such as dmcontrol (hart selection), command and data0–11

(instruction and return data), and sbaddress/sbdata (system bus interface). This centralized register interface serves as an ideal monitoring point for enforcing fine-grained access control over debug operations. For instance, write protection on sbaddress can restrict unauthorized access to SE-connected buses [15].

B. Debug Register Protection Unit (DRPU)

With the identification of the debug register interface as a critical enforcement point, we propose a hardware module—Debug Register Protection Unit (DRPU)—to implement and enforce security policies for secure debug. The DRPU is integrated with the RISC-V debug architecture and monitors accesses to the debug module's register interface. The DRPU has three primary functions: (1) Certificate Channel Support: We introduce a new authentication opcode alongside the conventional DMI read/write operations. This opcode redirects data from the JTAG register to a certificate buffer memory. The debugger certificate is shifted into the system using this opcode, and the DRPU manages access to the buffer based on address decoding. (2) Access Control Enforcement: After certificate validation, access permissions are configured via firmware running on the secure processor. The DRPU enforces these restrictions during conventional debug operations. (3) Status Reporting: The DRPU provides real-time feedback to the debugger, indicating authentication status and the state of the debug infrastructure. In our prototype, we implemented three core policies: blocking access to SE-related harts, restricting system bus access to SE-protected addresses, and preventing external access to watermark-related registers. Developers can extend this model to enforce additional policies tailored to their SoC's threat model. Our implementation is based on the open-source CVA6 core, a 6-stage, single-issue, in-order RISC-V CPU (RV32IM). The certificate validation is handled by secure firmware within the SE. The minimal area, power, and timing overhead introduced by the DRPU and certificate buffer are summarized in Table IV-A, demonstrating the efficiency of our secure debug solution.

V. CONCLUSION

This work addresses the critical challenge of secure hardware-assisted debugging in SoCs that integrate Security Engines (SEs). While debug interfaces are essential for development and diagnostics, they pose significant security risks if left unprotected—especially in open architectures like RISC-V. We presented a layered framework that uses certificate-based authentication and fine-grained privilege control, enforced through a Debug Register Protection Unit (DRPU) integrated into the RISC-V debug path. Our hardware prototype demonstrates the feasibility of enforcing access control without modifying the RISC-V debug specification. Experimental results show minimal area and power overhead, validating the practicality of our approach.

979-8-3315-9813-6/25 $31.00 © 2025 IEEE

References

[1] Sk Subidh Ali, Ozgur Sinanoglu, and Ramesh Karri. "Test-mode-only scan attack using the boundary scan chain". In: *2014 19th IEEE European Test Symposium (ETS)*. 2014.

[2] ARM Limited. *Security technology: Building a secure system using TrustZone® technology*. Online. 2008. URL: http://infocenter.arm.com/help/index.jsp?topic=/com.arm.doc.prd29-genc-009492c/index.html.

[3] Subodha Charles, Vincent Bindschaedler, and Prabhat Mishra. "Digital Watermarking for Detecting Malicious Intellectual Property Cores in NoC Architectures". In: *IEEE Transactions on Very Large Scale Integration (VLSI) Systems* (2022).

[4] "IEEE Standard for Test Access Port and Boundary-Scan Architecture". In: *IEEE Std 1149.1-2013 (Revision of IEEE Std 1149.1-2001)* (2013).

[5] Intel. *Intel® Software Guard Extensions programming reference*. Online. Accessed on: Aug. 9, 2019. 2014. URL: https://software.intel.com/sites/default/files/managed/48/88/329298-002.pdf.

[6] Patrick Jauernig, Ahmad-Reza Sadeghi, and Emmanuel Stapf. "Trusted Execution Environments: Properties, Applications, and Challenges". In: *IEEE Security Privacy* (2020).

[7] Sudeendra Kumar et al. "PUF-based secure test wrapper for SoC testing". In: *2018 IEEE Computer Society Annual Symposium on VLSI (ISVLSI)*. IEEE. 2018, pp. 672–677.

[8] Vinay BY Kumar et al. "Towards designing a secure RISC-V system-on-chip: ITUS". In: *Journal of Hardware and Systems Security* (2020).

[9] Dayeol Lee et al. "Keystone: An Open Framework for Architecting Trusted Execution Environments". In: *Proceedings of the Fifteenth European Conference on Computer Systems*. EuroSys '20. 2020.

[10] Kuen-Jong Lee, Zheng-Yao Lu, and Shih-Chun Yeh. "A Secure JTAG Wrapper for SoC Testing and Debugging". In: *IEEE Access* (2022).

[11] Zhenyu Ning and Fengwei Zhang. "Understanding the Security of ARM Debugging Features". In: *2019 IEEE Symposium on Security and Privacy (SP)*. 2019.

[12] Xuanle Ren et al. "IC Protection Against JTAG-Based Attacks". In: *IEEE Transactions on Computer-Aided Design of Integrated Circuits and Systems* (2019).

[13] Md Sami Ul Islam Sami et al. "Invited: End-to-End Secure SoC Lifecycle Management". In: *2021 58th ACM/IEEE Design Automation Conference (DAC)*. 2021.

[14] Alan Sguigna. "Mitigating JTAG as an Attack Surface". In: *2019 IEEE AUTOTESTCON*. 2019.

[15] "SiFive, RISC-V External Debug Support 0.13, 2018." In: ().

[16] Emanuele Valea et al. "A Survey on Security Threats and Countermeasures in IEEE Test Standards". In: *IEEE Design & Test* (2019).

[17] Emanuele Valea et al. "Encryption-Based Secure JTAG". In: *2019 IEEE 22nd International Symposium on Design and Diagnostics of Electronic Circuits Systems (DDECS)*. 2019.

[18] Sebastian Vasile, David Oswald, and Tom Chothia. "Breaking all the things—A systematic survey of firmware extraction techniques for IoT devices". In: Springer. 2019.

[19] Weizheng Wang et al. "Ensuring Cryptography Chips Security by Preventing Scan-Based Side-Channel Attacks With Improved DFT Architecture". In: *IEEE Transactions on Systems, Man, and Cybernetics: Systems* (2022).

Towards Full Integration of a Three-Level Flying Capacitor Converter Control in a Mixed-Signal ASIC

Nelson Salvador, *Student Member, IEEE*, Francisca Donoso, *Student Member, IEEE*,
Jorge Marín, *Member, IEEE*, Victor Grimblatt, *Senior Member, IEEE*, Christian A. Rojas, *Senior Member, IEEE*.

Abstract—This work presents the evolution of a Three-Level Flying Capacitor Converter (3L-FCC) ASIC design from early-stage modeling and simulation towards full integration in a mixed-signal ASIC. As an intermediate step, the power converter core was first implemented in an analog ASIC, complemented by a System-on-Chip (SoC) integrating an embedded CPU for the digital control loop and FPGA fabric for the modulation scheme. This hybrid platform enabled functional validation and post-silicon testing of the analog stage before migrating the remaining digital control and phase-shifted pulse-width modulation (PS PWM) logic into the ASIC. In the current stage of the project, the control architecture is being migrated to a hardware description language (HDL) implementation, with the goal of enabling full integration into a standard digital design flow for mixed-signal ASICs. An FPGA-based prototype is being used to verify functionality and assess timing performance prior to full integration, providing a validation path towards future monolithic implementation.

Index Terms—Three-Level Flying Capacitor Converter (3L-FCC), Power Management Integrated Circuit (PMIC), Mixed-Signal ASIC Integration, FPGA-Based Prototyping, HDL-Based Digital Control.

I. INTRODUCTION

The Three-Level Flying Capacitor Converter (3L-FCC) architecture has shown promising performance as a core component for fully integrated power conversion systems. It has been successfully employed in diverse applications such as interleaved DC-DC converters [1] and hybrid switched-capacitor DC-DC converters [2], as well as in power-efficient class-D audio amplifiers for industrial use [3]. In particular, the 3L-FCC topology stands out for high-efficiency and compact power conversion, making it well suited for integrated systems with stringent size and power constraints.

Fully integrating the 3L-FCC's control and power stages into a single mixed-signal ASIC presents significant design challenges, including the need for tight coordination between analog and digital components as well as strict timing requirements. Achieving this level of integration offers clear advantages such as reducing the system footprint, improving energy efficiency, and increasing reliability. These benefits are especially important for applications with strict power,

size, and latency demands such as space systems and portable electronics.

Our research group has developed a single ASIC prototype of the 3L-FCC which implements only the analog power stage. Validation and characterization of this ASIC were performed using a Xilinx Zynq SoC where the control algorithm, a proportional-integral control loop, and phase-shifted PWM signals were executed across the processing system and programmable logic. Although this setup provided flexibility for testing, the implementation of the control algorithm in a high-level programming language limited the potential for a fully integrated solution. To address these limitations, this work focuses on migrating the control functions from a CPU-based implementation to a fully HDL based design. This transition facilitates compatibility with standard digital design flows. The proposed HDL architecture has been prototyped on an FPGA to verify functionality and timing performance, paving the way towards a monolithic mixed-signal ASIC implementation.

II. PREVIOUS ITERATIONS AND SYSTEM DEVELOPMENT

Figure 1 summarizes the evolution of the 3L-FCC system development, tracing the path from initial simulations to the current HDL-based closed-loop prototype. This timeline provides a visual overview of the key milestones and improvements achieved throughout the design process. The work started with system-level simulations of the 3L-FCC. With fabrication access to the SKY130 process, the design, simulation, and implementation of the converter were presented in [4] using Open-Source tools. After fabrication of the ASIC, initial tests with the 3L-FCC IC were conducted in an open-loop configuration [5], during which the system lacked automated control over the flying capacitors and required manual adjustments to regulate the output voltage under varying load conditions. Although this approach provided valuable insights into the ASIC's behavior, it also revealed the limitations of operating without a closed-loop feedback control system.

To overcome these limitations, the control strategy was transitioned to a closed-loop scheme, automating regulation and significantly enhancing system reliability. Subsequently, the 3L-FCC ASIC was validated on a SoC platform, where the FPGA fabric generated PS-PWM signals while the embedded CPU executed the control algorithm. This SoC-based solution first validated the control algorithm and its parameters, enabling dynamic control and comprehensive functional testing of the converter. However, reliance on external processing resources

---

This work was supported in part by ANID + Fondecyt de Exploración 2025 + 13250147, ANID + FONDECYT Initiation Research Project 11240947, ANID/BASAL/AFB240002.

N. Salvador, F. Donoso, J. Marín and C. A. Rojas are with the Electronics Department, Universidad Técnica Federico Santa María and with the Advanced Center of Electrical and Electronics Engineering (AC3E), Valparaíso, Chile.

V. Grimblatt is with Synopsys Chile, Santiago, Chile.

979-8-3315-9813-6/25 $31.00 © 2025 IEEE

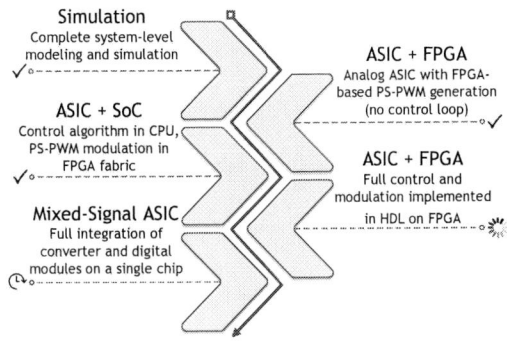

Fig. 1. Evolution timeline of the 3L-FCC system development.

TABLE I
FPGA RESOURCE UTILIZATION.

Resource	Utilization	Available
LUT	2381	63400
FF	1322	126800
Max Frequency	67 MHz	

limited the potential for full integration, motivating the migration of the control functions into an HDL-based design. This progression reflects the goal of a fully integrated mixed-signal ASIC combining the analog power stage and digital control logic. The current focus is on completing the HDL prototyping for subsequent integration within a standard ASIC design flow.

III. HDL-BASED ARCHITECTURE AND FPGA PROTOTYPING

The proposed HDL architecture fully implements the digital control scheme for the 3L-FCC, including the PI control loop, dead-time generation, and the PS-PWM. This modular design supports future integration with the analog power stage in a mixed-signal ASIC. The PI controller uses fixed-point arithmetic to reduce resource usage and ensure compatibility with open-source ASIC flows, unlike floating-point IP from the HLS tool. Dead-time generation delays outputs to prevent short circuits. PS-PWM facilitates dynamic switching adjustments to maintain flying capacitor voltages near reference values via switching state redundancies [6], ensuring stability and efficiency.

The HDL design is deployed on a Xilinx Nexys A7 FPGA to verify functionality and assess timing limits before ASIC integration. A Shift Register block is used to preconfigure system parameters using only two signals. Additionally, a Timer Controller block synchronizes the controller operation with the desired switching period and coordinates timing with the ADC sampling. Figure 2 shows the modular HDL design with the fixed-point PI controller. Functional validation and

timing verification on FPGA are ongoing. Although physical system validation is pending, current results indicate a clear path to full integration.

Table I summarizes the FPGA resource utilization and timing results, indicating low resource usage and timing margins sufficient to meet system requirements, thus supporting the design feasibility. It should be noted that the data in the table was obtained without using FPGA DSP blocks. Depending on the target ASIC PDK, DSP-like cells may not be available, so this scenario is assumed for the estimation of basic logical resources. Consequently, the resources reported are not necessarily the most area or performance efficient for the FPGA, but they provide a meaningful reference for the corresponding ASIC implementation, where optimized usage of standard cells and potentially higher maximum frequency are expected.

An important aspect remaining to be addressed is the design and integration of the analog-to-digital converter (ADC) for voltage sensing in the fully integrated ASIC. The choice of ADC architecture and its interface with the digital control logic will be critical to ensure accurate measurement, low latency, and efficient power consumption in the monolithic solution.

IV. CONCLUSION AND FUTURE WORK

This work presents the evolutionary path towards full integration of the 3L-FCC control in a mixed-signal ASIC, highlighting the transition from analog designs and hybrid platforms to an HDL solution validated on FPGA. Future efforts focus on monolithic implementation and conducting experimental tests to certify the performance and robustness of the integrated system.

REFERENCES

[1] J.-Y. Lee, G.-S. Kim, K.-I. Oh, and D. Baek, "Fully integrated low-ripple switched-capacitor dc–dc converter with parallel low-dropout regulator," *Electronics*, vol. 8, no. 1, 2019. [Online]. Available: https://www.mdpi.com/2079-9292/8/1/98

[2] Z. Xia and J. T. Stauth, "A cascaded hybrid switched-capacitor dc–dc converter capable of fast self startup for usb power delivery," *IEEE Journal of Solid-State Circuits*, vol. 57, no. 6, pp. 1854–1864, 2022.

[3] M. Høyerby, J. K. Jakobsen, J. Midtgaard, and T. H. Hansen, "A 2 × 70 w monolithic five-level class-d audio power amplifier in 180 nm bcd," *IEEE Journal of Solid-State Circuits*, vol. 51, no. 12, pp. 2819–2829, 2016.

[4] J. Marin, J. Gak, A. Cortes, N. Calarco, A. Oliva, E. Lindstrom, M. Miguez, A. Falcón, N. Osterman, and C. A. Rojas, "Integrated three-level flying capacitor dc-dc buck converter for cubesat applications," in *2023 Argentine Conference on Electronics (CAE)*. IEEE, 2023, pp. 90–95.

[5] J. Marin, J. Gak, C. A. Rojas, A. H. Wilson-Veas, N. Calarco, M. Miguez, A. R. Oliva, and N. Salvador, "Open-source multilevel converter power ic design and test," *IEEE Design & Test*, pp. 1–1, 2024.

[6] C. Feng, J. Liang, and V. G. Agelidis, "Modified phase-shifted pwm control for flying capacitor multilevel converters," *IEEE Transactions on Power Electronics*, vol. 22, no. 1, pp. 178–185, 2007.

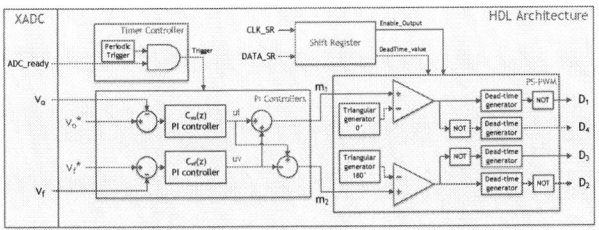

Fig. 2. High-level block diagram of the proposed HDL-based architecture.

Design of a 32nm Ultra-Low Power 6T SRAM Cell Analyzing Emerging Technologies such as FinFET, TFET, and CNFET for energy-efficient applications

Jesús González Huarancca, Carlos Silva Cárdenas(Advisor)
Faculty of Science and Engineering
Research Group on Microelectronic
Pontifical Catholic University of Peru
Lima, Peru
jgonzalezh@pucp.edu.pe, csilva@pucp.edu.pe

Abstract—**This work presents the design, simulation, and comparative analysis of a 6-transistor (6T) static random-access memory (SRAM) cell optimized for ultra-low-power operation. Emerging device technologies—Fin Field-Effect Transistor (FinFET), Tunnel Field-Effect Transistor (TFET), and Carbon Nanotube Field-Effect Transistor (CNFET)—are evaluated individually and in hybrid configurations to overcome the limitations of conventional CMOS SRAM cells. Device-level models are calibrated to ensure fair comparison, matching leakage currents and parasitic capacitances. Using NGSPICE and HSPICE simulations at subthreshold supply voltages (0.6–0.8 V), we measure static, dynamic, and total power consumption for each configuration. Results show that the FinFET–TFET hybrid achieves the lowest total power dissipation, combining the low leakage of TFETs with the high speed and channel control of FinFETs, making it suitable for energy-constrained applications such as IoT and implantable devices.**

Index Terms—**Ultra-low-power SRAM, FinFET, TFET, CNFET, hybrid technology, low-voltage design, power consumption.**

I. INTRODUCTION

Six-transistor (6T) SRAM cells are fundamental in high-performance applications due to their simplicity and operational efficiency [1]. However, their energy consumption, particularly during read and write operations, remains a significant challenge in embedded and portable systems where energy efficiency is critical. Recently, emerging device technologies have been introduced to overcome these limitations. FinFETs, with their three-dimensional structure, provide superior electrostatic control and reduced leakage, enabling stable operation at lower voltages [1], [2]. CNFETs also show high carrier mobility and notable leakage reduction, allowing the design of SRAM cells with improved static and dynamic power performance [3]. TFETs, exploiting band-to-band tunneling, enable ultra-low voltage operation, and recent architectures such as disturb-free 10T cells [4] and hybrid 8T TFET–MOSFET designs [5] have demonstrated significant gains in static noise margin and power efficiency. This research focuses on the design of 6T SRAM cells using FinFET, TFET, and CNFET technologies, with the primary goal of reducing energy consumption. Through calibrated simulations in NGSPICE and

HSPICE, the 6T architecture is evaluated in terms of static and dynamic power, showing notable improvements in energy efficiency without compromising stability.

II. METHODOLOGY

The work focuses on the design and evaluation of 6-transistor (6T) SRAM cells at 32nm optimized to reduce energy consumption, using CMOS, FinFET, CNFET, and TFET technologies. An experimental methodology based on circuit simulations was used to directly compare the impact of each technology on consumption, while maintaining the basic functionality of the cell. The CMOS and FinFET models were taken from the literature with parameters for static and dynamic analysis [6], the CNFET was modeled according to the proposal from the University of Minnesota, which considers mobility, parasitic capacitances, and leakage current [7], and the TFET was implemented using a model based on TCAD simulations for ultra-low voltage operation [8]. Detailed structures and characteristics of each technology can be found in the cited references [6]–[8]

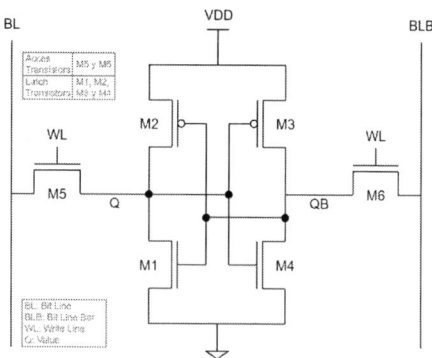

Fig. 1: Schematic of a 6T SRAM cell

The cell designs were created in Xschem, generating the netlists that were then simulated in HSPICE, allowing the extraction of consumption and stability metrics under subthreshold conditions (0.6 V to 0.8 V).

TABLE I: Proposed cases with the mentioned technologies

Case	Latch Transistors	Access Transistors
1	NMOS	NMOS
2	NFinFET	NFinFET
3	NCNFET	NCNFET
4	NTFT	NTFET
5	NCNFET	NTFET
6	NCNFET	NMOS
7	NMOS	NTFET
8	NFinFET	NCNFET
9	NFinFET	NMOS

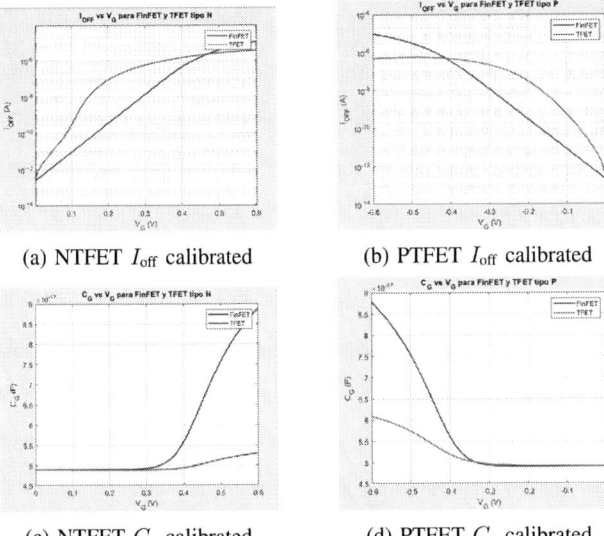

(a) NTFET I_{off} calibrated   (b) PTFET I_{off} calibrated

(c) NTFET C_g calibrated   (d) PTFET C_g calibrated

Fig. 2: TFET model calibration

The device models are calibrated primarily for leakage currents (I_{off}) and parasitic capacitances ($C_g = C_{gs} + C_{gd}$), using the Verilog-A TFET model. The calibration employs three key curves: CGS, CGD, and I_{off}, which are critical for modeling device behavior as seen in previous studies[9]. However, parameters such as ON/OFF resistors, leakage inductance, and voltage thresholds will be considered in subsequent studies to enhance model accuracy and provide a more comprehensive evaluation of power dissipation in SRAM cells.

III. RESULTS

By designing both the individual and hybrid cells, static, dynamic, and total consumption results were obtained.

TABLE II: Static, dynamic, and total power consumption

technology	VDD (V)	P_{sta} (W)	P_{dyn} (W)	P_{tot} (W)
CMOS	0.8	5.67E-10	2.70E-05	2.64E-07
CNFET	0.6	2.12E-10	4.53E-06	1.26E-08
FinFET	0.6	**2.80E-12**	2.79E-06	9.49E-09
TFET	0.6	4.19E-09	1.64E-06	2.66E-08
TFET-CMOS	0.6	1.87E-09	1.67E-06	5.89E-08
FinFET-TFET	0.6	1.69E-10	**1.10E-06**	**6.11E-09**
FinFET-CNFET	0.6	4.99E-11	3.25E-06	1.17E-08
FinFET-CMOS	0.6	4.81E-12	3.80E-06	4.46E-08
CNFET-TFET	0.6	3.04E-10	6.67E-06	3.29E-08
CNFET-CMOS	0.8	2.14E-10	2.86E-05	1.16E-07

The cell current was measured in an idle state, with static power calculated from current and supply voltage (0.6 V and 0.8 V). Dynamic consumption was evaluated by applying write and read transitions. To determine total consumption, a 60 ns interval was simulated, capturing both static and dynamic operations. This interval ensures scalability (repetitive switching behavior) and computational efficiency, providing a sufficient analysis of power consumption for the given low-voltage operation. Although it may be narrow for some IoT applications, it is adequate for the scope of this study.

IV. CONCLUSIONS

The evaluation of 6T SRAM cells using FinFET, TFET, and CNFET technologies confirmed that hybrid approaches, particularly the FinFET–TFET combination, can significantly reduce total power consumption while maintaining stability. These results reinforce the potential of emerging device technologies to improve energy efficiency in memory design.

However, this study was limited to a narrow voltage range and single-cell analysis, without exploring variability through corner simulations or the scalability of large arrays and peripheral circuits. Extending the analysis to include these aspects, along with updated benchmarking against recent device models, will provide a more comprehensive validation of hybrid SRAM designs for practical, ultra-low power applications.

REFERENCES

[1] V. Kumar and V. K. Tomar, "A Comparative Performance Analysis of 6T, 7T and 8T SRAM Cells in 18nm FinFET Technology," Proc. Int. Conf. Power Electronics & IoT Applications in Renewable Energy and its Control (PARC), Mathura, India, pp. 329–334, Feb. 2020, doi: 10.1109/ICSPCRE62303.2024.10675165.

[2] P. Bikki, A. S. V. L. Sai Poojitha, A. N. Sree, and B. S. L. Chowdary, "Design and Implementation of FinFET 12-T SRAM cell for low power Applications," Proc. 3rd Int. Conf. Intelligent Technologies (CONIT), Karnataka, India, pp. 1–5, Jun. 2023, doi: 10.1109/CONIT59222.2023.10205733.

[3] L. Saranya, H. Sayyed, P. Tanwar, N. S. Navaneethan, S. Manikandan, and R. J. Venkatesh, "Power Optimization in 6T-SRAM Cell using Carbon Nanotube Field-Effect Transistors (CNTFET)," Proc. 7th Int. Conf. Electronics, Communication and Aerospace Technology (ICECA), pp. 118–122, 2023, doi: 10.1109/ICECA58529.2023.10395589.

[4] Z. Lin, L. Li, X. Wu, C. Peng, W. Lu, and Q. Zhao, "Half-Select Disturb-Free 10T Tunnel FET SRAM Cell With Improved Noise Margin and Low Power Consumption," IEEE Trans. Circuits Syst. II: Express Briefs, vol. 68, no. 7, pp. 2628–2632, Jul. 2021, doi: 10.1109/TC-SII.2021.3057678.

[5] Y. Zhou, R. Jia, R. Zhao, K. Wang, Q. Huang, and R. Huang, "A Novel 8T TFET-MOSFET Hybrid SRAM Cell for Ultra-Low Power and Computing In-Memory Applications," Proc. IEEE Int. Symp. Circuits and Systems (ISCAS), pp. 1–5, 2022, doi: 10.1109/CSTIC64481.2025.11017934

[6] S. Lin, Y. B. Kim y F. Lombardi, "Stanford CNFET model," Stanford Nanoelectronics Group. Disponible en: https://nano.stanford.edu/downloads/stanford-cnfet-model.

[7] University of Minnesota, "Predictive Technology Model (PTM)." Disponible en: https://mec.umn.edu/ptm.

[8] S. Strangio, F. Settino, P. Palestri, M. Lanuzza, F. Crupi, D. Esseni y L. Selmi, "Digital and analog TFET circuits: Design and benchmark," Solid-State Electronics, vol. 146, pp. 50–65, 2018, doi: 10.1016/j.sse.2018.05.003.

[9] A. Arevalo, R. Liautard, D. Romero, L. Trojman, and L.-M. Procel, "New insight for next generation SRAM: Tunnel FET versus Fin-FET for different topologies," in Proc. 32nd Symp. Integr. Circuits Syst. Design (SBCCI), Sao Paulo, Brazil, Aug. 2019, pp. 1–6. doi: 10.1145/3338852.3339871.

979-8-3315-9813-6/25 $31.00 © 2025 IEEE

A Novel CMOS Time Register

Johnatan Felipe Silva Garcia, Dalton Martini Colombo, Kamal El-Sankary, Mahsa Zareie
Email: johnatanfelipe@ufmg.br, daltonmc@ufmg.br, kamal.el-sankary@dal.ca, Mahsa.Zareie@dal.ca

Abstract—**This work presents a time register designed for time-domain signal processing, employing a delay line in a ring configuration topology combined with a digital counter to achieve a high dynamic range. The proposed architecture supports both addition and subtraction operations, offering a practical framework for understanding the implementation of such functionalities in the time domain. The register is capable of efficiently processing signals with durations ranging from picoseconds to milliseconds and demonstrating feasibility for physical implementation.**

Index Terms—**time register, time domain, delay line.**

I. INTRODUCTION

The continuous miniaturization of semiconductor devices has improved efficiency and processing speed; however, it also introduces significant challenges for analog circuits. A key limitation arises from the reduction in supply voltage, which in turn degrades the signal-to-noise ratio (SNR) and dynamic range. Time-domain signaling offers a promising solution by representing signal amplitude through pulse width, thereby ensuring compatibility with digital processing and improving scalability. In this approach, time—an inherently analog parameter—is digitally encoded, yielding multiple performance and integration benefits. This processing paradigm is known as Time-Mode Signal Processing (TMSP) [1]–[3]. Within TMSP architectures, the time register plays a pivotal role, as it enables the storage and retrieval of time-domain signals. Furthermore, it serves as a versatile building block capable of performing fundamental operations such as addition, subtraction, amplification, and filtering in the time domain [3]–[6].

II. TIME DOMAIN

In the time domain, the amplitude of an analog signal is encoded in the pulse width of a time signal, rather than the voltage level, as shown in Fig. 1. These time-domain signals have only two distinct values, allowing them to be interpreted similarly to digital signals. Consequently, circuits operating in the time domain are, in essence, digital [7].

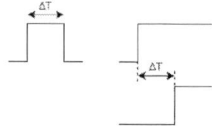

Fig. 1: Time Domain Representation [7].

III. TIME REGISTER

The time register used in this work has three inputs—Tin, Trigger, and Tsub—and one output, Tout. Tin applies the pulse width of a time-domain signal for storage or addition, Trigger retrieves the stored signal, and Tsub applies the signal for subtraction. The resulting time-domain signal is available at Tout.

To facilitate a better understanding of the circuit's operation, it can be compared to a stopwatch equipped with three control buttons, as depicted in the Fig. 2. The first button resets the stopwatch to zero. The second button, Tsub, initiates counting in the reverse (counterclockwise) direction, while the third button, Tin, starts counting in the forward (clockwise) direction. In this analogy, forward counting corresponds to addition in the time register, whereas reverse counting corresponds to subtraction.

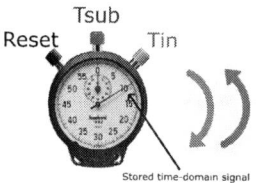

Fig. 2: Stopwatch illustrating the analogy to a time register.

To store a time-domain signal, the forward (clockwise) counting mode is activated for the duration of the input pulse. When the pulse ends, the value displayed on the counter corresponds to the stored time-domain signal. As presented in Fig. 3.

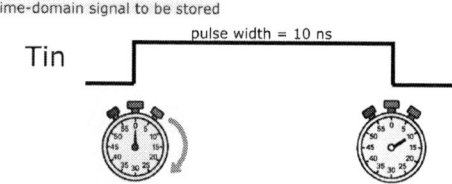

Fig. 3: Conceptual analogy between a stopwatch and a time register to store a time-domain signal.

To perform addition operations, the time-domain signal is applied while the forward (clockwise) counting mode is active. Each time an input pulse is applied, the value displayed on the counter increases proportionally to the duration of the time-domain signal. This is illustrated in Fig. 4.

For subtraction operations, the stopwatch must first contain a stored value. The subtraction is then performed by apply-

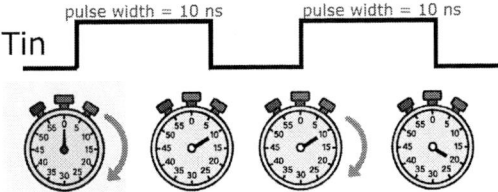

Fig. 4: Stopwatch illustrating the analogy to a time register performing addition operations.

ing the time-domain signal to the reverse (counterclockwise) stopwatch input, which decreases the stopwatch value by an amount proportional to the duration of the applied signal. As shown in Fig. 5.

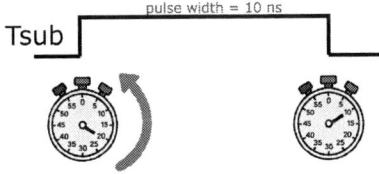

Fig. 5: Stopwatch illustrating the analogy to a time register performing subtraction operations.

IV. CIRCUIT OF TIME REGISTER AND THE SIMULATION RESULTS

The simplified circuit of the proposed time register, which performs addition and subtraction operations, is presented in Fig. 6.

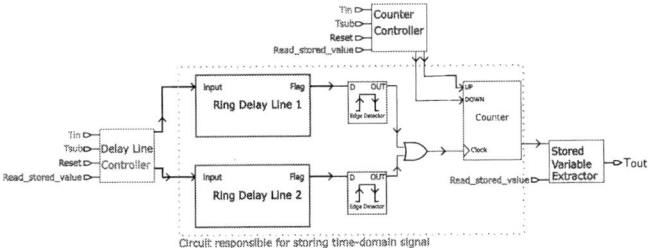

Fig. 6: Simplified circuit of Time Register.

The Table. I shows the result from an addition and subtraction operations. The proposed circuit was implemented using the AMS 350nm technology process with a supply voltage of VDD = 3.3V and simulated in Cadence Virtuoso. The ring delay line was 12 delay cells in cascaded unitary and a 4-bit counter its used.

The tests performed on the proposed time register with addition and subtraction operations demonstrate that the circuit functions correctly and exhibits low error rates across all operations conducted.

V. CONCLUSION

This work demonstrates that the proposed novel time register can perform both addition and subtraction operations in

Table I: Result of the operations on the Time Register

T_a (ns)	T_b (ns)	T_{SUB} (ns)	Expected (ns)	T_{out} (ns)	Error (%)
5.00	5.00	0.00	10.00	11.23	12.31
25.00	25.00	0.00	50.00	50.91	1.82
45.00	45.00	0.00	90.00	90.80	0.89
65.00	65.00	0.00	130.00	131.28	0.99
120.00	50.00	0.00	170.00	170.79	0.47
150.00	80.00	0.00	230.00	231.19	0.51
210.00	140.00	0.00	340.00	338.70	−3.22
340.00	0.00	300.00	40.00	39.23	−1.91
280.00	0.00	250.00	30.00	30.96	3.21
220.00	0.00	200.00	20.00	20.27	1.37
160.00	0.00	150.00	10.00	8.65	−13.43
120.00	50.00	10.00	160.00	161.59	0.99
150.00	80.00	60.00	170.00	170.22	0.12
180.00	110.00	110.00	180.00	181.09	0.60
210.00	140.00	160.00	190.00	191.08	0.57

the time-domain with high accuracy. It preserves the well-known advantages of time-mode circuits while offering superior performance over traditional time register designs [8]. A key benefit of the proposed approach is that its layout area remains constant, except for the counter size, which can be adjusted to increase the dynamic range.

VI. ACKNOWLEDGMENTS

This study was financed in part by Instituto SerraPilheira (grant number Serra – process 2211-42117), in part by Fundação de Amparo à Pesquisa de Minas Gerais (FAPEMIG, APQ-05837-23), in part by "National Council for Scientific and Technological Development – CNPq (process 404467/2024-5), and in part by Coordenação de Aperfeiçoamento de Pessoal de Nível Superior - Brasil (CAPES) - Finance Code 001.

REFERENCES

[1] G. W. Roberts and M. Ali-Bakhshian, "A Brief Introduction to Time-to-Digital and Digital-to-Time Converters," in IEEE Transactions on Circuits and Systems II: Express Briefs, vol. 57, no. 3, pp. 153-157, March 2010, doi: 10.1109/TCSII.2010.2043382.

[2] Zhu G, Yuan F, Khan G (2013) Time-Mode Approach for Mixed Analog-Digital Signal Processing. J Elec Electron 2: e109. doi:10.4172/2167-101X.1000e109.

[3] Asada, Kunihiro & Nakura, Toru & Iizuka, Tetsuya & Ikeda, Makoto. (2018). Time-domain approach for analog circuits in deep sub-micron LSI. IEICE Electronics Express. 15. 20182001-20182001. 10.1587/elex.15.20182001.

[4] Yuan, F. (2018), Design techniques of all-digital arithmetic units for time-mode signal processing. IET Circuits Devices Syst., 12: 753-763. https://doi.org/10.1049/iet-cds.2017.0327.

[5] Ali-Bakhshian, M.. (2011). Digital storage, addition and subtraction of time-mode variables. Electronics Letters. 47. 910-911. 10.1049/el.2011.1406.

[6] Panetas-Felouris, O., & Vlassis, S. (2022). A 3rd-Order FIR Filter Implementation Based on Time-Mode Signal Processing. Electronics, 11(6), 902. https://doi.org/10.3390/electronics11060902.

[7] P. S. Locatelli, D. M. Colombo and K. El-Sankary, "Time-Domain Multiply–Accumulate Unit," in IEEE Transactions on Very Large Scale Integration (VLSI) Systems, vol. 31, no. 6, pp. 762-775, June 2023.

[8] K. S. Kim, Y. H. Kim, W. S. Yu, and S. H. Cho, "A 7 bit, 3.75 ps resolution two-step time-to-digital converter in 65 nm CMOS using pulse-train time amplifier," IEEE J. Solid-State Circuits, vol. 48, no. 4, pp. 1009–1017, Apr. 2013.

Hybrid Lightweight Soft Error Mitigation Techniques for Edge Devices

Author: Jonas Gava
UFRGS
Porto Alegre, Brazil
jfgava@inf.ufrgs.br

Advisor: Ricardo Reis
UFRGS
Porto Alegre, Brazil
reis@inf.ufrgs.br

Co-advisor: Luciano Ost
Loughborough University
Loughborough, UK
l.ost@lboro.ac.uk

Abstract—The increasing deployment of artificial intelligence (AI) at the edge, particularly convolutional neural networks (CNNs) in resource-constrained devices, has created new challenges for ensuring system reliability and safety. Market analysts project a 21% annual growth rate in the edge AI market size over the next five years. These devices are being used in safety-critical applications such as autonomous vehicles, industrial control systems, and medical devices, where malfunctions due to radiation-induced soft errors can have severe consequences, ranging from degraded performance to life-threatening situations. Soft errors, caused by energetic particles, can corrupt data and instructions, resulting in unpredictable system behaviour. To meet safety standards in these domains, reliability engineers must proactively explore and implement efficient mitigation solutions during the initial design cycle.

Index Terms—soft error, reliability, mitigation techniques, edge devices

I. INTRODUCTION

The proliferation of connected Internet of Things (IoT) edge devices is rapidly increasing, with a growing number incorporating machine learning (ML) and artificial intelligence (AI) models, such as Convolutional Neural Networks (CNNs), for various applications [1]. These CNN models are playing increasingly critical roles in safety-sensitive autonomous systems, making their reliability paramount. This includes addressing the limitations imposed on these architectures from radiation issues [2]. However, ensuring the robustness and resilience of CNNs against radiation-induced soft errors, such as Single-Event Upsets (SEUs), is particularly challenging in resource-constrained edge environments, where traditional mitigation strategies become unsuitable.

While preceding works have extensively investigated soft error vulnerabilities in CNNs implemented on various platforms [3]–[5], including exploring fault-aware training and replication-based mitigation, challenges persist. To solve this, our contribution is:

- **Proposal and Implementation of RAT (Register Allocation Technique):** A lightweight, compiler-based mitigation technique implemented using the LLVM infrastructure, designed to reduce the soft error vulnerability of applications running on resource-constrained devices.
- **Proposal and Implementation of MATE (Memory Allocation Technique):** A memory allocation technique that protects critical data stored in memory.

- **Integration and Validation within the SOFIA Framework:** Integration of the mitigation techniques into the SOFIA framework [6] for automated soft error analysis and mitigation, enabling comprehensive evaluation across different architectures and software stacks.
- **Experimental Validation through Fault Injection and Neutron Radiation Testing:** Rigorous validation of RAT and MATE through extensive fault injection campaigns using simulation [5] and accelerated neutron radiation testing [7], [8] on representative embedded platforms.

II. METHODOLOGY

This work employed a multi-faceted approach to evaluate the soft error resilience of edge computing systems, combining simulation-based fault injection with real-world neutron irradiation experiments. Simulation-based assessments were conducted using the SOFIA framework [6]. SOFIA facilitates fault injection by emulating single-bit upsets in processor registers and memory, enabling the systematic analysis of error propagation and application behaviour under a wide range of fault conditions. To complement the simulation studies and provide a realistic assessment of soft error effects, neutron irradiation tests were carried out at the Laboratoire de Physique Subatomique et de Cosmologie (LPSC) and the Institut Laue-Langevin (ILL). These experiments involved exposing target systems, running representative edge applications, to controlled neutron fluxes and monitoring for error events.

III. PROPOSED MITIGATION TECHNIQUES

This work makes significant contributions in the area of soft error mitigation for resource-constrained embedded systems, with a particular focus on enhancing the reliability of CNN-based applications. The key contributions are centered around the development and validation of two novel techniques: RAT and MATE, and their integration within the SOFIA framework.

A. RAT Contributions:

1. Proposal and implementation of a lightweight, compiler-based mitigation technique called Register Allocation Technique (RAT) [9]. RAT restricts the number of available registers for specific functions to minimise the number of vulnerable registers. Unlike techniques that require

code redundancy, RAT focuses on limiting the processor's exposed area to radiation-induced soft errors.

2. RAT validation for applications running Linux on top of a multi-core processor through fault injection [6], [9].

3. RAT validation for CNN applications running on resource-constrained devices through fault injection [5].

4. RAT validation under neutron radiation for a 7-layer CNN application [7]. Experiments conducted at neutron radiation facilities confirmed that RAT can reduce the number of critical faults, exhibiting its effectiveness in mitigating radiation-induced errors.

Experiments demonstrate that hardened bare-metal applications lacking external dependencies exhibit notable soft error reliability improvements due to complete code access. Conversely, code protection is less effective for typical Linux applications. For ML algorithms, partial TMR offers similar improvements to full TMR, but with up to 50% less performance overhead. CNN application results indicate that replication techniques may be unsuitable for resource-constrained platforms, necessitating lightweight alternatives. Applying RAT to a 7-layer CNN inference model on a resource-constrained device under neutron radiation improved reliability against critical faults by 50%, with minor increases in memory usage and execution time (approximately 30%).

B. MATE Contributions:

1. Proposal and implementation of MATE [8], a novel semi-automated, compiler-based memory allocation technique. MATE facilitates assigning the most vulnerable data to any memory region equipped with parity check or error correction code (ECC) mechanisms, enhancing resilience against soft errors in resource-constrained devices.

2. First experimental assessment of the impact of allocating CNN weights and biases in the SRAM parity checking area on the soft error reliability of an inference CNN model running on a commercial off-the-shelf (COTS) Arm Cortex-M4 microprocessor under 14-MeV neutron radiation.

MATE is a novel semi-automated compiler-based memory allocation technique implemented in LLVM 12.0.1, designed to enhance the soft error reliability of CNNs operating in resource-constrained IoT devices. MATE enables the allocation of the most vulnerable data to memory regions protected by parity check or error correction code (ECC) mechanisms. Authors in [8] present the first experimental assessment of this approach, evaluating the impact of allocating weights and biases to an SRAM parity checking area on a CNN inference model running on a commercial off-the-shelf (COTS) Arm Cortex-M4 microprocessor under 14-MeV neutron radiation. Experimental results, obtained from over 80,000 runs per case-study scenario, demonstrate that MATE can achieve up to a 90% reduction in critical Silent Data Corruption (SDC) events with negligible runtime overhead (approximately 0.3%) and minimal energy consumption overhead (approximately 4.4-4.7%). The findings suggest that MATE offers a viable, low-cost, and rapidly deployable method for improving the reliability of CNN inference models in safety-critical embedded systems without requiring extensive expertise in CNN internals.

IV. CONCLUSION AND FUTURE WORK

This research demonstrated the viability of lightweight and hybrid system-level approaches for mitigating soft errors in resource-constrained edge devices. By combining the Register Allocation Technique (RAT) and Memory Allocation Technique (MATE) with the SOFIA framework, this work provides a practical and efficient solution for enhancing the reliability of edge computing systems without compromising performance or energy efficiency.

Numerous avenues for future work stem from this research. One promising direction involves the development of a machine learning (ML) model to predict the optimal soft error mitigation technique for a given application and target architecture. By leveraging application characteristics such as function profiles, register usage, and instruction profiles, this ML model could automate the selection of the most effective mitigation strategy, minimising development effort and maximising system reliability.

Another key area for future investigation involves validating these findings in real-world deployments. This includes integrating a CNN-based application into a drone platform and assessing its soft error resilience under realistic operating conditions. Furthermore, future work should focus on porting the developed mitigation techniques (RAT and MATE) to newer LLVM compiler versions (21.0+ and beyond).

REFERENCES

[1] Z. L. Zaid Saeb Sabri, "Low-cost intelligent surveillance system based on fast CNN," *PeerJ. Computer science*, vol. 7, no. e402, pp. 1–23, 2021.

[2] C. Slayman, "JEDEC Standards on Measurement and Reporting of Alpha Particle and Terrestrial Cosmic Ray Induced Soft Errors," in *Soft Errors in Modern Electronic Systems*, Aug. 2011, pp. 55–76.

[3] H.-B. Wang *et al.*, "Impact of Single-Event Upsets on Convolutional Neural Networks in Xilinx Zynq FPGAs," *IEEE Transactions on Nuclear Science*, vol. 68, no. 4, pp. 394–401, April 2021.

[4] F. F. Dos Santos *et al.*, "Characterizing a Neutron-Induced Fault Model for Deep Neural Networks," *IEEE Transactions on Nuclear Science*, vol. 70, no. 4, pp. 370–380, Apr. 2023.

[5] G. Abich *et al.*, "Applying lightweight soft error mitigation techniques to embedded mixed precision deep neural networks," *IEEE Transactions on Circuits and Systems I: Regular Papers*, vol. 68, no. 11, pp. 4772–4782, 2021.

[6] J. Gava *et al.*, "Sofia: An automated framework for early soft error assessment, identification, and mitigation," *Journal of Systems Architecture*, vol. 131, p. 102710, 2022.

[7] J. Gava, A. Hanneman, G. Abich, R. Garibotti, S. Cuenca-Asensi, R. P. Bastos, R. Reis, and L. Ost, "A lightweight mitigation technique for resource-constrained devices executing DNN inference models under neutron radiation," *IEEE Transactions on Nuclear Science*, vol. 70, no. 8, pp. 1625–1633, 2023.

[8] J. Gava *et al.*, "The Impact of Allocating Weights and Biases in SRAM Parity Checking Areas on the Soft Error Reliability of CNN Inference Models," in *Conference on Radiation Effects on Components and Systems (RADECS 2024)*, 2024.

[9] J. Gava, R. Reis, and L. Ost, "Rat: A lightweight architecture independent system-level soft error mitigation technique," in *IFIP/IEEE International Conference on Very Large Scale Integration-System on a Chip.* Springer, 2020, pp. 235–253.

Cross-Layer Approximate Hardware Design of Interpolation Filters for Fractional Motion Estimation in Versatile Video Coding

Author: Rafael da Silva
UFRGS
Porto Alegre, Brazil
rsilva@inf.ufrgs.br

Advisor: Ricardo Reis
UFRGS
Porto Alegre, Brazil
reis@inf.ufrgs.br

Co-advisor: Mateus Grellert
UFRGS
Porto Alegre, Brazil
mateus.grellert@inf.ufrgs.br

Abstract—The fact that we can stream video on multiple devices in our homes, on the go using mobile devices, or even while video chatting across the globe, even with low bandwidth, is owed to video coding. Versatile Video Coding (VVC) is the latest video coding standard, which introduces a range of innovative tools. One example is adopting an alternative fractional motion estimation (FME) filter that is part of the Advanced Motion Vector Resolution (AMVR) extension. This work introduces a low-power hardware architecture accelerator specifically designed for Fractional Motion Estimation (FME) with support for the AMVR extension of VVC.

Index Terms—Versatile Video Coding, Approximate Computing, Cross-layer, Approximate Adders, VLSI.

I. INTRODUCTION

Motion Estimation (ME) and Motion Compensation (MC) are essential to reduce temporal redundancy in VVC. ME accounts for about 65% of the total encoder complexity [4], driven by advanced models such as Fractional Motion Estimation (FME) and Affine Motion Estimation (AME), which enable sub-pixel accuracy. FME, which is responsible for interpolating integer samples from reference frames, is particularly computationally and power-intensive.

The current version of the VVC standard incorporates the Adaptive Motion Vector Resolution (AMVR) scheme, which operates at the Coding Unit (CU) level [1], allowing for more precise motion vector representation by adapting the resolution of Motion Vector Differences (MVD) based on the specific requirements of the video content. The concept of a switched alternative Gaussian interpolation filter for the half-pixel position can be considered an AMVR extension. This approach allows different luma half-pixel interpolation filters to be selected at the CU level, enhancing motion estimation accuracy and flexibility. In AMVR mode, an alternative half-pixel interpolation filter is permitted, specifically designed for integration with the AMVR scheme [3].

With this in mind, the main goal of this thesis is to explore techniques of approximate computing in the context of video coding, more specifically, hardware design for FME interpolation, as this stage is particularly critical due to the data dependencies involved. This trade-off must also consider image quality, where a tolerable reduction in accuracy may

be acceptable in exchange for significant hardware efficiency gains. To accomplish this claim, the following goals were defined:

- Propose cross-layer approximate architectures at different abstraction levels (i) reduction of coefficients, (ii) approximate operators, and (iii) hierarchical gate-level pruning. And with that, to ease the design space exploration and analyses of trade-offs;
- Evaluate and explore the most suitable approximate adders for inclusion in the interpolation filters, aiming to optimize performance, area, and power efficiency;
- Provide an interpolation architecture with multiple versions, varying in approximation levels and precision, to enable evaluation across different points in the power-quality trade-off.

A. Thesis Contributions

Based on the goals, several explorations have been made to support the thesis claim. Along with data gating, exploring approximate computation is one of the main strategies to mitigate power, energy, and timing issues, as they are designed solely for specific applications, such as Application-Specific Integrated Circuits (ASICs) and Field-Programmable Gate Arrays (FPGAs). Initially, we developed a baseline architecture with precise filters to study the acquired achievements and compare them to the state-of-the-art architecture. This architecture enables the analysis of power, area, and accuracy trade-offs. This architecture was extended to include the approximation (i) reduction of filter coefficients and (ii) approximation of arithmetic operators. As a case study, a set of approximate adders at different approximation levels was analyzed to search for the best combination that can be adapted to the best results in terms of area, memory, and power consumption.

The contributions acquired from this work so far include the following:

- Novel VLSI dedicated architecture that supports both regular filters and the alternative filter in parallel, controlled by the data gating technique;

- A set of approximate interpolation filters, 4T, 4TAxA, 2T, and 2TAxA, significantly reduces power dissipation and area in the VLSI circuit.

II. PROPOSAL DEVELOPMENT

This work develops a hardware architecture for the VVC FME interpolation stage, incorporating cross-layer approximation techniques to reduce computational complexity. The proposed Multifilter Interpolation Unit (MIU) supports the standard set of FIR filters and an alternative Gaussian filter, which is dynamically selected based on the input motion vector.

We first proposed this architecture in [2], because this architecture was designed to support any VVC interpolation filter and 8×8 up to 128×128 sized blocks. This proposal is the adaptation of the Lower-Part-OR Adder (LOA) adder in the baseline and the 4T (4-tap) approximate filter, and the second is a 4TxA (4-tap + approximate adder) architecture proposal. In other words, it is a cross-layer that comes from coefficient reduction combined with an approximate adder. Both proposals have a precise alternative Gaussian filter.

In the work [5], we propose exploring the alternative Gaussian filter with this approximate 4T, proposing an approximate version of the alternative filter for 4-tap, forming an approximate 4T/4T architecture (4-tap regular filters and 4-tap alternative filter).

A cross-layer methodology is applied at both the software and hardware levels to reduce filtering complexity. In the software stage, approximations are simulated within the VTM reference encoder to evaluate their impact on the coding quality. In the hardware stage, the architecture integrates analytical models of approximate adder (AA) topologies with different approximation levels (k), representing the percentage of approximate bits. Two reduced complexity versions of the regular luma interpolation filters are proposed: a 4-tap [7] and a 2-tap [6] configuration.

Approximate computing (AxC) principles are applied to arithmetic units to trade computational accuracy for area, power, and speed gains. Given the prevalence of arithmetic operations in VLSI designs, such strategies offer significant efficiency improvements. The evaluated adders include Approximate Adder 2 (AA2), Copy Adder, Error-Tolerant Adder I (ETA-I), Lower-Part-OR Adder (LOA), Lower-Part-XOR Adder (LXOA), and Truncation Adder.

The experimental results demonstrated that the proposed designs achieved significant hardware savings. Depending on the configuration and approximation level, power reductions ranged from 69.20% to 96.19%, while area savings reached up to 91.13%, all while maintaining acceptable video reconstruction quality (e.g., $SSIM \geq 0.88$, which can be interpreted as a fair perception of image quality according to [8]). These findings validate the potential of cross-layer AxC as a viable design paradigm for high-performance video encoders in energy-constrained environments such as mobile and edge devices.

III. CONCLUSION

A key contribution of this research was the design and evaluation of multiple interpolation filter architectures that incorporate approximation at different levels of abstraction. The work progressed from precise baseline implementations to include approximations such as tap reduction (e.g., from 6-tap to 4-tap and 2-tap configurations), and the integration of approximate arithmetic units such as LOA, ETA-I, and LXOA. These elements were combined to create hybrid architectures—such as the 4TAxA, 2TAxA/4T, and 2TAxA/2TAxA—which were systematically analyzed in terms of power consumption, area, and video quality metrics (e.g., SSIM and PSNR), using an ASIC synthesis flow.

For future work, contributing to the work already developed in the PhD and expanding upon the techniques introduced so far involves exploring gate-level approximations through indiscriminate pruning of synthesized netlists. A recent tool, AxLS (available online), enables such gate-level pruning. This tool operates on synthesized netlists generated from standard-cell libraries, accompanied by a file describing the switching activity of each gate. AxLS supports two main pruning techniques:

- *inOuts*, which removes inputs specified by the user, thus eliminating dependent gates;
- *probprun*, automatically prunes gates with the lowest switching activity, based on specified thresholds.

Pruning at the netlist level can result in approximations that are difficult to obtain in earlier design stages, primarily because designers have limited control over how synthesis tools translate RTL descriptions into gate-level implementations.

REFERENCES

[1] J. Chen, Y. Ye, and S. Kim. "Algorithm description for versatile video coding and test model 5 (VTM 5)". In: *JVET-M1002, Joint Video Exploration Team (JVET)*. 14th Meeting: Geneva, CH: Joint Video Experts Team (JVET), Document JVET-N1002, 19-27 March 2019.

[2] R. Da Silva, M. Grellert, and R. Reis. "An energy-efficient interpolation unit targeting VVC encoders with approximate adder". In: *2023 36th SBC/SBMicro/IEEE/ACM Symposium on Integrated Circuits and Systems Design (SBCCI)*. IEEE. 2023, pp. 1–5.

[3] A. Henkel, I. Zupancic, B. Bross, M. Winken, H. Schwarz, D. Marpe, et al. "Alternative half-sample interpolation filters for versatile video coding". In: *ICASSP 2020-2020 IEEE International Conference on Acoustics, Speech and Signal Processing (ICASSP)*. IEEE. 2020, pp. 2053–2057.

[4] F. Pakdaman, M. A. Adelimanesh, M. Gabbouj, and M. R. Hashemi. "Complexity analysis of next-generation VVC encoding and decoding". In: *2020 IEEE International Conference on Image Processing (ICIP)*. IEEE. 2020, pp. 3134–3138.

[5] R. da Silva, M. Grellert, and R. Reis. "Energy-Efficient Design of Approximate VVC Interpolation Filters Units". In: *2024 IEEE Computer Society Annual Symposium on VLSI (ISVLSI)*. 2024, pp. 63–68.

[6] R. da Silva, P. T. Pereira, M. Grellert, and R. Reis. "Cross-Layer Approximate Design of Low-Power Fractional Motion Estimation Accelerators for VVC". In: *IEEE Transactions on Very Large Scale Integration (VLSI) Systems* (2025).

[7] R. da Silva, P. T. Pereira, M. Grellert, and R. Reis. "Energy-Efficient Interpolation Filter Design for VVC Encoders Using Cross-Layer Approximate Computing". In: *Journal of Integrated Circuits and Systems* 19.3 (2024), pp. 1–11.

[8] T. Zinner, O. Hohlfeld, O. Abboud, and T. Hoßfeld. "Impact of frame rate and resolution on objective QoE metrics". In: *2010 second international workshop on quality of multimedia experience (QoMEX)*. IEEE. 2010, pp. 29–34.

A Case Study on the Migration of a High-Level PI Controller to ASIC-Compatible HDL Representation

Francisca Donoso, *Student Member, IEEE*, Nelson Salvador, *Student Member, IEEE*,
Jorge Marin, *Member, IEEE*, Christian A. Rojas, *Senior Member, IEEE*, Gonzalo Carvajal, *Member, IEEE*.

Abstract—This article presents a structured methodology for migrating digital control algorithms developed in high-level languages into synthesizable HDL suitable for ASIC implementation. A proportional–integral controller for a three-level flying capacitor converter is used as a representative case study to demonstrate the design flow. The methodology encompasses initial modeling using floating-point arithmetic, conversion to fixed-point representation, generation of reference models for validation, translation to C/C++, and high-level synthesis for HDL generation. Preliminary results consider co-simulation to ensure behavioral consistency between the high-level description and the generated HDL. The HDL was successfully synthesized using the open-source LibreLane tool-chain targeting the SkyWater SKY130 PDK. Ongoing work includes FPGA-based prototyping and detailed characterization of the ASIC. The proposed approach establishes a verified and systematic pathway for transitioning high-level control logic to ASIC-ready hardware implementations, preserving the original algorithmic behavior across abstraction levels.

Index Terms—Fixed-Point Control, HDL Generation, High-Level Synthesis, ASIC Design Flow, FPGA Prototyping.

I. INTRODUCTION

The migration of high-level digital control algorithms to hardware description languages (HDLs) is essential for achieving efficient and reliable implementations on FPGAs and ASICs [1]. Early-stage development often relies on tools like MATLAB/Simulink due to their flexibility and rapid iteration, but these environments typically use floating-point arithmetic, which is costly in hardware [2].

Fixed-point arithmetic offers a practical bridge between algorithm design and hardware realization. It enables significant reductions in resource usage and better integration within constrained digital systems. When using high-level synthesis (HLS) tools such as AMD's Vitis HLS, adopting fixed-point representations also avoids reliance on proprietary, non–open-source floating-point IP blocks, ensuring compatibility with open-source ASIC design flows.

This work presents a case study on the migrating of a proportional–integral (PI) controller for a three-level flying capacitor converter (3L-FCC) [3] from a high-level, floating-point model to a fixed-point HDL implementation. The design flow emphasizes functional equivalence across abstraction levels and prepares the system for FPGA prototyping and eventual ASIC integration.

II. DESCRIPTION OF THE CASE STUDY

The case study considers a three-level flying capacitor converter (3L-FCC) regulated by two decoupled PI controllers.

This work was supported by ANID Fondecyt de Exploración 13250147 and by AC3E under project ANID/AFB240002.

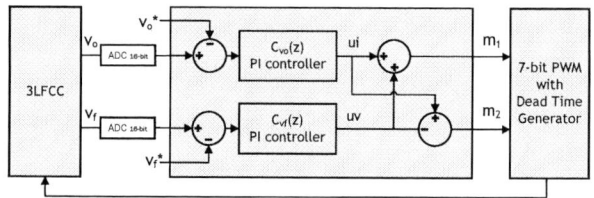

Fig. 1. High-level block diagram of the complete control system, including ADC interfacing and PWM modulation with deadtime.

The control scheme, including its inputs and outputs, is illustrated in Figure 1. The details of the 3L-FCC module and the requirements of the controller can be found in [3]. The module processes the measured output voltage (v_o) and the flying capacitor voltage (v_f), together with their corresponding reference values (v_o^* and v_f^*). Each PI block operates independently: the first drives the output voltage to follow its reference, while the second ensures proper voltage balancing across the flying capacitor. Both controllers incorporate anti-windup mechanisms and output saturations, which prevent integrator accumulation and ensure stable operation under different conditions. These computations produce two modulation indices (m_1 and m_2), which serve as duty-cycle commands for the power stage switches.

III. FIXED-POINT CONTROLLER DESIGN

The controller inputs and outputs are encoded in fixed-point formats selected to match the hardware constraints of the planned test setup. Specifically, the input type (*in_type*) corresponds to a 16-bit representation consistent with the resolution of the ADC, ensuring direct mapping from measurements without additional rescaling. The controller outputs are defined as 7-bit values (*out_type*), aligned with the modulation scheme used in the simulation model.

Internal computations share a common fixed-point type (*intern_type*), defined as a 32-bit signed format with 1 sign bit, 8 integer bits, and 23 fractional bits. This intentionally overprovisioned configuration minimizes quantization effects in the initial implementation while leaving headroom for later optimization. Table I summarizes the fixed-point configuration of the main data types. In addition to word length and fractional allocation, specific rounding and overflow rules were defined to ensure deterministic behavior. Rounding-to-nearest avoids bias accumulation in repetitive operations, while saturation on overflow prevents wrap-around effects that could destabilize the closed-loop system. These choices contribute to keeping

979-8-3315-9813-6/25 $31.00 © 2025 IEEE

TABLE I
FIXED-POINT CONFIGURATION FOR DIFFERENT DATA TYPES

Parameter	in_type	intern_type	out_type
Signed	no	yes	no
Word Length	16	32	7
Fractional Bits	0	23	0
Rounding Mode	round	round	round
Overflow Action	saturate	saturate	saturate

the controller response predictable and deterministic across the operating range, although alternative configurations may also be viable depending on the design constraints.

IV. PROPOSED DESIGN FLOW

The proposed design flow follows a structured sequence that begins with a high-level functional floating-point description of the control algorithm and concludes with HDL files suitable for FPGA prototyping or ASIC integration. The details for each step of the design flow are described below:

1) **High-Level Floating-Point Model:** The starting point is a C-based control implementation inside PLECS[1], operating in closed loop with the converter plant model. All computations are performed in floating-point arithmetic, allowing rapid validation of the control algorithm and plant interaction.

2) **Floating-Point Equivalent:** An equivalent floating-point MATLAB implementation of the controller was developed and validated against a trusted baseline model obtained from PLECS. This step ensures that the MATLAB model reproduces the exact behavior of the original PLECS/C implementation. Using MATLAB's tools for fixed-point representation, a version emulating the controller's behavior in fixed-point arithmetic is created within MATLAB, where preliminary word length, fractional precision, rounding mode, and overflow action, among other parameters, were configured.

3) **Fixed-Point Verification:** The fixed-point MATLAB implementation is embedded as a MATLAB Function block within a Simulink model, connected to the PLECS Blockset plant. This enables closed-loop simulation with the selected fixed-point settings and confirms that the plant remains properly controlled. Feedback between Steps 2 and 3 allows iterative tuning of the fixed-point configuration.

4) **Migration to HLS:** The fixed-point configuration obtained in Step 3 is manually transferred to Vitis HLS by defining equivalent data types from its fixed-point library, ensuring that all word length and precision parameters are preserved. A reference-based verification is then performed between the MATLAB fixed-point implementation and the Vitis HLS model, confirming functional equivalence. Once validated, Vitis HLS is used to run simulation, synthesis, and co-simulation within

[1] PLECS (Piecewise Linear Electrical Circuit Simulation) is a tool for system-level simulation of power electronic circuits.

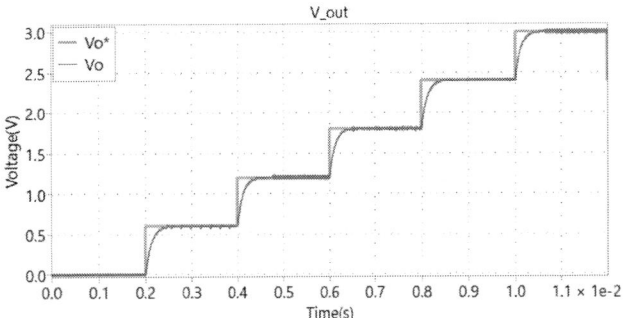

Fig. 2. Simulated output voltage of the converter and its reference under the fixed-point controller.

the AMD/Xilinx environment, producing HDL files for FPGA prototyping or ASIC integration.

The use of these software environments throughout the flow enables rapid iteration and verification of fixed-point configurations in a software-in-the-loop (SIL) context before committing to hardware synthesis.

V. CURRENT RESULTS AND VALIDATION STATUS

Validation has been performed through simulations in PLECS, MATLAB, and Vitis HLS, showing consistency with a trusted reference model. An example of this behavior is illustrated in Figure 2, where the converter output voltage is compared against its reference under the fixed-point controller. FPGA prototyping is in progress to evaluate closed-loop performance in hardware. As an additional step, the design was synthesized with the open-source LibreLane flow [4] using the SkyWater SKY130 PDK. The implementation achieved an instance area of $145\,927\ \mu m^2$, meeting timing requirements for a 10 ns (100 MHz) target clock in typical conditions.

VI. CONCLUSION AND FUTURE WORK

The case study demonstrates that the proposed methodology enables the migration of high-level controllers into HDL suitable for chip design flows. Using the SKY130 PDK, the synthesized implementation reached an area of $145\,927\ \mu m^2$ while satisfying timing at a 100 MHz clock target. These results highlight the applicability of the approach for practical chip implementation. Future work will extend validation to FPGA hardware and explore fixed-point optimizations to further reduce area while maintaining performance, as well as integration with the analog power stage in a mixed-signal ASIC.

REFERENCES

[1] D. Navarro, Lucía, L. A. Barragán, I. Urriza, and Jiménez, "High-level synthesis for accelerating the fpga implementation of computationally demanding control algorithms for power converters," *IEEE Transactions on Industrial Informatics*, vol. 9, no. 3, pp. 1371–1379, 2013.

[2] J. B. Romaine, T. I. Ashley, and M. P. Martín, "Digital fixed-point low powered area efficient function estimation for implantable devices," *IEEE Access*, vol. 10, pp. 70 793–70 805, 2022.

[3] R. L. Carmona, I. A. Acosta, F. J. Vargas, and C. A. Rojas, "On the multivariable control of a three-level flying capacitor buck dc-dc converter," in *2024 IEEE ANDESCON*, 2024, pp. 1–6.

[4] M. Shalan and T. Edwards, "Building openlane: A 130nm openroad-based tapeout-proven flow : Invited paper," in *2020 IEEE/ACM International Conference On Computer Aided Design (ICCAD)*, 2020, pp. 1–6.

Transistor Placement for Automatic Cell Layout Generation on Advanced Nodes: A Review

Vitor H. Fuerstenau, and Ricardo Reis
UFRGS, Porto Alegre, Brazil
Email: vhsfmaciel@inf.ufrgs.br, reis@inf.ufrgs.br

Abstract—**As technology nodes advance, design rules restrictions are becoming increasingly complex due to physical limitations. This has made automatic cell layout generation a crucial and rapidly evolving area of study. A core stage of the layout design flow is the placement of transistors, which directly impacts the intracell routing and overall layout characteristics. This work presents a concise review of the different approaches that tackle the transistor placement problem to meet cell design goals.**

Index Terms—**cell design automation, layout generation, standard cells, transistor placement, EDA, VLSI**

I. Introduction

In digital Very-Large-Scale Integration (VLSI) design, standard cells serve as the atomic units of logic implementation, with their predefined layouts and timing models enabling automated place-and-route (PnR) tools to construct higher-level functional blocks within a chip. Given their role, its key attributes—including placement efficiency, cell area, pin accessibility, and routing congestion—directly influence the overall performance, power efficiency, and area of the final synthesized project. Optimization in the design of individual cells emerges as a way for enhancing the quality and efficiency of the entire chip, while also enabling more efficient architecture exploration in Design-Technology Co-Optimization (DTCO) flows [1].

Advanced nodes, such as FinFETs and Gate-All-Around FETs, requires strict design rules to be obeyed, and manually creating layouts presents a highly laborious and time-consuming challenge. Cell synthesis tools address this problem [2], but the design of more complex standard cells—which are often the most critical for achieving high-performance and power-efficient designs—typically demands significant computational resources.

The placement of transistors is the foundation step of the layout design, directly affecting frameworks efficiency. Consequently, understanding and exploring this problem is essential to progress towards techniques that quickly generate optimized designs.

II. Transistor Placement

A. The Problem

Given a CMOS circuit, the transistor placement consists of organizing its devices in a certain order on a limited space, where the problem can be of one or two dimensions (single- or multi-height layouts).

Key techniques to spare routing and area in transistor placement are diffusion abutment of devices that share a same net as well as gate alignment of PMOS and NMOS transistors which possess the same gate net, since the transistors terminals positions determines the initial routing points and minimum layout size, impacting on pin accessibility, required number of metal tracks, vias and parasitic resistance.

B. Cell Synthesis Frameworks

Cell layout generation researches can be split into two different approaches, sequential and simultaneous transistor place-and-route. Simultaneous flow consists of concurrently optimizing transistor placement and internal cell routing, while in the sequential flow these are done in independent stages that can interact with each other.

Table I shows the state-of-the-art frameworks for sub 10nm nodes and its respective characteristics. The only simultaneous framework, SP&R [3], frames the placement problem as a multi-objective constraint satisfaction, leveraging an SMT solver to model it into a set of mathematical rules regarding transistor positions, diffusion breaks and diffusion heights, and those equations are appended to routing and pin allocation clauses to be computed by the solver.

The biggest barrier of simultaneous approach is that exponential computation is required to contemplate the search space due to its NP-complete order and high number of variables. As illustrated in Table I, a scan-flop cell with 36 FETs and a predictive technology node-which, when compared to real technologies, only considers a small set of basic design rules-can take up to 9 hours to run. Even considering search-space reduction techniques, the complexity of simultaneous methods restricts their use to simple cells while also introduce the risk of generating suboptimal layouts or even prematurely discarding superior solutions.

Sequential approach is the predominantly one in recent researches, due to the fact that splitting the flow into NP-complete problems with less variables reduces the computational challenge. However, simply generating a placement that minimizes area and attempting to route it afterward is inefficient, due to the risk that the configuration might not be routable, wasting computation. Besides, the final result could be significantly inferior to that achieved by a simultaneous optimization flow, which considers multiple objectives. When considering multi-objective optimization, placement design space becomes exceedingly large, making exhaustive exploration impractical and requiring predictive or pruning methodologies.

TABLE I
CELL SYNTHESIS TOOL CHARACTERISTICS. RUNTIME CONSIDERING PLACEMENT.

Tool	PnR approach	Tech node	Max Runtime	Multi-Height	#FETs
SP&R [3]	Simultaneous	7nm Predictive	≈ 9hr (SDFFQSx1)	No	36
BonnCell [4]	Sequential	7nm Real	≈ 8hr (SDFFQSx1)	Yes	44
CSyn-fp [5]	Sequential	7nm Predictive	≈ 4s (SDFL/SDFH)	No	32
Nvidia [6]	Sequential	7nm Real	≈ 27hr (Flop51/Latch7)	No	114
iTPlace [7]	Sequential	7nm Predictive	N/A	No	32
NCTUCell [8]	Sequential	7nm Predictive	≈ 85s (SDF/ICG)	No	36
MCell [9]	Sequential	7nm Predictive	≈ 81min (MBFF4-bits)	Yes	104

The known sequential frameworks to work with real 7nm technology, considering whole complex design rule decks, are BonnCell [4] and Nvidia NVCell [6]. The first tool places transistors from left to right with a feedback system, where placement is partial and incremental, with routing being used to test whether the partial solution is routable with the objective of pruning infeasible solutions early. However, it is highly time consuming to avail candidates, where for sequential cells, on average, 75% of the framework running time is spent on transistor placement. As for Nvidia last work, it is based on the optimization algorithm Simulated Annealing with a pin density aware congestion metric, leveraging a transformer model based clustering methodology to help the algorithm to converge to an optimal solution, addressing up to 114 devices. While both tools achieve compact layout solutions, running time is concerning when compared to other studies with predictive technology. Additionally, BonnCell can not find adequate solutions for multi-height and NVCell do not consider this feature in its research.

The tools CSyn-fp [5], iTPlace [7], NCTUCell [8] and MCell [9], are designed to generate layouts for predictive advanced technologies, each using a different strategy. CSyn-fp [5] combines a search tree-based design space exploration algorithm with a set of effective speeding up techniques that can create relatively intricate sequential cell layouts in 4 seconds without losing optimality, also incorporating a in-cell routability estimation metric to evaluate and select the best placement solutions. NCTUCell introduces a routability-aware transistor placement algorithm that considers design rule constraints during the placement phase. By integrating a dynamic programming-based approach with implicit grid adjustments, it aims to generate placements that are inherently more routable. This preemptive consideration of routing constraints helps in reducing the likelihood of encountering routing issues later in the design process.

Similarly to CSyn, iTPlace generates a list of candidate placements using a search tree based on a reward function, but it incorporates a machine learning based on router behavior data to filter out unroutable placements, enabling to focus computational resources on the most promising placement options. This is confirmed by its better delay performance score when compared to the predictive 7nm library and NCTUCell. Finally, MCell [9] is the only one that focuses on multi-row cell layouts, employing an A*-based placement algorithm to optimize intra-row and inter-row connections, trying to increase placement routability by minimizing an objective function.

The challenge for assessing predictive technology tools placement algorithms efficiency lies on the lack of transistors, design rules and characterized cells, being inconclusive if the same techniques would be more successful than [4], [6] considering the same restrict design constraints. Nonetheless, combining speed-up techniques with routability estimation indicates a promising research path for higher performance, while the lack of works that explore multi-height cells is also an opportunity to stress search tree based algorithms.

III. CONCLUSION

This work presented the transistor placement, a foundational step in layout automation that profoundly impacts design, and cell synthesis tools approaches to it. We discussed the problem and surveyed various frameworks algorithms, highlighting their strengths and limitations, comparing its characteristics. As technology evolves and design rules increase in quantity and complexity, the challenge of placing transistors will remain a critical area for innovation, continuously driving the need for sophisticated solutions.

REFERENCES

[1] T. Jhaveri, V. Rovner, L. Liebmann, L. Pileggi, A. J. Strojwas, and J. D. Hibbeler, "Co-Optimization of Circuits, Layout and Lithography for Predictive Technology Scaling Beyond Gratings," *IEEE Transactions on Computer-Aided Design of Integrated Circuits and Systems*, vol. 29, no. 4, pp. 509–527, 2010.

[2] J. Zhou, Z. Luo, Z. Li, H. You, M. He and Z. Zhao, "A Survey of Standard Cell Layout Design," 2025 International Symposium of Electronics Design Automation (ISEDA), Hong Kong, China, 2025, pp. 554-559, doi: 10.1109/ISEDA65950.2025.11101049.

[3] D. Lee et al., "SP&R: SMT-Based Simultaneous Place-and-Route for Standard Cell Synthesis of Advanced Nodes", in IEEE Transactions on Computer-Aided Design of Integrated Circuits and Systems, vol. 40, no. 10, pp. 2142-2155, Oct. 2021, doi: 10.1109/TCAD.2020.3037885.

[4] P. Van Cleeff, S. Hougardy, J. Silvanus, and T. Werner, "BonnCell: Automatic Cell Layout in the 7-nm Era," IEEE Transactions on Computer-Aided Design of Integrated Circuits and Systems, vol. 39, no. 10, pp. 2872–2885, Oct. 2020. doi: 10.1109/TCAD.2019.2962782.

[5] K. Baek and T. Kim, "CSyn-fp: Standard Cell Synthesis of Advanced Nodes With Simultaneous Transistor Folding and Placement," in IEEE Transactions on Computer-Aided Design of Integrated Circuits and Systems, vol. 43, no. 2, pp. 627-640, Feb. 2024, doi: 10.1109/TCAD.2023.3320631.

[6] Chia-Tung Ho, Ajay Chandna, David Guan, Alvin Ho, Minsoo Kim, Yaguang Li, and Haoxing Ren. 2024. Novel Transformer Model Based Clustering Method for Standard Cell Design Automation. In Proceedings of the 2024 International Symposium on Physical Design (ISPD '24). Association for Computing Machinery, New York, NY, USA, 195–203.

[7] T.-C. Lee, C.-Y. Yang, and Y.-L. Li, "iTPlace: Machine Learning-Based Delay-Aware Transistor Placement for Standard Cell Synthesis," in 2020 IEEE/ACM International Conference On Computer Aided Design (ICCAD), 2020, pp. 1–8.

[8] Y.-L. Li, S.-T. Lin, S. Nishizawa, H.-Y. Su, M.-J. Fong, O. Chen, and H. Onodera, "NCTUcell: A DDA- and Delay-Aware Cell Library Generator for FinFET Structure With Implicitly Adjustable Grid Map," IEEE Transactions on Computer-Aided Design of Integrated Circuits and Systems, vol. 41, no. 12, pp. 5568–5581, 2022.

[9] Y.-L. Li, S.-T. Lin, S. Nishizawa, and H. Onodera, "MCell: Multi-Row Cell Layout Synthesis with Resource Constrained MAX-SAT Based Detailed Routing," in 2020 IEEE/ACM International Conference On Computer Aided Design (ICCAD), 2020, pp. 1–8.

AUTHOR INDEX

Acharya, R. 100
Alarcón, L.P. 157
Anghel, L. 36
Astudillo, E. 105
Atienza, D. 81
Awais, M. 109
Aysu, A. 114
Baungarten-Leon, E.I. 26
Benini, L. 16
Beque, H. 95
Bezzam, I. 56
Biswas, R. 100
Bosio, A. 46
Boston, A. 71, 90
Bozzini, C. 123
Brito, J.P.M. 95
Caon, M. 81
Capulong, T.R.G. 157
Cárdenas, C.S. 178
Cardoso, R. 1
Carloni, L. 71
Carvajal, G. 186
Cauquil, A. 41, 51
Challagundla, D. 56
Cheng, Y. 31
Cisternas, J. 66
Coignus, J. 41
Colombo, D.M. 148, 180
Conti, F. 16
Cornett, E. 114
Costa, M.V. 1
Da Silva, R. 184
Darne, B. 46
Davidson, J. 90
Deleruyelle, D. 41, 46, 51
Di Pendina, G. 36
Donoso, F. 176, 186
Dos Santos, A.V.C. 133
Duan, J. 100
EduardoHolguin, E.G. 105
El-Sankary, K. 148, 180
Ella, M. 6
Farahmandi, F. 61, 171
Farias, R.C.G.D. 143
Farina, C. 123
Filsinger, M. 46, 51
Fuerstenau, V.H. 188
Gabbay, F. 6

Gaillardon, P.-E. 71, 90
Garcia, J.F.S. 148, 180
Gava, J. 182
Giuffrida, L. 81
Gómez, J. 16
Gomez, J. 66
Grellert, M. 184
Grimblatt, V. 176
Herman, K. 119
Huarancca, J.G. 178
Hunt, W. 16
Islam, R. 56
Jaramillo-Toral, U. 26
Jaramillo-Toral, E. 26
Jove, D.M.R. 166
Juruena, K.M.H. 157
Karn, R.R. 11
Kavun, E.B. 114
Kokkiligadda, K. 85
Labrak, L. 41
Larsson-Edefors, P. 153
Li, Z. 31
Liu, C. 16
Magalhães, J.P.P. 128
Magesh, R. 114
Marchand, C. 41, 46, 51
Marcon, C. 133, 161
Margala, M. 76, 138
Marín, J. 95, 176
Marin, J. 186
Martina, M. 81
Martins, J.B. 143
Masera, G. 81
McCarthy, K.G. 21
Mitchell, A. 1
Mohammadi, H.G. 109
Montanares, M. 21
Moraes, F.G. 133
Moraes, F. 161
Narayanan, V. 100
Navarro, D. 51
Ni, K. 100
Nisar, A. 36
Nunes, W.A. 133
Ortega-Cisneros, S. 26
Ost, L. 182
Osterno, M. 161
O'Connor, I. 1, 41, 46, 51

Padilha, W.	143	Villegas, N.	16
Palma, V.A.	21	Vilquin, B.	46
Panigrahi, P.	100	Vivet, P.	41
Patel, J.	85	Wang, Y.	31
Pendyala, S.	114	Warning, D.	119
Pimenta, T.C.	128	Weber, A.	143
Platzner, M.	109	Welch, J.	66
Pravadelli, G.	123	Xu, N.	31
Prócel, L.M.	105	Yamaguchi, Y.	166
Pronsat, R.	41	Yan, T.	31
Queiroz, M.G.D.	1	Zapata, H.E.M.	26
Rahman, T.	61	Zareie, M.	148, 180
Ramadan, F.	6		
Ramos, E.	143		
Reis, R.	143, 182, 184, 188		
Rojas, C.A.	176, 186		
Romanini, S.	16		
Rosales, M.E.R.	157		
Rossa, G.D.M.	95		
Rossi, D.	16		
Saha, S.	61		
Saha, S.K.	61, 171		
Sahruri, A.	138		
Salgado, G.M.	21		
Salvador, N.	176, 186		
Salvo, B.D.	16		
Sarwar, S.S.	16		
Scherer, M.	16		
Schiavone, P.D.	81		
Seyoum, B.	71		
Silva, D.L.F.	128		
Silveira, J.	161		
Sinanoglu, O.	11		
Singh, V.	85		
Snelgrove, A.	90		
Stockham, S.	90		
Sugiura, K.	166		
Sundarapu, V.K.V.	56		
Tandon, M.	85		
Tehranipoor, M.	61, 171		
Tenace, V.	90		
Thorawade, T.	85		
Tinti, I.	95		
Turetta, C.	123		
Udeji, U.L.	76		
Vafaei, A.	171		
Varasteh, M.	123		
Venkitaraman, V.	85		

9798331598136